# AI and Deep Learning in Biometric Security

# Artificial Intelligence (AI): Elementary to Advanced Practices

*Series Editors:*
Vijender Kumar Solanki, Zhongyu (Joan) Lu, and Valentina E. Balas

In the emerging smart city technology and industries, the role of artificial intelligence is getting more prominent. This AI book series will aim to cover the latest AI work, which will help the naïve user to get support to solve existing problems, and for the experienced AI practitioners, it will shed light on new avenues in the AI domains. The series will cover the recent work carried out in AI and associated domains and it will also cover a broad scope of application areas such as biometric security, Pattern recognition, NLP, Expert Systems, Machine Learning, Block-Chain, and Big Data. The work domain of AI is quite deep, so it will be covering the latest trends that are evolving with the concepts of AI, and it will be helpful to those who are new to the field, practitioners, students, and researchers to gain some new insights.

### Cyber Defense Mechanisms
Security, Privacy, and Challenges
*Gautam Kumar, Dinesh Kumar Saini, and Nguyen Ha Huy Cuong*

### Artificial Intelligence Trends for Data Analytics Using Machine Learning and Deep Learning Approaches
*K. Gayathri Devi, Mamata Rath, and Nguyen Thi Dieu Linh*

### Transforming Management Using Artificial Intelligence Techniques
*Vikas Garg and Rashmi Agrawal*

### AI and Deep Learning in Biometric Security
Trends, Potential, and Challenges
*Gaurav Jaswal, Vivek Kanhangad, and Raghavendra Ramachandra*

For more information on this series, please visit: https://www.crcpress.com/
Artificial-Intelligence-AI-Elementary-to-Advanced-Practices/book-series/
CRCAIEAP

# AI and Deep Learning in Biometric Security

## Trends, Potential, and Challenges

Edited by
Gaurav Jaswal, Vivek Kanhangad, and
Raghavendra Ramachandra

CRC Press
Taylor & Francis Group
Boca Raton London New York

CRC Press is an imprint of the
Taylor & Francis Group, an **informa** business

First edition published 2021
by CRC Press
6000 Broken Sound Parkway NW, Suite 300, Boca Raton, FL 33487-2742

and by CRC Press
2 Park Square, Milton Park, Abingdon, Oxon, OX14 4RN

*Library of Congress Cataloging-in-Publication Data*
Names: Jaswal, Gaurav, editor. | Kanhangad, Vivek, editor. |
Ramachandra, Raghavendra, editor.
Title: AI and deep learning in biometric security : trends, potential, and
challenges / edited by Gaurav Jaswal, Vivek Kanhangad, and Raghavendra Ramachandra.
Description: First edition. | Boca Raton, FL : CRC Press, 2021. |
Series: Artificial intelligence (AI) : elementary to advanced practices |
Includes bibliographical references and index.
Identifiers: LCCN 2020032531 (print) | LCCN 2020032532 (ebook) |
ISBN 9780367422448 (hardback) | ISBN 9781003003489 (ebook)
Subjects: LCSH: Biometric identification. | Artificial intelligence.
Classification: LCC TK7882.B56 A53 2021 (print) | LCC TK7882.B56 (ebook) |
DDC 006.2/48—dc23
LC record available at https://lccn.loc.gov/2020032531
LC ebook record available at https://lccn.loc.gov/2020032532

ISBN: 978-0-367-42244-8 (hbk)
ISBN: 978-0-367-67251-5 (pbk)
ISBN: 978-1-003-00348-9 (ebk)

Typeset in Times
by codeMantra

# Contents

Preface.................................................................................................................vii
Editors................................................................................................................ix
Contributors ......................................................................................................xi

**Chapter 1** Deep Learning-Based Hyperspectral Multimodal Biometric
Authentication System Using Palmprint and Dorsal Hand Vein ......... 1

*Shuping Zhao, Wei Nie, and Bob Zhang*

**Chapter 2** Cancelable Biometrics for Template Protection: Future
Directives with Deep Learning.........................................................23

*Avantika Singh, Gaurav Jaswal, and Aditya Nigam*

**Chapter 3** On Training Generative Adversarial Network for Enhancement
of Latent Fingerprints.........................................................................51

*Indu Joshi, Adithya Anand, Sumantra Dutta Roy, and
Prem Kumar Kalra*

**Chapter 4** DeepFake Face Video Detection Using Hybrid Deep Residual
Networks and LSTM Architecture......................................................81

*Semih Yavuzkiliç, Zahid Akhtar, Abdulkadir Sengür, and
Kamran Siddique*

**Chapter 5** Multi-spectral Short-Wave Infrared Sensors and Convolutional
Neural Networks for Biometric Presentation Attack Detection....... 105

*Marta Gomez-Barrero, Ruben Tolosana, Jascha Kolberg, and
Christoph Busch*

**Chapter 6** AI-Based Approach for Person Identification Using ECG
Biometric ....................................................................................... 133

*Amit Kaul, A.S. Arora, and Sushil Chauhan*

**Chapter 7** Cancelable Biometric Systems from Research to Reality:
The Road Less Travelled................................................................. 155

*Harkeerat Kaur and Pritee Khanna*

**Chapter 8**   Gender Classification under Eyeglass Occluded Ocular Region:
An Extensive Study Using Multi-spectral Imaging ........................ 175

*Narayan Vetrekar, Raghavendra Ramachandra,*
*Kiran Raja, and R. S. Gad*

**Chapter 9**   Investigation of the Fingernail Plate for Biometric
Authentication using Deep Neural Networks ................................ 205

*Surabhi Hom Choudhury, Amioy Kumar, and*
*Shahedul Haque Laskar*

**Chapter 10** Fraud Attack Detection in Remote Verification Systems for
Non-enrolled Users ........................................................ 239

*Ignacio Viedma, Sebastian Gonzalez, Ricardo Navarro, and*
*Juan Tapia*

**Chapter 11** Indexing on Biometric Databases ................................... 257

*Geetika Arora, Jagdiah C. Joshi, Karunesh K. Gupta, and*
*Kamlesh Tiwari*

**Chapter 12** Iris Segmentation in the Wild Using Encoder-Decoder-Based
Deep Learning Techniques ................................................. 283

*Shreshth Saini, Divij Gupta, Ranjeet Ranjan Jha,*
*Gaurav Jaswal, and Aditya Nigam*

**Chapter 13** PPG-Based Biometric Recognition: Opportunities with
Machine and Deep Learning ............................................... 313

*Amit Kaul and Akhil Walia*

**Chapter 14** Current Trends of Machine Learning Techniques in Biometrics
and its Applications ........................................................ 333

*B. S. Maaya and T. Asha*

**Index** ...................................................................... 361

# Preface

With the growth of data and the increasing awareness about the sensitivity of personal information, people have started to treat their privacy more seriously. Biometric systems have now significantly improved person identification and verification, playing an important role in personal, national, and global security. The recently evolved deep neural networks (DNN) learn the hierarchical features at intermediate layers automatically from the data and have shown many inspiring results for biometric applications. With this motivation, the text offers a showcase of cutting-edge research on the use of DNN in face, nail, finger knuckle, iris, ECG, palm print, fingerprint, vein, and medical biometric systems, and hence focuses on two parts: "Biometrics" and "Deep Learning for Biometrics".

This text highlights original case studies to solve real-world problems on biometric security and presents a broad overview of advanced deep learning architectures for learning domain-specific feature representation for biometrics-related tasks. The book aims to provide an in-depth overview of the recent advancements in the domain of biometric security using artificial intelligence (AI) and deep learning techniques, enabling readers to gain a deeper insight into the technological background of this domain. The text acts as a platform for the decision on the use of advanced architectures of convolutional neural networks, generative adversarial networks, autoencoders, recurrent convolutional neural networks, and graph convolution neural networks for various biometric security tasks such as indexing, gender classification, recognition in the wild, spoofing attacks/liveness detection, quality analysis, ROI segmentation, cross-sensor matching, and domain adaptation. In the text, feasibility studies on medical modalities (ECG, EEG, PPG) have been investigated using AI and deep learning. This book also examines the potential and future perspectives of AI and deep learning towards biometric template protection and multi-spectral biometrics. Overall, the reference provides better readability to readers through its chapter organisation and contains fourteen chapters only.

This text/reference is an edited volume by prominent academic researchers and industry professionals in the area of AI and biometric security. It will be essential reading for prospective undergraduate/postgraduate students, young researchers, and technology aspirants who are willing to research in the field of AI and biometric security.

**Gaurav Jaswal**
**Vivek Kanhangad**
**Raghavendra Ramachandra**

MATLAB® is a registered trademark of The MathWorks, Inc. For product information,
Please contact:

The MathWorks, Inc.
3 Apple Hill Drive
Natick, MA 01760-2098 USA
Tel: 508-647-7000
Fax: 508-647-7001
E-mail: info@mathworks.com
Web: www.mathworks.com

# Editors

**Dr. Gaurav Jaswal** is currently working as post-doctoral researcher at Indian Institute of Technology Delhi, India since January 2020. Before this, he served as Project Scientist (Electrical Engineering) at National Agri-Food Biotechnology Institute Mohali, India. He was research associate at School of Computing and Electrical Engineering, Indian Institute of Technology Mandi, India. He received MTech and PhD degrees in Electrical Engineering from National Institute of Technology Hamirpur in 2018. His research interests are in the areas of multimodal biometrics, medical imaging, and deep learning. He regularly reviews papers for various international journals including *IEEE Transactions on Information Forensics and Security* (TIFS), *IEEE Transactions on Biometrics, Behavior, and Identity Science* (T-BIOM), *IET Biometrics*, and *Pattern Recognition Letters*.

**Dr. Vivek Kanhangad** is currently working as associate professor, Department of Electrical Engineering, Indian Institute of Technology Indore, India since February 2012. Prior to this, he was visiting assistant professor, International Institute of Information Technology Bangalore, India (June 2010–December 2012). He received PhD from the Hong Kong Polytechnic University in 2010. Prior to joining Hong Kong PolyU, he received MTech degree in Electrical Engineering from Indian Institute of Technology Delhi in 2006 and worked for Motorola India Electronics Ltd, Bangalore for a while. His research interests are in the overlapping areas of digital signal and image processing, pattern recognition with a focus on biometrics and biomedical applications. He regularly reviews papers for various international journals including *IEEE Transactions on Information Forensics and Security* (TIFS), *IEEE Transactions on Cybernetics*, *IEEE Transactions on Human-Machine Systems*, and Elsevier journals – *Pattern Recognition* and *Pattern Recognition Letters*.

**Dr. Raghavendra Ramachandra** is currently working as a professor in Department of Information Security and Communication Technology (IIK). He is a member of Norwegian Biometrics Laboratory (http://nislab.no/biometrics_lab) at NTNU Gjøvik. He received B.E. (Electronics and Communication) from University of Mysore, India; MTech (Digital Electronics and Advanced Communication Systems) from Visvesvaraya Technological University, India; and PhD (Computer Science with specialisation of Pattern Recognition and Image Processing) from the University of Mysore, India, and Telcom SudParis, France. His research interest includes pattern recognition, image and video analytics, biometrics, human behaviour analysis, video surveillance, health biometrics, and smartphone authentication.

# Contributors

**Zahid Akhtar**
Department of Computer Science
University of Memphis
Memphis, Tennessee

**Adithya Anand**
Indian Institute of Technology Delhi
Delhi, India

**A.S. Arora**
Department of Electrical &
 Instrumentation Engineering
Sant Longowal Institute of Engineering
 and Technology
Longowal, India

**Geetika Arora**
Department of Computer Science and
 Information Systems
Birla Institute of Technology and
 Science Pilani
Pilani, India

**T. Asha**
Department of CSE
Banglore Institute of Technology
Bengaluru, India

**Christoph Busch**
da/sec – Biometrics and Internet
 Security Research Group
Hochschule Darmstadt
Darmstadt, Germany

**Sushil Chauhan**
Department of Electrical Engineering
National Institute of Technology
 Hamirpur
Hamirpur, India

**Surabhi Hom Choudhury**
Department of Electronics &
 Instrumentation Engineering
National Institute of Technology Silchar
Silchar, India

**R.S. Gad**
Department of Electronics
Goa University
Taleigao-Plateau, India

**Marta Gomez-Barrero**
Fakultät Wirtschaft
Hochschule Ansbach
Ansbach, Germany

**Sebastian Gonzalez**
R+D
TOC Biometrics Labs
Santiago, Chile

**Divij Gupta**
Department of Electrical Engineering
Indian Institute of Technology Jodhpur
Jodhpur, India

**Karunesh K. Gupta**
Department of Electrical and
 Electronics Engineering
Birla Institute of Technology and
 Science Pilani
Pilani, India

**Gaurav Jaswal**
Department of Electrical Engineering
Indian Institute of Technology Delhi
Delhi, India

**Ranjeet Ranjan Jha**
School of Computing and Electrical
    Engineering
Indian Institute of Technology Mandi
Mandi, India

**Indu Joshi**
Indian Institute of Technology Delhi
Delhi, India

**Jagdiah C. Joshi**
Department of Electrical and
    Electronics Engineering
Birla Institute of Technology and
    Science Pilani
Pilani, India

**Prem Kumar Kalra**
Indian Institute of Technology Delhi
Delhi, India

**Amit Kaul**
Department of Electrical Engineering
National Institute of Technology
    Hamirpur
Hamirpur, India

**Harkeerat Kaur**
Indian Institute of Technology Jammu

**Pritee Khanna**
PDPM Indian Institute of Information
    Technology, Design and
    Manufacturing, Jabalpur
Jabalpur, India

**Jascha Kolberg**
da/sec – Biometrics and Internet
    Security Research Group
Hochschule Darmstadt
Darmstadt, Germany

**Amioy Kumar**
Client Computing Group
Intel Corporation Bangalore
Bangalore, India

**Shahedul Haque Laskar**
Department of Electronics &
    Instrumentation Engineering
National Institute of Technology Silchar
Silchar, India

**B.S. Maaya**
Department of CSE
Banglore Institute of Technology
Bengaluru, India

**Ricardo Navarro**
R+D
TOC Biometrics Labs
Santiago, Chile

**Wei Nie**
Department of Computer and
    Information Science
University of Macau
Macau, China

**Aditya Nigam**
School of Computing and Electrical
    Engineering
Indian Institute of Technology Mandi
Mandi, India

**Kiran Raja**
Norwegian Biometrics Laboratory
Norwegian University of Science and
    Technology (NTNU)
Gjøvik, Norway

**Raghavendra Ramachandra**
Norwegian Biometrics Laboratory
Norwegian University of Science and
    Technology (NTNU)
Gjøvik, Norway

**Sumantra Dutta Roy**
Indian Institute of Technology Delhi
Delhi, India

**Shreshth Saini**
Department of Electrical Engineering
Indian Institute of Technology Jodhpur
Jodhpur, India

**Abdulkadir Sengür**
Department of Electrical and
   Electronics Engineering
Fırat University
Elazig, Turkey

**Kamran Siddique**
Department of Information and
   Communication Technology
Xiamen University Malaysia
Sepang, Malaysia

**Avantika Singh**
School of Computing and Electrical
   Engineering
Indian Institute of Technology Mandi
Mandi, India

**Juan Tapia**
R+D
TOC Biometrics Labs
Santiago, Chile

**Kamlesh Tiwari**
Department of Computer Science and
   Information Systems
Birla Institute of Technology and
   Science Pilani
Pilani, India

**Ruben Tolosana**
Biometrics and Data Pattern Analytics –
   BiDA Lab
Universidad Autonoma de Madrid
Madrid, Spain

**Narayan Vetrekar**
Department of Electronics
Goa University
Taleigao-Plateau, India

**Ignacio Viedma**
R+D
TOC Biometrics Labs
Santiago, Chile

**Akhil Walia**
Department of Electrical Engineering
National Institute of Technology
   Hamirpur
Hamirpur, India

**Semih Yavuzkiliç**
Department of Electrical and
   Electronics Engineering
Fırat University
Elazig, Turkey

**Bob Zhang**
Department of Computer and
   Information Science
University of Macau
Macau, China

**Shuping Zhao**
Department of Computer and
   Information Science
University of Macau
Macau, China

# 1 Deep Learning-Based Hyperspectral Multimodal Biometric Authentication System Using Palmprint and Dorsal Hand Vein

*Shuping Zhao, Wei Nie, and Bob Zhang*
University of Macau

## CONTENTS

1.1 Introduction ................................................................................................. 1
1.2 Device Design ............................................................................................. 5
1.3 System Implementation ............................................................................. 6
    1.3.1 ROI Extraction ................................................................................. 6
        1.3.1.1 Hyperspectral Palmprint ROI Extraction ........................... 6
        1.3.1.2 Hyperspectral Dorsal Hand Vein ROI Extraction ............... 8
    1.3.2 Feature Extraction ......................................................................... 10
    1.3.3 Feature Fusion and Matching ........................................................ 13
1.4 Experiments ............................................................................................. 13
    1.4.1 Multimodal Hyperspectral Palmprint and Dorsal Hand Vein Dataset ..... 14
    1.4.2 Optimal Pattern and Band Selection ............................................. 14
    1.4.3 Multimodal Identification ............................................................. 17
    1.4.4 Multimodal Verification ................................................................ 17
    1.4.5 Computational Complexity Analysis ............................................. 18
1.5 Conclusions .............................................................................................. 19
Acknowledgements ........................................................................................... 19
References ......................................................................................................... 19

## 1.1 INTRODUCTION

Biometric recognition system has been widely used in the construction of a smart society. Many types of biometric systems, including face, iris, palmprint, palm vein, dorsal hand vein, and fingerprint, currently exist in security authentication. Palmprint

1

recognition system is a kind of reliable authentication technology, due to the fact that palmprint has stable and rich characteristics, such as textures, local orientation features, and lines. In addition, a palmprint is user-friendly and cannot be easily captured by a hidden camera device without cooperation from the users. However, palmprint images captured using a conventional camera cannot be used in liveness detection. Palm vein is a good remedy for the weakness of palmprint acquired using a near-infrared (NIR) camera. The vein pattern is the vessel network underneath human skin. It can successfully protect against spoofing attacks and impersonation. Similar to palm vein, dorsal hand vein also has stable vein structures that do not change with age. Besides vein networks, some related characteristics to palmprint such as textures and local direction features can also be acquired.

Up to now, palmprint and dorsal hand vein-based recognition methods have achieved competitive performances in the literature. Huang et al. [1] put forward a method for robust principal line detection from the palmprint image, even if the image contained long wrinkles. Guo et al. [2] presented a binary palmprint direction encoding schedule for multiple orientation representation. Sun et al. [3] presented a framework to achieve three orthogonal line ordinal codes. Zhao et al. [4] constructed a deep neural network for palmprint feature extraction, where a convolutional neural network (CNN)-stack was constructed for hyperspectral palmprint recognition. Jia et al. presented palmprint-oriented lines in [5]. Khan et al. [6] applied the principle component analysis (PCA) to achieve a low-dimensionality feature in dorsal hand vein recognition. Khan et al. [7] obtained a low-dimensionality feature representation with Cholesky decomposition in dorsal hand vein recognition. Lee et al. [8] encoded multiple orientations using an adaptive two-dimensional (2D) Gabor filter in dorsal hand vein feature extraction.

The palmprint and dorsal hand vein recognition is usually carried out by conventional and deep learning-based methods. The conventional methods need to design a filter to extract the corresponding feature, i.e., local direction, local line, principal line, and texture. These hand-crafted algorithms usually require rich prior knowledge based on the specific application scenario. PalmCode [9] encoded palmprint features on a fixed direction by using a Gabor filter. Competitive code [10] extracted the dominant direction feature by using six Gabor filters. Xu et al. [11] encoded a competitive code aiming to achieve the accurate palmprint dominant orientation. Fei et al. [12] detected the apparent direction from the palmprint image. In addition, Huang et al. [13] put forward a centroid-based circular key-point grid (CCKG) pattern in dorsal hand vein recognition, which extracts local features based on key-points detection. Deep learning-based algorithms require a mass of training data to train the parameters in the deep convolutional neural network (DCNN). Afterwards, the optimal DCNN can be utilised for classification or convolution feature extraction. However, a mass of training data is usually unavailable for a palmprint or dorsal hand vein recognition task. Especially, the transfer learning technology with DCNN supports an approach that a pretrained DCNN can be fine-tuned with a few training samples for classification in a specific application. Zhao et al. [14] proposed a deep discriminative representation method, which extracted palmprint features from deep discriminative convolutional networks (DDCNs). DDCNs contain a pretrained DCNN and a set of lightened CNNs corresponding to the global and local patches segmented from the palmprint image. Wan et al. [15] trained the VGG depth CNN to extract dorsal hand vein features and used the logistic regression for identification. Deep learning-based methods can be widely applied in generic application scenarios.

Increasing research studies have moved to the area of hyperspectral imagery technology in the past decades. Contrary to the traditional imagery technology, not only skin texture but also vascular networks are imaged using the designed hyperspectral imagery system with the specific spectrum setup. In the phase of imaging palmprint or dorsal hand combined hyperspectral technology, more discriminative information from the palmprint or dorsal hand image can be captured achieving a high recognition rate. With more than 60 bands covered in hyperspectral palmprint, the three-dimensional (3D) feature was extracted through 3D Gabor filters [16]. Due to the redundant data, hyperspectral palmprint authentication improved but not remarkably when every spectral data were considered in the feature extraction phase. Based on band combination, Shen et al. [17] clustered typical bands in hyperspectral palmprint images for authentication, which performed better compared with in Ref. [16], while Guo et al. [18] applied an approach of $k$ means algorithm for representative band selection in hyperspectral palmprint database to improve performance. What's more, the band clustering method can decrease computation and increase efficiency in hyperspectral biometrics. As is known, dorsal hand vein and palmprint are concentrated in one hand, which makes it more convenient to collect these two different modalities simultaneously. Based on this observation, the combination of hyperspectral palmprint and dorsal hand biometrics is developed to meet a higher security requirement and to guarantee an exceptional recognition performance. In addition, unimodal biometrics recognition based on a single trait easily suffers from spoofing and other attacks as stated in the literature [19,20]. Table 1.1 illustrates the survey of the current multimodal biometric recognition algorithms. First, it is observed from this table that palmprint and dorsal hand vein have been fused before [21]. However, Ref. [21] and the other methods in Table 1.1 used only two single-spectrum images (one for each modality) to improve the recognition performance.

Different from the literature in Table 1.1, this work will study and implement the merging hyperspectral palmprint feature into dorsal hand vein feature to develop a novel hyperspectral multimodal biometric authentication system, which is demonstrated by a flow diagram (refer to Figure 1.1). A hyperspectral acquisition device was utilised for collecting hyperspectral palmprint and dorsal hand images. Then, region of interest (ROI) is detected from hyperspectral palmprint, and dorsal hand images resulted in two corresponding ROI cubes. After ROI extraction, the optimal feature pattern, i.e., local binary pattern (LBP) [22], local derivative pattern (LDP) [9], 2D-Gabor [2], and deep convolutional feature (DCF) [23], is selected for the palmprint and dorsal hand vein, correspondingly. In the pattern selection procedure, each image in the ROI cube is extracted and its features are used in recognition. Thus, the pattern and band which can achieve the highest recognition are treated as the optimal pattern and band for hyperspectral palmprint and dorsal hand images. Afterwards, the feature corresponding to the optimal pattern from palmprint on the optimal band and the feature concerning to the optimal pattern from dorsal hand vein on the optimal band are merged as one feature vector. At last, this fused multimodal feature vector is directly used in matching with the 1-NN classifier.

## TABLE 1.1
## The Survey of Multimodal Biometric Recognition Algorithms

| Literature | Algorithms | Modalities | Features | Year |
|---|---|---|---|---|
| [19] | Concatenation | Palmprint and hand-geometry | Line features; hand lengths and widths | 2003 |
| [20] | Combined face-plus-ear image | Face and ear | PCA | 2003 |
| [24] | Concatenation | Face and hand | PCA, linear discriminant analysis (LDA) and 9-byte features | 2005 |
| [25] | Concatenation | Face and palmprint | 2D-Gabor PCA | 2007 |
| [26] | Concatenation | Fingerprint and face | Minutia features | 2007 |
| [27] | Concatenation | Side face and gait | PCA | 2008 |
| [28] | Fusion | Palmprint and fingerprint | Discrete cosine transforms | 2012 |
| [29] | Fusion | Profile face and ear | Speeded up robust features (SURF) | 2013 |
| [30] | Concatenation | Palmprint and fingerprint | Bank of 2D-Gabor | 2014 |
| [31] | Weighted concatenation | Face and ear | PCA | 2015 |
| [32] | Feature level | Iris, face and fingerprint | Group sparse representation-based classifier (GSRC) | 2016 |
| [21] | Score level | Palmprint and dorsal hand vein | Mean and average absolute deviation (AAD) features | 2016 |
| [33] | Bayesian decision fusion | Face and ear | CNN features | 2017 |
| [34] | Score level | Finger-vein and finger shape | CNN features | 2018 |
| [35] | Concatenation | Face and ear | CNN features | 2017 |

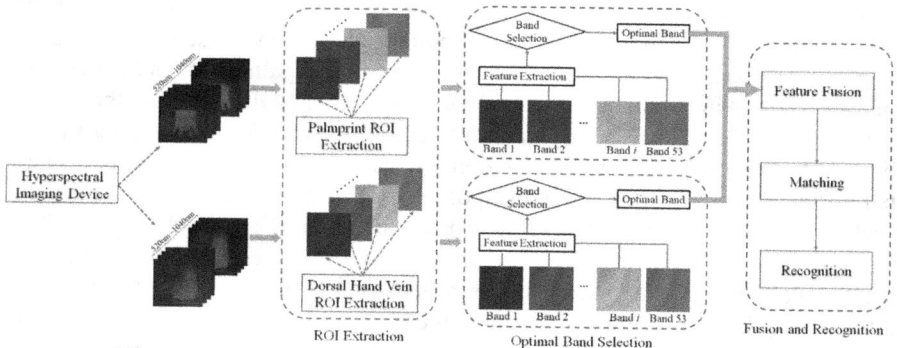

**FIGURE 1.1** The flowchart of the designed system-merged hyperspectral palmprint feature with dorsal hand feature.

The major contributions in the chapter are briefly introduced as follows:

1. A novel real-time hyperspectral multimodal biometric authentication system is conceived. It captures hyperspectral hand images by the proposed hyperspectral imaging acquisition device under 53 spectrums in the range of 520–1040 nm with intervals of 10 nm.
2. We collected a big multimodal dataset containing hyperspectral palmprint and dorsal hand images using the designed device. More information about this dataset can be found in Section 1.4.1.

The remaining work is organised as follows. In Section 1.2, the designed capture device is introduced. Following this, the designed system is illustrated in Section 1.3, including ROI and feature extraction as well as multimodal fusion and matching. Extensive experiments and analysis are included in Section 1.4, while Section 1.5 concludes the proposed system.

## 1.2 DEVICE DESIGN

The hyperspectral imaging acquisition system consists of two halogen lamps made by Osram, Inc., one charge coupled device (CCD) camera produced by Cooke, Inc., and one liquid crystal tunable filter manufactured by Meadowlark, Inc. The cost of the setup is approximately USD 6,500.00. The prototype of this acquisition system is illustrated in Figure 1.2. The CCD camera is placed in the middle with one halogen lamp on either side. The halogen lamps produce both visible light and NIR with spectra ranging from 520 to 1,040 nm. The light from the two halogen lamps irradiates on the palm or dorsal hand, and then reflects to the camera sensor for capturing images. A tunable filter is settled ahead of the camera lens and allows a single band to pass through its settings. To obtain stable spectral images, 10 nm is set as the spectral distance in the tunable filter. Therefore, this hyperspectral

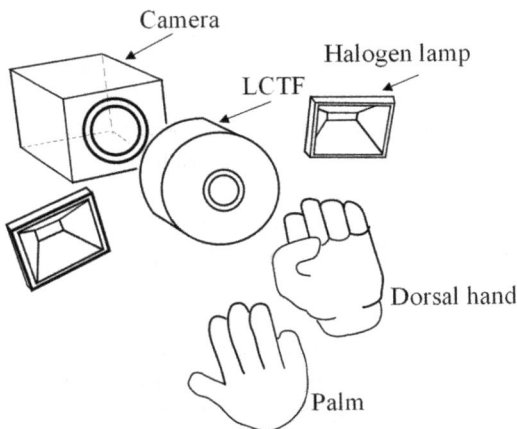

**FIGURE 1.2**  Schematic of our designed hyperspectral imaging device.

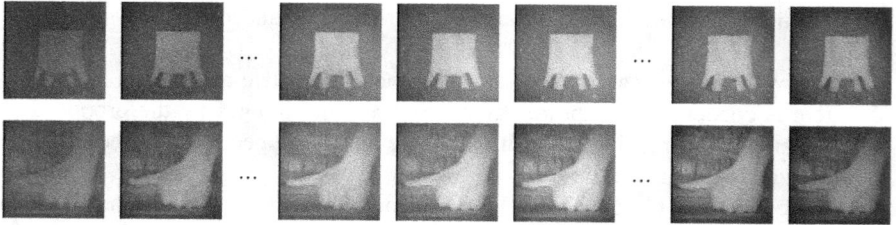

**FIGURE 1.3**   Hyperspectral palm (the upper row) and dorsal hand (the lower row) samples.

imagery acquisition system contains 53 bands in the range of 520–1,040 nm with 10 nm intervals.

Each volunteer was asked to grasp a prop making a fist when capturing his\her dorsal hand images. Contrary to an open hand, a closed dorsal hand makes the vascular network more visible achieving discriminant feature exaction. For the palmprint, each individual placed his/her hand on a plate with pegs to somewhat fix their hand, while a cutout was made to expose the palm. Examples of hyperspectral palm and dorsal hand images captured using the designed apparatus are shown in Figure 1.3.

## 1.3   SYSTEM IMPLEMENTATION

First, the ROI detection algorithms for hyperspectral palmprint images and dorsal hand vein images are introduced, respectively. Afterwards, several widely used patterns are presented for feature extraction. At last, a feature fusion strategy is proposed for multimodal recognition of hyperspectral palmprint and dorsal hand vein.

### 1.3.1   ROI EXTRACTION

#### 1.3.1.1   Hyperspectral Palmprint ROI Extraction

It is necessary to conduct ROI extraction from the palm image, due to the fact that the location of the ROI will influence the effectiveness of the extracted feature and the recognition performance. Here, we adaptively and reliably detect an ROI from the original palm image, which contains rich and stable characteristics. This step also makes the discriminative characteristics of palmprint separable from the background that contains noise and interference information. In this system, we used the hyperspectral palmprint ROI extraction method, which is based on our previous work in Ref. [36] (refer to Figure 1.4):

1. Image Enhancement: A Laplacian operator with eight neighbourhoods [37] is utilised for sharpness improvement of the original palmprint image. Afterwards, the image quality is much enhanced and will be beneficial for further preprocessing in the next steps (refer to Figure 1.4a and b). The utilised Laplacian operator is defined as follows:

$$\begin{bmatrix} 0 & -1 & 0 \\ -1 & 5 & -1 \\ 0 & -1 & 0 \end{bmatrix}$$

2. Binarisation: The Niblack [38] algorithm is a binarisation method which adaptively and locally computes the threshold of the image by performing a convolution. We first transform the enhanced palmprint image into grey-scale. Then, a 2D median filter is utilised for noise reduction. In Ref. [39], it has been proved that a 2D median filter can achieve a better performance on denoising in the hyperspectral images. Lastly, we obtain the binary palm-print image using the Niblack method [38] (refer to Figure 1.4b and c).

3. Palm Detection: Given the binarisation palmprint image (refer to Figure 1.4c), we initially locate the tips of the fingers (a–d) and valleys (e–h) of the palm by conducting the method in Ref. [40]. Afterwards, we detect the maximum inscribed circle ($T$) of the palm to find the centre of the palm (see Figure 1.4c). Therefore, the location of the maximum inscribed circle in the enhanced image can be achieved (refer to Figure 1.4d). To acquire pixels from the background, four external tangent circles of $T$ are located as $B_1$, $B_2$, $B_3$, and $B_4$ (see Figure 1.4c), which are on the vertical and horizontal directions. We define the radius of $T$ as $R$; thus, the radiuses of $B_1$, $B_2$, $B_3$, and $B_4$ are defined as $0.5 \times R$, $0.5 \times R$, $0.5 \times R$, and $0.3 \times R$, correspondingly. Here, pixels in $T$ are randomly selected as the positive data, and pixels in $B_1$, $B_2$, $B_3$, and $B_4$ are randomly selected as the negative data. Afterwards, the positive data and the negative data are put into the SVM to segment the palm from the background (see Figure 1.4e).

4. Contour Detection and ROI Extraction: Given the detected palm image (refer to Figure 1.4e), the Canny operator is utilised to achieve the boundary of the palm in the original image. Then, the boundaries named GAP$_1$ and GAP$_2$ between the forefinger and second finger and the fourth finger

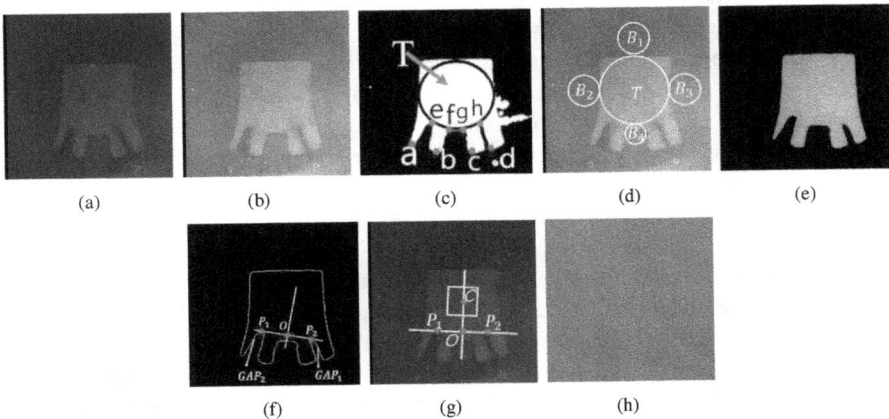

FIGURE 1.4   Steps of hyperspectral palmprint ROI extraction [30].

and little finger are obtained using the method in Ref. [41] (see Figure 1.4f), respectively. A line can then be drawn through one point in $GAP_1$ and another point in $GAP_2$, simultaneously. Then, we can define the two key points $P_1$ and $P_2$ when all points in $GAP_1$ and $GAP_2$ are below the line (see Figure 1.4f). Afterwards, a coordinate system is constructed based on $P_1$ and $P_2$ that the midpoint of line $P_1-P_2$ is defined as the origin $O$ and a vertical line with $P_1-P_2$ passing $O$ is defined as the $x$-axis (see Figure 1.4g). At last, a sub-image with a size of $128 \times 128$ in the palm centre is separated from the image using the constructed coordinate system, where $OC = \frac{3}{4} P_1 P_2$ (as seen in Figure 1.4g and h).

### 1.3.1.2   Hyperspectral Dorsal Hand Vein ROI Extraction

In the dorsal hand image, ROI indicates to the area that simply includes the vein part applied to extract feature. Dorsal hand vein images gathered through the acquisition device covers much redundant information such as a complicated background, the wrist, and the thumb. The unnecessary information can be eliminated by cropping the ROI from the collected image. The ROI not only maintains the vein structure with noise decreased but also reduces the computation cost, which can improve the recognition performance. The procedures of hyperspectral dorsal hand vein ROI extraction are presented in the following, which is adapted from our earlier study in Ref. [42] (refer to Figure 1.6):

1. **Pinky Knuckle Point Detection**: Based on a dorsal hand in the closed state (refer to Figure 1.6a), bulges at joints of the fingers and the boundary of a dorsal hand can be taken into consideration when locating the ROI. Here, the ROI can be extracted by locating one invariant point combined with a line of the profile of the dorsal hand. To this end, the template (refer to Figure 1.5) is constructed to search the point on a pinky knuckle. Based on a correlation operation between a template and a dorsal hand image, the maximal response (see the red point denoted in Figure 1.6f) can be found as the invariant point of the pinky knuckle.
2. **Dorsal Hand Profile Location**: The binarisation of a dorsal hand vein image was required for foreground segmenting from its background (refer to Figure 1.6b). Then, morphological opening and closing operations were applied to eliminate minor holes and remove tinny protrusions in the contour of the image (refer to Figure 1.6c). From the largest connected area (refer to Figure 1.6d), a profile of a dorsal hand (refer to Figure 1.6e) can be found by a boundary through single pixel-wise searching.
3. **Key Line Determination**: A circle was drawn with its centre at the point of the detected pinky knuckle, where the two crossing dots between the circle and the dorsal hand profile are located (refer to Figure 1.6f). A point was found concerning a lower area of a dorsal hand, which is connected with the pinky knuckle formed a closely horizontal line. Another point was searched regarding a higher reign of a dorsal hand, which is connected with the point on the pinky knuckle produced a closely vertical line. Here,

| 1 | 1 | 1 | 0 | 0 | 0 | 0 | 0 | 0 | 0 | 0 | 0 | 0 | 0 | 0 | 0 | 0 |
|---|---|---|---|---|---|---|---|---|---|---|---|---|---|---|---|---|
| 1 | 1 | 1 | 0 | 0 | 0 | 0 | 0 | 0 | 0 | 0 | 0 | 0 | 0 | 0 | 0 | 0 |
| 1 | 1 | 1 | 0 | 0 | 0 | 0 | 0 | 0 | 0 | 0 | 0 | 0 | 0 | 0 | 0 | 0 |
| 1 | 1 | 1 | 1 | 0 | 0 | 0 | 0 | 0 | 0 | 0 | 0 | 0 | 0 | 0 | 0 | 0 |
| 1 | 1 | 1 | 1 | 1 | 0 | 0 | 0 | 0 | 0 | 0 | 0 | 0 | 0 | 0 | 0 | 0 |
| 1 | 1 | 1 | 1 | 1 | 1 | 0 | 0 | 0 | 0 | 0 | 0 | 0 | 0 | 0 | 0 | 0 |
| 1 | 1 | 1 | 1 | 1 | 1 | 1 | 0 | 0 | 0 | 0 | 0 | 0 | 0 | 0 | 0 | 0 |
| 1 | 1 | 1 | 1 | 1 | 1 | 1 | 1 | 0 | 0 | 0 | 0 | 0 | 0 | 0 | 0 | 0 |
| 1 | 1 | 1 | 1 | 1 | 1 | 1 | 1 | 1 | 0 | 0 | 0 | 0 | 0 | 0 | 0 | 0 |
| 0 | 0 | 0 | 0 | 0 | 0 | 0 | 0 | 0 | 0 | 0 | 0 | 0 | 0 | 0 | 0 | 0 |
| 0 | 0 | 0 | 0 | 0 | 0 | 0 | 0 | 0 | 0 | 0 | 0 | 0 | 0 | 0 | 0 | 0 |
| 0 | 0 | 0 | 0 | 0 | 0 | 0 | 0 | 0 | 0 | 0 | 0 | 0 | 0 | 0 | 0 | 0 |
| 0 | 0 | 0 | 0 | 0 | 0 | 0 | 0 | 0 | 0 | 0 | 0 | 0 | 0 | 0 | 0 | 0 |
| 0 | 0 | 0 | 0 | 0 | 0 | 0 | 0 | 0 | 0 | 0 | 0 | 0 | 0 | 0 | 0 | 0 |
| 0 | 0 | 0 | 0 | 0 | 0 | 0 | 0 | 0 | 0 | 0 | 0 | 0 | 0 | 0 | 0 | 0 |
| 0 | 0 | 0 | 0 | 0 | 0 | 0 | 0 | 0 | 0 | 0 | 0 | 0 | 0 | 0 | 0 | 0 |
| 0 | 0 | 0 | 0 | 0 | 0 | 0 | 0 | 0 | 0 | 0 | 0 | 0 | 0 | 0 | 0 | 0 |

**FIGURE 1.5** The template to locate the pinky knuckle.

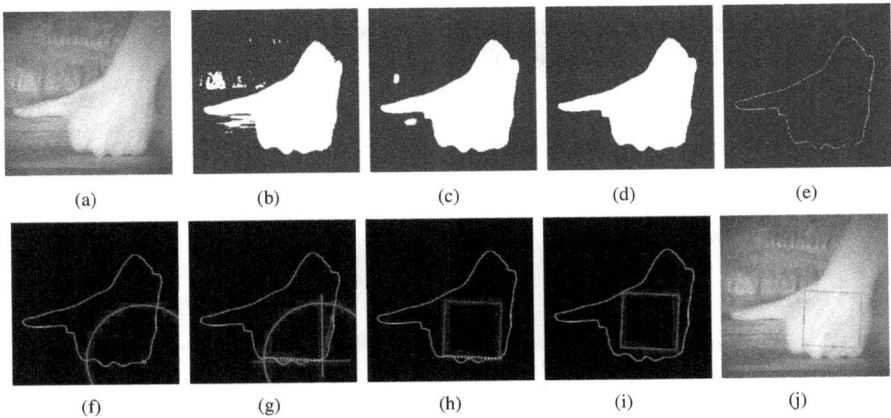

**FIGURE 1.6** The steps of hyperspectral dorsal hand vein ROI extraction [42].

we chose a horizontal line or vertical line in place of an edge of an ROI
(refer to Figure 1.6g).

4. **ROI Extraction**: Finally, with the pinky knuckle point detected and one
key line drawn, the other three edges of the ROI are determined (refer to
Figure 1.6h). Due to the insufficient vein information in margin of a dorsal
hand image, the ROI is moved a few pixels to the up and right to achieve
rich vascular features (refer to Figure 1.6i). The experiments showed that
this method is robust and adaptive at locating the ROI precisely for hyperspectral dorsal hand image (refer to Figure 1.6j).

## 1.3.2 FEATURE EXTRACTION

LBP is an effective and widely used texture feature descriptor [22] in biometric recognition. Not only does LBP obtain a better performance in many applications, it
is also computationally simplistic [43]. Compared with LBP, LDP was proposed
as a high-order texture encoding scheme for local patterns. Furthermore, LDP can
extract the derivative direction variation information of each pixel in the image.
The 2D-Gabor filter is sensitive to orientations, making it the most promising in
the extraction of local palmprint and dorsal hand vein [2,5]. Otherwise, DCNN has
obtained significant performances in image classification [44]. DCNN has powerful ability of abstract and impact feature representation by executing several non-
linear convolutional layers. Usually, abundant training data are necessary to train
the parameters in the DCNN. Particularly, the derived characteristics from a certain
layer can be utilised as the DCF for biometric authentication [10,44].

In this subsection, the classical feature extractors including LBP, LDP, 2D-Gabor,
and DCNN are introduced as follows. Each will be utilised for the hyperspectral
palmprint and dorsal hand vein ROIs (refer to Sections 1.4.2–1.4.4).

1. LBP: Texture has been proved an effective pattern in biometric recognition
[9] due to its rich local characteristics. Given an ROI image, the key step to
transform a pixel into the LBP code is to binarise its neighbouring eight pixels that the value of the centre pixel is chosen as the threshold. Afterwards,
each pixel can be encoded as follows:

$$\text{LBP}_{L,C} = \sum_{d=0}^{d-1} S(v - v_d) \times 2^d \tag{1.1}$$

$$S(x) = \begin{cases} 0, \, x > 0 \\ 1, \, x \leq 0 \end{cases} \tag{1.2}$$

where $v$ is the value of the pixel at the location $(L, C)$ in the image, and $v_d$ is
the value of the $d$th neighbour pixel. Finally, a LBP vector can be generated
by using a histogram for all the encoded values. It is shown in Figure 1.7 that
we can define the LBP descriptor with a variety of sizes ($\text{LBP}_{d,r}$), where $d$
denotes the number of neighbour adjacent points and $r$ denotes the radius.

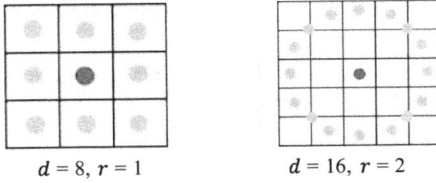

$$d = 8, r = 1 \qquad\qquad d = 16, r = 2$$

**FIGURE 1.7**  LBP neighbourhood sizes.

2. LDP: The LDP [9] is utilised to encode the local direction pattern. Given the ROI image $I(Z)$, we define its first-order derivatives on different orientations as $I'_{\partial}(Z)$, where $\partial = 0°, 45°, 90°$, and $135°$. Here, we assume that $Z_0$ is a point in $I(Z)$, and $Z_i$ ($i = 1, ..., 8$) (see Figure 1.8) denotes the $i$th neighbour pixel. Therefore, the first-order derivatives of $Z_0$ is calculated as follows:

$$I'_{0°} = I(Z_0) - I(Z_4) \tag{1.3}$$

$$I'_{45°} = I(Z_0) - I(Z_3) \tag{1.4}$$

$$I'_{90°} = I(Z_0) - I(Z_2) \tag{1.5}$$

$$I'_{135°} = I(Z_0) - I(Z_1) \tag{1.6}$$

The second-order derivative of $Z_0$ on $\partial$ ($\partial = 0°, 45°, 90°$, and $135°$) can be described as follows:

$$\mathrm{LDP}^2_{\partial}(Z_0) = \{f(I'_{\partial}(Z_0), I'_{\partial}(Z_1)), ..., f(I'_{\partial}(Z_0), I'_{\partial}(Z_8))\} \tag{1.7}$$

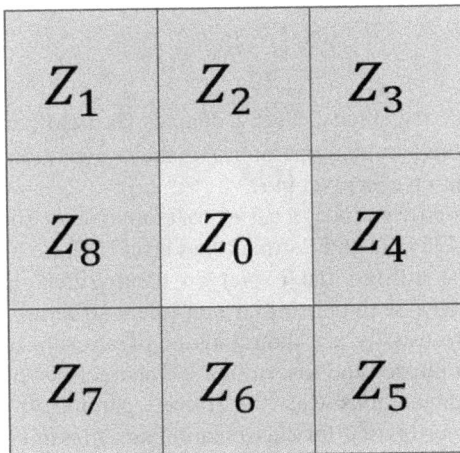

**FIGURE 1.8**  Surrounding pixels around the centre point $Z_0$.

where $f(.,.)$ is an equation on binary transformation:

$$f\left(I'_\partial(Z_0),\ I'_\partial(Z_i)\right)=\begin{cases}0,\ \text{if}\ I'_\partial(Z_0)\cdot I'_\partial(Z_i)>0\\1,\ \text{if}\ I'_\partial(Z_0)\cdot I'_\partial(Z_i)\le 0\end{cases}\tag{1.8}$$

At last, a 32-bit feature vector can be generated as follows on different orientations:

$$\text{LDP}^2(Z_0)=\left\{\text{LDP}_\partial^2(Z_0)|\partial=0°,\ 45°,\ 90°\ \text{and}\ 135°\right\}\tag{1.9}$$

3. 2D-Gabor: Due to the fact that it has a good 2D spectral specificity property, the 2D-Gabor filter is frequently exploited in orientation feature extraction [2,5]. The 2D-Gabor is presented as follows:

$$G\left(x,\ y,\ \varphi,\ \mu,\ \sigma\right)=\frac{1}{2\pi\sigma^2}\exp\left\{-\frac{x^2+y^2}{2\sigma^2}\right\}\exp\left\{2\pi i(\mu x\cos\varphi+\mu y\sin\varphi)\right\}\tag{1.10}$$

where $i=\sqrt{-1}$, $\mu$ presents the frequency of the sinusoidal wave, $\varphi$ denotes the direction, and $\sigma$ denotes the standard deviation. Usually, a 2D-Gabor filter bank contains a set of filters on $n$ orientations with the same scale. The orientation $\varphi_j$ is obtained as follows:

$$\varphi_j=\frac{\pi(j-1)}{n}\ j=1,\ 2,\ \dots,\ n.\tag{1.11}$$

Then, the convolution of the 2D-Gabor filter is conducted on the palmprint image to obtain a line response as follows:

$$r_j=\left(I*G\left(\varphi_j\right)\right)_{(x,\ y)}\tag{1.12}$$

where $I$ denotes the image, $G(\varphi_j)$ denotes the real part of the filter on $\varphi_j$, "*" is the convolutional operator, $r$ is the convolution result, and $(x,\ y)$ denotes the position of a pixel in $I$.

4. DCF: DCNN usually includes a variety of components, such as pooling [11], convolution, ReLU [12], and Softmax-loss layer [23], as shown in Figure 1.9. LeCun [11] first utilised the LeNet on handwritten digit classification. Since then, DCNNs with the similar non-linear structure have been widely used [23]. There usually are thousands of parameters in different layers. Therefore, high impact and discriminative characteristics can be obtained after several convolutions with the trained parameters. Particularly, the Softmax-loss layer is used for classification as a classifier in DCNN. Here, we ignore the Softmax-loss layer and extract discriminative features as DCFs directly from the second FC layer (see Figure 1.9) of the DCNN.

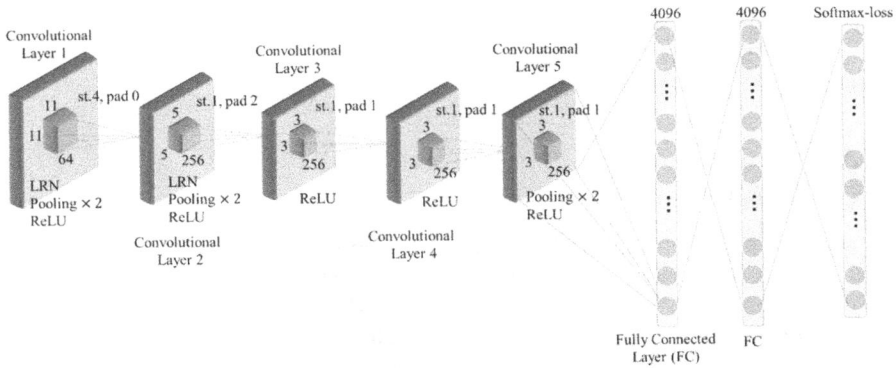

**FIGURE 1.9** The architecture of DCNN for VGG-F [45].

### 1.3.3 FEATURE FUSION AND MATCHING

In the feature extraction phase (refer to Section 1.3.2), different features, including LBP, LDP, 2D-Gabor, and DCF, can be applied to each ROI image in the palmprint and dorsal hand vein cubes, respectively. If all images from the different bands are fused for recognition, it will be costly and time consuming. Consequently, we selected the optimal bands with respect to the types of features achieving the best recognition results on palmprint and dorsal hand vein, respectively.

Let $F_{palm} = [f_1, f_2, ..., f_i, ..., f_n] \in R^{d \times n}$ and $P_{dhv} = [p_1, p_2, ..., p_s, ..., p_n] \in R^{d \times n}$ denote the hyperspectral palmprint features and hyperspectral dorsal hand vein features, respectively, where $f_i$ is the feature vector for the $i$th band palmprint image, $p_s$ is the feature vector for the $s$th band dorsal hand vein ROI, $d$ denotes the dimensionality of the feature, and $n$ denotes the number of spectrums. Afterwards, the optimal features can be fused as follows:

$$W = \left[ O\left(F_{palm}\right); O\left(P_{dhv}\right) \right] \tag{1.13}$$

where $O(\cdot)$ is the selection of the optimal feature from $F_{palm}$ or $P_{dhv}$, with the selected feature vector obtaining the highest recognition accuracy. Specifically, $W$ is to be concatenated with the optimal $O\left(F_{palm}\right)$ and the optimal $O\left(P_{dhv}\right)$.

After feature fusion, we use the "Euclidean" distance for the final matching:

$$\text{dist}\left(X, Y\right) = \sqrt{\sum_{i=1}^{d} \left(x_i - y_i\right)^2} \tag{1.14}$$

where $X$ and $Y$ are features extracted from two objects.

## 1.4 EXPERIMENTS

First, we briefly introduce the collected hyperspectral palmprint and dorsal hand vein dataset. Then, the optimal band and pattern selection are performed on

different modalities, respectively. Afterwards, multimodal identification and verification results are presented, correspondingly. At last, the time consumption of the
proposed method is analysed.

### 1.4.1 MULTIMODAL HYPERSPECTRAL PALMPRINT AND DORSAL HAND VEIN DATASET

We constructed a hyperspectral palmprint and dorsal hand vein dataset captured
from the same volunteers utilising the proposed hyperspectral imaging device (refer
to Section 1.2). As mentioned in Section 1.2, the device can acquire hyperspectral
images covering a spectrum range of 520–1,040 nm with 10 nm intervals, which
means that the images on 53 different spectrums can be obtained. The dataset was
acquired from 209 persons, and each volunteer was required to provide both left and
right hands for imaging. This dataset contains two sessions which were acquired
with intervals about 30 days. In each session, a volunteer was requested to capture
both their left and right hands a total of five times. Therefore, this dataset totally
includes 443,080 (209 subjects × 5 samples × 2 objects × 53 bands × 2 sessions ×
2 modalities) images. Some original and ROI samples from one object are shown in
Figures 1.10 and 1.11, respectively.

### 1.4.2 OPTIMAL PATTERN AND BAND SELECTION

To obtain the best performance in recognition using multimodal features, we should
select the best bands for palmprint and dorsal hand vein, respectively, in which the
image contains rich and clear information and can derive the most discriminative
features. For different feature patterns, including LBP, LDP, 2D-Gabor, and DCF,
we aim to choose the optimal pattern and band for palmprint recognition and dorsal
hand vein recognition, respectively. For every experiment, each algorithm was conducted 10 times. Finally, the mean accuracy of recognition rate was calculated as the
performance evaluation:

$$ARR = \frac{\text{number of correctly classifised samples}}{\text{total number of samples}} \qquad (1.15)$$

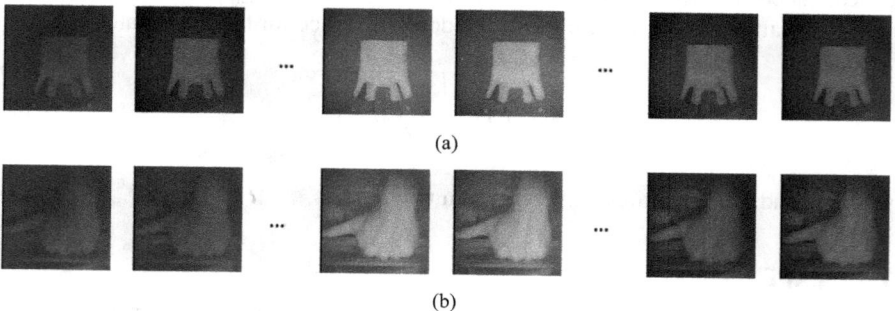

(a)

(b)

FIGURE 1.10   Hyperspectral palmprint (a) and dorsal hand vein (b) samples.

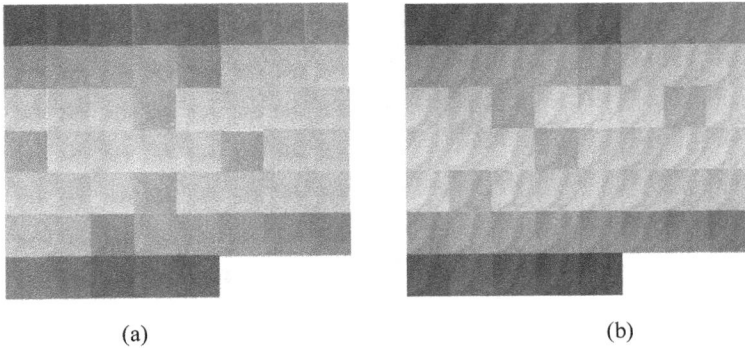

(a)                                        (b)

**FIGURE 1.11**   The hyperspectral palmprint and dorsal hand vein ROIs coming from the same individual. (a) denotes palmprint ROI samples and (b) denotes dorsal hand vein ROI samples. From left to right, top to down, the band increases from 520 to 1040 nm with 10 nm intervals.

In this work, the experiments were implemented using MATLAB 2015a on a CPU 3.40 with RAM 16.0 GB running Windows 10.

When we extracted LBP and LDP features from the ROI image, each image was segmented into 16 non-overlapping sub-images with the same size of 32 × 32. Afterwards, the LBP or LDP features were extracted from each sub-image and further to be concatenated to one feature vector. As for 2D-Gabor, we defined a bank with five scales on eight directions. Otherwise, we applied VGG-F for DCF extraction with the DCF derived from the 19th layer of VGG-F. At last, the nearest neighbour (1-NN) was chosen for identification and verification. Figures 1.12 and 1.13 show the identification rates of different patterns on each band of the hyperspectral palmprint and dorsal hand vein cubes, respectively.

From Figure 1.12a, one can see that LBP achieved the highest ARR (98.09%) on the 44th band corresponding to 950 nm. LDP (Figure 1.12b) obtained the highest ARR (94.74%) on the 41th band corresponding to 930 nm. 2D-Gabor (Figure 1.12c) achieved the highest ARR (76.08%) on the 37th band corresponding to 880 nm. DCF (Figure 1.12d) obtained the highest ARR (97.89%) on the 21th band corresponding to 730 nm. As for the dorsal hand vein results presented in Figure 1.13, one can see that LBP (Figure 1.13a) achieved the highest ARR on the 38th band corresponding to 880 nm with 92.20%. LDP (Figure 1.13b) achieved the highest ARR of 97.00% on the 52th band corresponding to 1030 nm. 2D-Gabor (Figure 1.13c) achieved the highest ARR on the 40th band corresponding to 900 nm with 88.20%. DCF (Figure 1.13d) obtained the highest ARR on the 26th band corresponding to 780 nm with 92.20%.

Both Figures 1.12 and 1.13 show that different patterns have their own corresponding optimal bands. For hyperspectral palmprint identification, LBP can achieve the highest ARR on 950 nm with 98.09%. On the other hand, in hyperspectral dorsal hand vein identification, LDP can achieve the highest ARR on 1,030 nm with 97.00%.

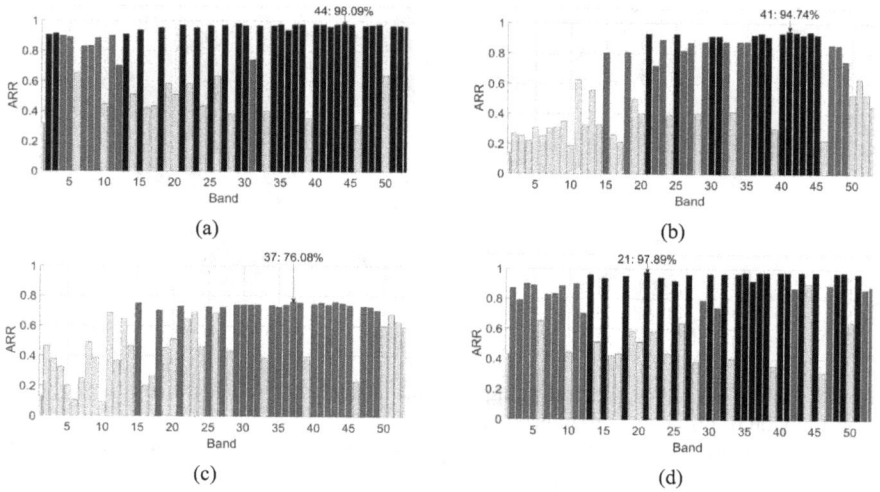

**FIGURE 1.12** ARRs of different patterns for each band of the hyperspectral palmprint cube: (a) LBP, (b) LDP, (c) 2D-Gabor, and (d) DCF. When the ARR $\geq 0.9$, the bar colour is black. With $0.7 \leq ARR < 0.9$, the bar colour is dark grey. When ARR $< 0.7$, the bar is coloured light grey.

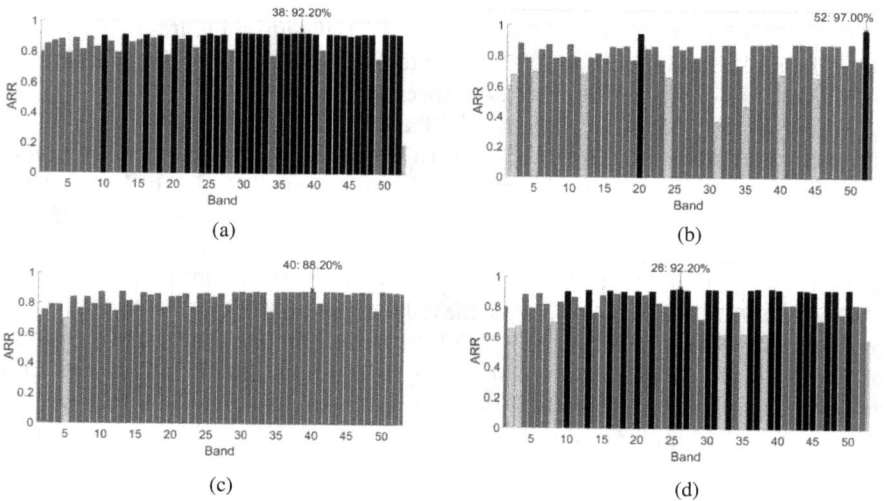

**FIGURE 1.13** ARRs of different patterns for each band of the hyperspectral dorsal hand vein cube: (a) LBP, (b) LDP, (c) 2D-Gabor, and (d) DCF. When the ARR $\geq 0.9$, the bar colour is black. With $0.7 \leq ARR < 0.9$, the bar colour is dark grey. When ARR $< 0.7$, the bar is coloured light grey.

### 1.4.3 Multimodal Identification

After pattern and band selection, we can obtain a multimodal pattern combination of $W_{m,n}$, where $m$ is the selected pattern for the palmprint and $n$ is the selected pattern for the dorsal hand vein. To select the optimal combinations of patterns for both the palmprint and dorsal hand vein, we tested every combination for identification in groups of two, including $W_{LBP,LBP}$, $W_{LBP,LDP}$, $W_{LBP,DCF}$, $W_{LDP,DCF}$, $W_{LDP,LDP}$, $W_{DCF,DCF}$, $W_{LBP,Gabor}$, $W_{LDP,Gabor}$, $W_{DCF,Gabor}$, and $W_{Gabor,Gabor}$. The performance of multimodal identification was evaluated once again using ARR.

Table 1.2 depicts the identification ARRs of different combinations for multimodal identification. From this table, it can be observed that $W_{LBP,LBP}$ achieved the highest ARR with 99.21% compared to the other features. As we know, LBP can achieve the highest ARR for palmprint recognition (refer to Section 1.4.2), while LDP obtained the highest ARR for the dorsal hand vein. In addition, from Table 1.2, we can see that multimodal identification produces a better identification performance than unimodal identification for either palmprint or dorsal hand vein.

### 1.4.4 Multimodal Verification

In addition to the identification results mentioned above, we performed verification as well. Verification is a one-to-one matching scheme to verify if the given two samples are from the same object or sharing the same label. The performance of multimodal verification was evaluated using equal error rate (EER) as follows:

$$FAR=\frac{NFA}{NIRA} \tag{1.16}$$

$$GAR = 1-\frac{NFR}{NGRA}\times100\% \tag{1.17}$$

**TABLE 1.2**
**Identification ARRS of Different Pattern Combinations for Multimodal Identification**

| Feature | Palmprint (nm) | Dorsal Hand Vein (nm) | ARR (%) |
|---|---|---|---|
| $W_{LBP,LBP}$ | 950 | 880 | 99.21 |
| $W_{LBP,LDP}$ | 950 | 1,030 | 98.87 |
| $W_{LBP,DCF}$ | 950 | 780 | 98.64 |
| $W_{LDP,DCF}$ | 930 | 780 | 99.10 |
| $W_{LDP,LDP}$ | 930 | 1,030 | 94.57 |
| $W_{DCF,DCF}$ | 730 | 780 | 98.39 |
| $W_{LBP,Gabor}$ | 950 | 900 | 97.96 |
| $W_{LDP,Gabor}$ | 930 | 900 | 95.32 |
| $W_{DCF,Gabor}$ | 730 | 900 | 96.97 |
| $W_{Gabor,Gabor}$ | 880 | 900 | 93.26 |

**TABLE 1.3**

**Multimodal Verification with Different Features**

| Features | Palmprint (nm) | Dorsal Hand Vein (nm) | EER (%) |
|---|---|---|---|
| $W_{\text{LBP, LBP}}$ | 950 | 880 | 0.006 |
| $W_{\text{LDP, LDP}}$ | 920 | 1,030 | 0.004 |
| $W_{\text{Gabor, Gabor}}$ | 880 | 910 | 0.032 |
| $W_{\text{DCF, DCF}}$ | 730 | 780 | 0.002 |

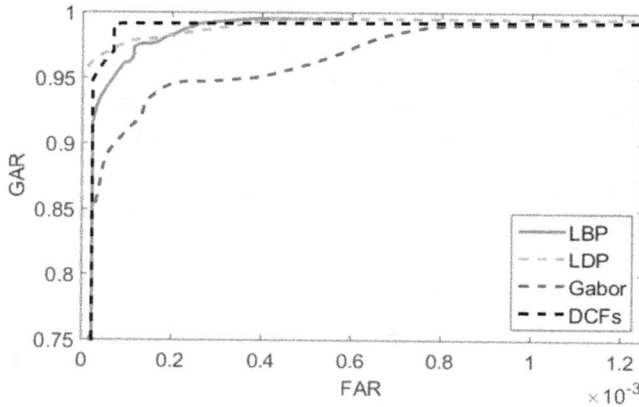

**FIGURE 1.14**   ROC curves of different features for verification.

where NGRA denotes the number of times of intra-class test, NIRA denotes the number of times of inter-class test, NFR presents the number of times of false rejections, and NFA presents the number of times of false acceptances. Therefore, we can obtain the EER, while FAR is equal to 1−GAR. Due to the fact that the same pattern has a better fusion property, as shown in the identification experiments, we conducted verification experiments using the patterns of $W_{\text{LBP, LBP}}$, $W_{\text{LDP, LDP}}$, $W_{\text{Gabor, Gabor}}$, and $W_{\text{DCF, DCF}}$. Table 1.3 illustrates the verification results. From Table 1.3, we can see that $W_{\text{DCF, DCF}}$ obtained the lowest EER of 0.002%. Figure 1.14 shows the ROC curves of GAR and FAR for the four combined patterns.

### 1.4.5   COMPUTATIONAL COMPLEXITY ANALYSIS

For computational complexity evaluation, we compared the computation costs of $W_{\text{LBP, LBP}}$, $W_{\text{LDP, LDP}}$, $W_{\text{Gabor, Gabor}}$, and $W_{\text{DCF, DCF}}$ due to the fact that these four fusion strategies have similar feature extraction, fusion, and matching procedures. We randomly selected 100 classes from the multimodal dataset in Section 1.4.1. For each pattern, the experiments were conducted five times with one test sample and the remaining data as the training samples. At last, we calculated the mean time as the time consumption. From Table 1.4, it can be seen that $W_{\text{LBP, LBP}}$ takes an average time

**TABLE 1.4**
**Time Cost of Different Kinds of Features**

| Methods | Feature Extraction (ms) | Matching (ms) |
|---|---|---|
| $W_{\text{LBP, LBP}}$ | 24.740 | 0.063 |
| $W_{\text{LDP, LDP}}$ | 39.113 | 0.071 |
| $W_{\text{Gabor, Gabor}}$ | 44.120 | 0.083 |
| $W_{\text{DCF, DCF}}$ | 57.655 | 0.122 |

of 24.740 ms for feature extraction, which is the quickest when compared with the other methods. Furthermore, the average matching time was 0.063 ms for $F_{\text{LBP, LBP}}$, which is the lowest simultaneously.

## 1.5 CONCLUSIONS

This chapter presented a novel multimodal biometric recognition system utilising palmprint and dorsal hand vein. A unique hyperspectral imaging device was developed that can capture an individual's palmprint or dorsal hand vein under 53 spectral bands. After the ROI was extracted from the two modalities, the different feature extractors were applied to each band. The optimal results in terms of the feature and its corresponding band were fused to perform multimodal identification and verification. Using this strategy, $W_{\text{LBP, LBP}}$ for multimodal hyperspectral recognition achieved the highest identification ARR of 99.21%. On the other hand, $W_{\text{DCF, DCF}}$ obtained the lowest EER of 0.002% for verification. Given its performance, the designed system can be implemented in a real-world application.

## ACKNOWLEDGEMENTS

This work was supported by the National Natural Science Foundation of China (grant number 61602540) and the University of Macau (file number MYRG2018–00053-FST).

## REFERENCES

1. Huang, D. S., Jia, W., and Zhang, D. 2008. Palmprint verification based on principal lines. *Pattern Recognition* 41(4): 1316–1328.
2. Guo, Z., Zhang, D., Zhang, L. et al. 2009. Palmprint verification using binary orientation co-occurrence vector. *Pattern Recognition Letters* 30(13): 1219–1227.
3. Sun, Z., Tan, T., Wang, Y., and Li, S. Z. 2005. Ordinal palmprint representation for personal identification. In *Proceedings of the IEEE Conference on Computer Vision and Pattern Recognition*, pp. 279–284.
4. Zhao, S., Zhang, B., and Chen, C. L. P. 2019. Joint deep convolutional feature representation for hyperspectral palmprint recognition. *Information Sciences* 489: 167–181.
5. Jia, W., Hu, R., Lei, Y., Zhao, Y., and Gui, J. 2013. Histogram of oriented lines for palmprint recognition. *IEEE Transactions on Systems, Man, and Cybernetics: Systems* 44(3): 385–395.

6. Khan, M.H., and Subramanian, R.K. 2009. Low dimensional representation of dorsal hand vein features using principle component analysis (PCA). *World Academy of Science, Engineering and Technology* 49: 1001–1007.
7. Khan, M.H., Subramanian, R.K., and Khan, N.A. 2009. Representation of hand dorsal vein features using a low dimensional representation integrating Cholesky decomposition. In *2nd International Congress on Image and Signal Processing*, pp. 1–6.
8. Lee, J.C., Lee, C.H., Hsu, C.B. et al. 2014. Dorsal hand vein recognition based on 2D Gabor filters. *The Imaging Science Journal* 62(3): 127–138.
9. Zhang, B., Gao, Y., Zhao, S., and Liu, J. 2009. Local derivative pattern versus local binary pattern: face recognition with high-order local pattern descriptor. *IEEE Transactions on Image Processing* 19(2): 533–544.
10. Cheng, J., Sun, Q. and Zhang, J. 2017. Supervised hashing with deep convolutional features for palmprint recognition. *Chinese Conference on Biometric Recognition*, Springer, pp. 259–268.
11. LeCun, Y., Bottou, L. and Bengio, Y. 1998. Gradient-based learning applied to document recognition. *Proceedings of the IEEE* 86: 2278–2324.
12. Odena, A., Olah, C., and Shlens, J. 2017. Conditional image synthesis with auxiliary classifier gans. *arXiv preprint arXiv*:1610.09585, (Jul). https://arxiv.org/pdf/1610.09585.pdf%5D.
13. Huang, D., Zhang, R., Yin, Y., and Wang, Y. 2017. Local feature approach to dorsal hand vein recognition by centroid-based circular key-point grid and fine-grained matching. *Image and Vision Computing* 1 (58): 266–277.
14. Krizhevsky, A., Sutskever, I., and Hinton, G.E. 2012. Imagenet classification with deep convolutional neural networks. *Advances in Neural Information Processing Systems*, 1097–1105.
15. Ioffe, S., and Szegedy, C. 2015. Batch normalization: accelerating deep network training by reducing internal covariate shift. *arXiv preprint arXiv*: 1502.03167.
16. Shen, L., Wu, W., Jia S. et al. 2014. Coding 3D Gabor features for hyperspectral palmprint recognition. *2014 International Conference on Medical Biometrics*, pp. 169–173.
17. Shen, L., Dai, Z., Jia, S. et al. 2015. Band selection for Gabor feature based hyperspectral palmprint recognition). *2015 International Conference on Biometrics*, pp. 416–421.
18. Guo, Z., Zhang, D., Zhang L. et al. 2012. Feature band selection for online multispectral palmprint recognition. *IEEE Transactions on Information Forensics and Security* 7(3): 1094–1099.
19. Kumar, A., Wong, D.C.M., Shen, H.C. et al. 2003. Personal verification using palmprint and hand geometry biometric. *Lecture Notes in Computer Science* 2688(1): 668–678.
20. Chang, K., Bowyer, K.W., Sarkar, S. et al. 2003. Comparison and combination of ear and face images in appearance-based biometrics. *IEEE Transactions on Pattern Analysis and Machine Intelligence* 25(9): 1160–1165.
21. Srivastava, S., Bhardwaj, S., and Bhargava, S. 2016. Fusion of palm-phalanges print with palmprint and dorsal hand vein. *Applied Soft Computing* 47: 12–20.
22. Ojala, T., Pietikainen, M., and Maenpaa, T. 2002. Multiresolution gray-scale and rotation invariant texture classification with local binary patterns. *IEEE Transactions on Pattern Analysis and Machine Intelligence* 24: 971–987.
23. Simonyan, K., and Zisserman, A. 2015. Very deep convolutional networks for large-scale image recognition. *arXiv preprint arXiv*:1409.1556, (Sep.). https://arxiv.org/pdf/1409.1556.pdf%20http://arxiv.org/abs/1409.1556.pdf.
24. Ross, A. 2005. Feature level fusion of hand and face biometrics. *The International Society for Optical Engineering* 5779: 196–204.
25. Yao, Y.F., Jing, X.Y., and Wong, H.S. 2007. *Letters: Face and Palmprint Feature Level Fusion for Single Sample Biometrics Recognition*. Amsterdam: Elsevier Science Publishers B. V.

26. Rattani, A., Kisku, D.R., Bicego, M. et al. 2007. Feature level fusion of face and finger-print biometrics. *IEEE International Conference on Biometrics*, pp. 1–6.
27. Zhou, X., and Bir, B. 2008. Feature fusion of side face and gait for video-based human identification. *Pattern Recognition* 41(3): 778–795.
28. Krishneswari, K., and Arumugam, S. 2012. Multimodal biometrics using feature fusion. *Journal of Computer Science* 8(3): 431.
29. Rathore, R., Prakash, S., and Gupta, P. 2013. Efficient human recognition system using ear and profile face. *Biometrics. IEEE Sixth International Conference on Theory, Applications and Systems*, pp. 1–6.
30. Chin, Y.J., Ong, T.S., Teoh, A.B.J. et al. 2014. Integrated biometrics template protection technique based on fingerprint and palmprint feature-level fusion. *Information Fusion* 18: 161–174.
31. Huang, Z., Liu, Y., and Li, X. 2015. An adaptive bimodal recognition framework using sparse coding for face and ear. *Pattern Recognition Letters* 53: 69–76.
32. Goswami, G., Mittal, P., and Majumdar, A. 2016. Group sparse representation based classification for multi-feature multimodal biometrics. *Information Fusion* 32: 3–12.
33. Fan, T.Y., Mu, Z.C., and Yang, R.Y. 2017. Multi-modality recognition of human face and ear based on deep learning. In *2017 International Conference on Wavelet Analysis and Pattern Recognition (ICWAPR)*, pp. 38–42.
34. Kim, W., Song, J.M., and Park, K.R. 2018. Multimodal biometric recognition based on convolutional neural network by the fusion of finger-vein and finger shape using near-infrared (NIR) camera sensor. *Sensors* 18(7): 2296.
35. Omara, I., Xiao, G., Amrani, M., Yan, Z., and Zuo, W. 2017. Deep features for efficient multi-biometric recognition with face and ear images. In *Ninth International Conference on Digital Image Processing (ICDIP 2017)*. International Society for Optics and Photonics 10420, p. 104200D.
36. Zhao, S., and Zhang, B. 2019. Robust and adaptive algorithm for hyperspectral palmprint region of interest extraction. *IET Biometrics* 8: 391–400.
37. Polesel, A., Ramponi, G., and Mathews, V.J. 2000. Image enhancement via adaptive unsharp masking. *IEEE Transactions on Image Processing* 9(3): 505–510.
38. Niblack, W. 1985. *An Introduction to Digital Image Processing*. Copenhagen, Denmark: Strandberg Publishing Company.
39. Khandizod, A.G., and Deshmukh, R.R. 2015. Comparative analysis of image enhancement technique for hyperspectral palmprint images. *International Journal of Computer Applications* 121: 30–35.
40. Zhu, L., and Zhang, S. 2010. Multimodal biometric identification system based on finger geometry, knuckle print and palm print. *Pattern Recognition Letters* 31(12): 1641–1649.
41. Zhang, D., and Kong, W.K. 2003. Online palmprint identification. *IEEE Transactions on Pattern Analysis and Machine Intelligence* 25(9): 1041–1050.
42. Nie, W., and Zhang, B. 2019. Robust and adaptive ROI extraction for hyperspectral dorsal hand vein images. *IET Computer Vision* 13(6): 595–604.
43. Ahonen, T., Hadid, A., and Pietikainen, M. 2006. Face description with local binary patterns: application to face recognition. *IEEE Transactions on Pattern Analysis and Machine Intelligence* 28: 2037–2041.
44. Chopra, S., Hadsell, R., and LeCun, Y. 2005. Learning a similarity metric discriminatively, with application to face verification. In *CVPR*, pp. 539–546.
45. Chatfield, K., Simonyan, K., Vedaldi, A., and Zisserman, A. 2014. Return of the devil in the details: delving deep into convolutional nets. *arXiv preprint arXiv*:1405.3531.

# 2 Cancelable Biometrics for Template Protection
## *Future Directives with Deep Learning*

*Avantika Singh*
IIT Mandi

*Gaurav Jaswal*
IIT Delhi

*Aditya Nigam*
IIT Mandi

## CONTENTS

2.1 Introduction ...................................................................................................24
2.2 Template Protection.......................................................................................25
    2.2.1 Consequences of Template Compromise.............................................25
    2.2.2 Template Protection Techniques..........................................................25
    2.2.3 Comparative Analysis between Template Protection Techniques......27
    2.2.4 Fundamental Requirements of Template Protection Techniques.......27
    2.2.5 Potential Attacks on Template Protection Techniques ......................28
2.3 Role of Deep Learning Approaches in Biometrics.......................................29
    2.3.1 Deep Learning in Face Recognition.....................................................29
    2.3.2 Deep Learning in Iris Recognition.......................................................30
    2.3.3 Deep Learning in Fingerprint Recognition .........................................30
    2.3.4 Deep Learning in Other Biometric Traits ............................................31
2.4 Related Work: Template Protection...............................................................31
    2.4.1 Biometric Encryption ...........................................................................31
    2.4.2 Biometric Cryptosystems......................................................................31
    2.4.3 Cancelable Biometrics ..........................................................................33
        2.4.3.1 Deep Learning-Based Cancelable Techniques ...................34
        2.4.3.2 Deep Learning versus Non-deep Learning Cancelable
                Techniques ............................................................................37
2.5 Performance Measures and Datasets in Cancelable Biometrics ................38
    2.5.1 Performance Measures for Non-invertibility Analysis........................38
    2.5.2 Performance Measures for Unlinkability Analysis .............................39

2.5.3    Performance Measures for System Usability Analysis .....................39
2.5.4    Performance Measures for Revocability Analysis ..........................40
2.5.5    Databases Used in Cancelable Biometrics.........................................40
2.6    Comparative Performance Analysis: Cancelable Biometrics......................41
2.7    Conclusions and Future Prospective of Deep Learning in Biometrics ..........42
References..............................................................................................43

## 2.1  INTRODUCTION

Biometric-based verification systems are quite common in the present-day world. In fact, application areas of biometrics are expanding day-by-day. They are used for a variety of purposes ranging from online banking, e-commerce, health-care, airport check-ins, border security, mobile phone unlocking, law enforcement, and many more. In the present-day scenario, it is worthy to say that biometric-based authentication systems are replacing traditional ones one by one to provide better security. Big technocrat giants like Google, Apple, Microsoft, Intel, and Samsung are investing a huge amount of money for implementing biometric authentication systems in their future products for better customer experience and satisfaction. Moreover, deployment of large biometric systems worldwide like Aadhaar (India), eKTP (Indonesia), and MyKad (Malaysia) surges the immediate need to secure biometric systems from potential security threats. Although biometric authentication-based systems can help in alleviating the problems associated with traditional systems, they are prone to inadvertent security lapses as well as to deliberate attacks that can result due to illegitimate intrusion or theft of sensitive biometric information. Figure 2.1 depicts the significant points of vulnerabilities on biometric recognition systems, as suggested by [72]. These vulnerability points, as suggested by [26], can be broadly classified into two categories as follows:

1. **Direct Vulnerability**: Here, the attacker attacks the sensing device by presenting the spoofed biometrics of the registered user. For mounting this type of attack, adversary requires no knowledge about the system. Furthermore,

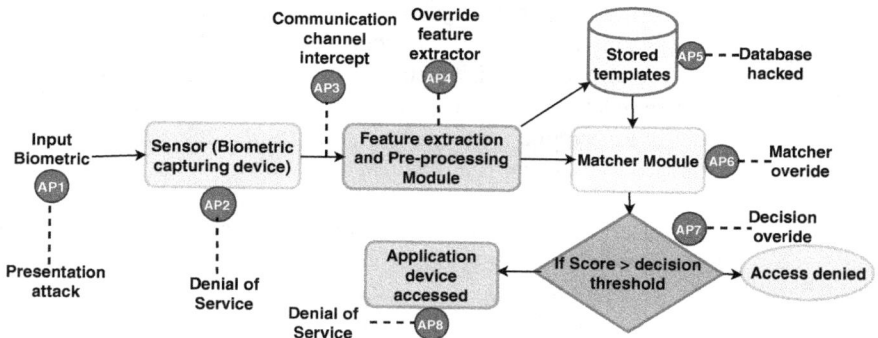

FIGURE 2.1   Major vulnerability points in a biometric-based system.

digital protection mechanisms like watermarking, encryption cannot be used here because this type of attack is carried out outside the system at the sensor level [26]. In Figure 2.1, this type of vulnerability is depicted as an attack point AP1.

2. **Indirect Vulnerability**: Here, the adversary needs to have an expertise knowledge about the internal working of the biometric system [38]. In Figure 2.1, this type of vulnerability is depicted as an attack point from AP2 to AP8. This type of attack mainly comprises manipulation of the database (either by altering a template or by deleting it), communication channel interception, or by bypassing the feature extractor and matcher module.

In recent years, several researchers have paid attention to address these vulnerabilities, but still, it is not fully solvable. Direct vulnerabilities are normally accessed by studying the physiological characteristics of biometric traits as carried out in liveliness detection, while indirect ones are addressed by securing the communication channel and databases. The focus of this chapter is to highlight the importance of template protection along with its techniques.

## 2.2   TEMPLATE PROTECTION

Biometric verification systems are based on the uniqueness of anatomical and observable patterns; however, the permanence of these features poses a challenge if it is stolen. Unlike conventional password-based systems, it cannot be revoked. Thus, gaining one's biometric information is regarded as a compromise of the user's privacy [60]. Even the EU General Data Protection Regulation 2016/679 [1] has defined biometric data as sensitive data. So, it is important and essential to secure biometric templates from adversarial attackers who can alter biometric templates for illegitimate access and fraudulent activities.

### 2.2.1   Consequences of Template Compromise

On gaining access to a person biometric template, an adversary can launch not only financial attack but can hamper a person's social life also by falsely plotting its biometric templates at crime scenes. Moreover, an intruder getting access to a template stored with least security can launch cross-domain linkage attacks. In the past, it was postulated [60] that biometric features can detect a certain type of medical condition in an individual. Furthermore, this information can be used to deny employment and insurance to subjects having a certain kind of medical disorder.

### 2.2.2   Template Protection Techniques

Broadly template protection techniques are classified under two main categories: hardware-based solutions and software-based solutions. The former one is a close recognition system [61] from which the biometric template is never transmitted and thus secured. Privaris PlusID [2] is one such example of hardware-based solution. Major limitations are that they are less flexible (need to be carried everywhere) and are expensive and prone to being lost like conventional credit cards. In the latter case (software-based solutions), biometric data are combined with some helper data to transform it into another form, and this resultant form is stored in the database rather than the original biometric template. Further, software-based template protection techniques can be divided into three subcategories as follows:

1. **Biometric Encryption**: In this type of technique, the biometric template is encrypted during the enrollment phase using a key; thus, an encrypted version of the biometric template is stored in the database. During authentication attempt, stored encrypted template is decrypted and matched with the query biometrics. On the basis of the key used, it can be further classified into two categories: (i) symmetric encryption (same key for encryption and decryption) and (ii) asymmetric encryption (different keys for encryption and decryption).

2. **Biometric Cryptosystems**: As the name suggests, biometric cryptosystem (BC) is an amalgam of two terms biometrics and cryptosystem. Designed specifically to take benefits from both like uniqueness and non-repudiation from biometrics and high security from cryptography [44]. Here, during the enrollment phase, the biometric template is associated with a key to obtain a secure sketch (which is stored in the database) while during authentication, query biometric is used to recover the original biometric template from the stored secure sketch. On the basis of the key used to generate secure sketch, it is mainly divided into two categories, as shown in Figure 2.2 and described below as: (a) Key binding-based cryptosystems (here, the cryptographic key is hidden within the enrolled biometric template using secret bit replacement algorithm. Fuzzy vault [40] and fuzzy commitment [41] are two popular examples of this category). (b) Key generation-based cryptosystems (here, the secure sketch is derived only from the biometric template while the cryptographic key is generated from the helper data and query biometric features. A fuzzy extractor is a popular example of this category).

3. **Cancelable Biometrics (CB)**: During the enrollment phase, a transformed version of the biometric template is stored in the database known as pseudo biometric identity (PBI), while during authentication query, biometric is again transformed to match with PBI. Based on the transformation functions, they are further classified into two subcategories: (a) non-invertible transformation-based and (b) salting-based approaches.

    a. **Non-invertible Transformation-Based CB**: Here, the transformation function is non-invertible in nature, major limitation performance, and security degradation if transformation function is stolen. Two popular approaches under this category are random projection-based transformations and geometric transformations.

    b. **Salting-Based Approaches**: Here, original biometric features are randomly permuted and convolved to generate transformed versions. GRAY-SALT, BIN-SALT, GRAY-COMBO, and BIN-COMBO [108] are some of the popular earlier works carried out under this category.

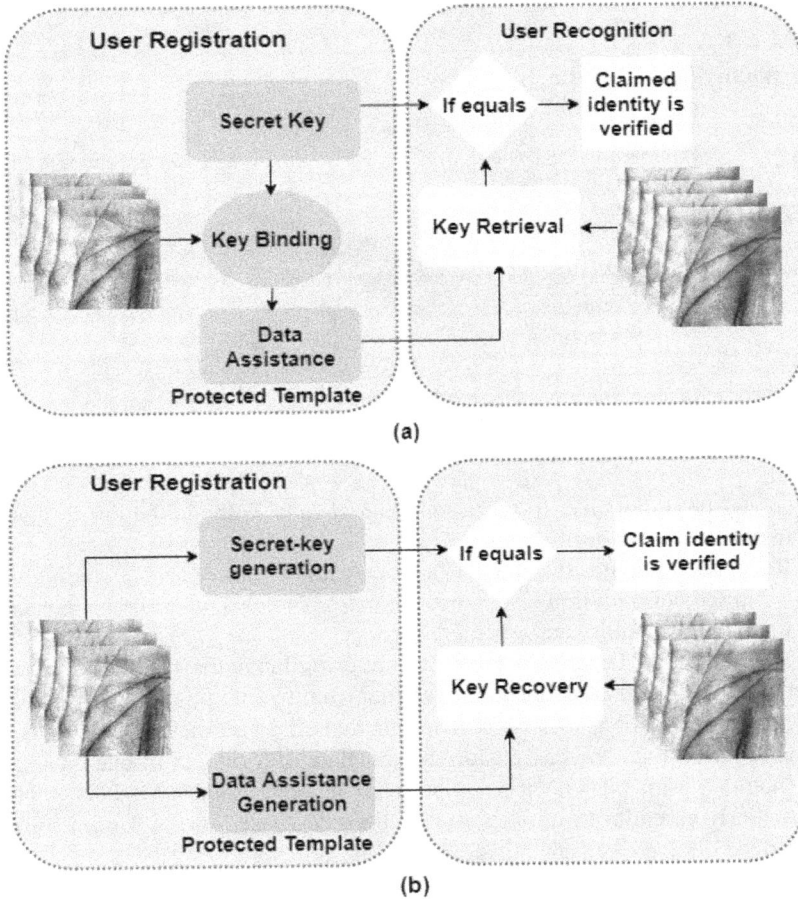

**FIGURE 2.2** Two variants of biometric cryptosystem: (a) Key binding scheme (b) key generation scheme.

### 2.2.3 COMPARATIVE ANALYSIS BETWEEN TEMPLATE PROTECTION TECHNIQUES

We have seen that a biometric template can be secured using any three of the software-based template protection techniques discussed in the previous section. Table 2.1 postulates major advantages and limitations of the above-mentioned techniques.

### 2.2.4 FUNDAMENTAL REQUIREMENTS OF TEMPLATE PROTECTION TECHNIQUES

1. **Non-Invertibility**: This property ensures non-invertibility of stored transformed template in the database. Mathematically, it is defined as: if $F_i$ is the original biometric template corresponding to $subject_i$ and $T_i$ is its transformed version stored in the database, then reconstruction of $F_i$ from $T_i$

**TABLE 2.1**

**Approach-Wise Advantages and Issues**

| Approach | Advantages | Issues |
|---|---|---|
| Encryption | Performance preservation | 1. Key management<br>2. During authentication original biometrics is accessible |
| Biometric cryptosystem | 1. Combines benefits of cryptography and biometrics<br>2. Secure key release mechanism based on biometrics | 1. Original biometrics is accessible after accept decision<br>2. Linkability |
| Cancelable biometrics | Original biometrics is never stored and thus not accessible | 1. Performance degradation<br>2. Weak security |

should be impossible. In short, mapping of $F_i$ to $T_i$ should be one to many instead of one to one.

2. **Revocability**: Since the number of biometrics associated with an individual is limited, it is required if somehow biometric is stolen, it should be replaced wisely. Revocability ensures this property.

3. **Unlinkability**: In today's world, we are using biometric-based authentication in a number of applications ranging from mobile unlocking to sophisticated applications like online banking and all. In all these applications, it is required that stored, transformed template of a subject in one database should not match with templates stored in other databases. This is particularly essential to limit cross-matching database attacks. Unlinkability among databases ensures this.

4. **System Usability**: The above three mentioned requirements, i.e., non-invertibility, revocability, and unlinkability are non-functional requirements of cancelable templates while system usability is a functional requirement which ensures that the system performance in terms of false acceptance rate (FAR) and false rejection rate (FRR) should not degrade while applying any kind of transformation to biometric templates in order to meet non-functional requirements. In fact, for an ideal protected biometric system all the four requirements should be met simultaneously, although it is difficult to achieve in reality.

### 2.2.5 Potential Attacks on Template Protection Techniques

Although protected biometric templates are more robust against different types of attacks as compared to the one without protection, they are vulnerable to some attacks. One of the major shortcomings of these protected biometric templates is that they are vulnerable to presentation attacks. In fact, some of the techniques have been specially fabricated to attack popular BCs and CB systems. In Table 2.2, template protection techniques along with their vulnerable attacks are discussed.

**TABLE 2.2**
**Security Attacks Against Biometric Templates**

| Approach | Possible Security Attacks |
|---|---|
| Biometric Encryption [94] | Hill climbing, substitution attack, attack via record multiplicity |
| **Biometric Cryptosystems** | |
| (i) Key binding scheme [32,40] | Attack on error-correcting codes, substitution attack, chaff elimination, ARM (attack via record multiplicity), |
| (ii) Key generation scheme [17,97] | Hill climbing, false acceptance attack, brute force attack |
| **Cancelable Biometrics** | |
| (i) Non-invertible transformations [73] | Overwriting final decision, ARM, substitution attack, linkage attack |
| (ii) Salting-based Approaches [28] | Stolen token attack, substitution attack, overwriting final decision, linkage attack, masquerading attack |

## 2.3   ROLE OF DEEP LEARNING APPROACHES IN BIOMETRICS

Deep learning-based models have shown outrageous performance on a variety of application areas like computer vision (self-driving cars, robotics), natural language processing (automatic text generation, automatic machine translation), biomedical, forensics, and many more in recent years. Even the biometric recognition system does not remain untouched to deep learning. In recent years, deep learning-based models are extensively used in the biometric domain to improve the accuracy of different recognition systems based on various biometric traits like face, iris, fingerprint, palmprint, gait, and many more. There is a wide range of applicability of deep learning concepts in the biometric domain that ranges from segmentation, authentication to generation of artificial real looking biometric samples. One of the fundamental problems with handcrafted features in the biometric domain is that they are highly trait-specific (like LBP, Gabor-based features well suited for the face, log Gabor features for iris) and require parameter tuning according to the dataset in consideration. On the other hand, deep learning-based methods provide an end-to-end learning framework that automates the process of learning the best feature representation irrespective of biometric trait and thus more universal. It should be noted that the tremendous success of deep learning mainly leverages the availability of large datasets. Non-availability of labeled voluminous datasets in the biometric domain except for face is a fundamental problem. To circumvent this situation, the transfer learning paradigm can be used to handle labeled data constraint. Thus, in the literature [84,92], many deep learning-based models for biometric authentication are based on transfer learning paradigm.

### 2.3.1   DEEP LEARNING IN FACE RECOGNITION

As compared to other vision community datasets, there is large intra-class variability and high inter-class similarity in biometric datasets; thus, the applicability of deep learning models in the biometric domain is quite challenging. To handle this situation, many researchers have proposed domain-specific loss functions that deal quite well with intra-class dissimilarity and inter-class similarity. For example, in the year 2016, Liu et al. [55] proposed an alternative of cross-entropy loss by the name large margin Softmax loss, and claimed that this loss explicitly enhanced inter-class separability and at the same time encouraged intra-class compactness between learned features. They experimentally verified their claim on the Labeled Faces in the Wild (LFW) (face dataset) and achieved good results. In another notable work, Liu [54] proposed hypersphere embedding for face recognition by the name SphereFace. This hypersphere embedding is constructed by proposing a novel loss function (angular softmax) that helps convolutional neural network (CNN) in learning angular discriminative features. CosFace [98], UniformFace [22], ArcFace [18], and AdaptiveFace [53] are popular deep learning-based frameworks for face recognition. It should be noted that deep learning-based frameworks in the biometric domain are mostly employed in the face recognition domain due to the availability of large labeled face datasets; Ref. [99] gives a detailed overview of various deep learning-based face recognition techniques.

### 2.3.2   DEEP LEARNING IN IRIS RECOGNITION

Deep learning-based framework is mostly used in the iris domain for iris recognition and spoof detection. Traditional iris recognition mainly comprises three major steps: (i) iris segmentation, (ii) iris normalisation, and (iii) iris feature extraction and matching. It is worth mentioning that earlier works in iris recognition using deep learning-based techniques mainly focus on the iris feature extraction part only [9,27]. Like in the work carried out by Minaee et al. [57,58], they have extracted discriminative iris features from normalised iris images using the transfer learning paradigm on VggNet and ResNet models. Recently, it has been pointed out by Ahmad et al. [4] that due to the highly discriminative learning capability of deep networks, outrageous results on iris recognition can be achieved by directly feeding segmented iris images to deep networks without normalising them. They have experimentally validated their claim on several challenging iris datasets (ND-0405, UbirisV2, and IITD). Very recently, a unified framework (deep learning-based) that detects, segments, and recognises iris images simultaneously without any need for pre-processing was proposed by Zhao et al. [107]. Recently, GANs are also used for augmenting iris datasets. It has been pointed out by researchers [50] that traditional iris augmentation techniques result in generating highly correlated samples, and thus, they are not as robust as compared to the one generated through generative models. Through the progress made in iris recognition, one can see how biometric researchers are adapting deep learning frameworks for challenging biometric problems.

### 2.3.3   DEEP LEARNING IN FINGERPRINT RECOGNITION

Deep learning-based approaches are widely used in fingerprint recognition. For example, FingerNet [91], a deep learning-based model for fingerprint minutiae detection, jointly performs tasks like feature extraction, segmentation, and orientation estimation; similarly in another work by Stajanovic et al. [86], CNNs are used for fingerprint ROI segmentation. There has also been some work for detecting finger-print spoofs using deep learning-based techniques [45,65].

### 2.3.4 Deep Learning in Other Biometric Traits

Deep learning-based frameworks are not so popular for recognising other biometric traits apart from the face, iris, and fingerprints due to the non-availability of labeled training datasets as well as due to their nonpopularity in comparison to face, iris, and fingerprints. But still, biometric researchers have used deep learning frameworks by utilising the concept of few shot learning where deep models can be trained by using few training samples like the one done in palmprint recognition [81]. In this work, features extracted through CNNs are represented as nodes in graph neural networks. Apart from that, siamese networks are also used for learning intra-class similarity and inter-class dissimilarity from small training datasets.

In addition, many works have been carried out so far [70,93] using deep learning-based techniques for mitigating presentation attacks in biometric traits, but prevent-ing adversarial attacks and template attacks through deep learning techniques is still in their infancy state. We will discuss more on mitigating template attacks based on the deep learning-based architecture in the following sections.

## 2.4 RELATED WORK: TEMPLATE PROTECTION

Here, in this section, we are discussing pioneer techniques in template protection, majorly focusing on template protection using CB. Furthermore, we are also illus-trating a comparative analysis between templates transformed via deep learning- and non-deep learning-based methods.

### 2.4.1 Biometric Encryption

Sahai and Waters in the year 2005 [79] were the first to develop biometric-based encryption systems. Since then, several systems have been developed. In one of the work [34], face data are protected using the Shamir secret sharing key. Here, in the first step, binarised face features are obtained using tokenised pseudo-random numbers. This binarised representation is known as FaceHash. Later FaceHash is protected via the Shamir secret sharing key. In another notable work, Bansal [8] gen-erates the key for the RSA algorithm using a matrix forged by fingerprints. Recently, a method is proposed by the name symmetric keyring encryption [47]. Here, in this method, biometric secret binding is carried out as fuzzy symmetric encryption.

### 2.4.2 Biometric Cryptosystems

BCs are majorly divided into two types: (i) key generation based and (ii) key binding based. In the case of key generation-based BCs during the enrollment phase, helper data are generated from biometric templates and these helper data are used to gener-ate keys. Here, in this case, generated keys along with helper data are stored in the database. Here, matching is performed by comparing keys stored in the database.

A major difficulty in these technique is to generate keys having high entropy values. The following section mentions some of the notable work in this field.

Pioneer work in this field is carried out by Yevgeniy Dodis [20] in the year 2008. He has proposed a fuzzy extractor for securing fingerprint templates. In another notable work [101], iris features are extracted through 2-D Gabor filters, and further Reed–Solomon error-correcting code is used along with the hash function to generate cipher key. This cipher key is used for encrypting and decrypting the iris features. In Ref. [104], the authors have proposed a fingerprint authentication technique based on a delaunay triangle-based fuzzy extractor. The major advantage of this technique is that it exploits the distinctive properties of delaunay triangulation net to attain robust features and use them further to achieve registration free matching. In Ref. [102], the authors proposed a near equivalent dual-layer structure check (NeDLSC) algorithm based on the minutiae local structure for generating secure fingerprint templates. An online voting system is proposed by [89]. This scheme is based on a fuzzy extractor for providing biometric-based authentication, which is paired with a secret password to provide add-on security to the voter. In another notable work [63], a biometric authentication protocol is proposed based on Kerberos. Here, for the first time, the fuzzy extractor is embedded in the Kerberos scheme. This proposed protocol is resilient against several attacks like man-in-middle and reply attacks. In one of the latest work [64], the authors utilised the Chebyshev polynomial in combination with a fuzzy extractor to protect face datasets. Due to the chaotic properties of the Chebyshev polynomial, it serves as a good candidate for designing cryptosystems.

In the case of key binding-based systems, fuzzy commitment [41] and fuzzy vault [40] are two popular categories. The former one is used to secure biometric templates that can be used as a binary vector-like iris. Here, in the case of fuzzy commitment during the enrollment phase, the binary represented feature vector is XORed with the binary representation of the error-correcting code (obtained from the key) to obtain helper data. While during biometric authentication, feature is XORed with helper data to obtain the error-corrected code, which further generates the key. The latter one is used to secure point set-based biometric features like fingerprint minutiae. Here, during enrollment, the biometric feature point is embedded in a finite field and is evaluated on a polynomial, which is generated by a key. In order to add biometric security points, its polynomials are further mixed with random points. While during authentication, query biometric is used to generate actual polynomial from the stored representation. Table 2.3 depicts some of the popular key binding-based BCs.

Unimodal biometric templates often suffer from inter-class variations and non-universality. In such situations, multi-biometric templates come as a rescue measure. Here, two or more modalities of the same person are used to increase the efficacy of the system. A fuzzy vault-based template protection method for fusing fingerprints and palmprints was first proposed by Brindha and Natarajan [10]. Later, in the same year, Nagar [62] proposed a template protection technique that combines the advantages of both fuzzy vault and fuzzy commitment. Here, the biometric features are fused feature-wise. Recently, a multimodal biometric authentication system is proposed that fuses feature vectors from fingerprints and palmprints based on fuzzy vault [88]. The proposed scheme exhibits potential results.

**TABLE 2.3**
**Key Binding-Based Biometric Cryptosystems**

| Author | Trait | Description | Advantages | Shortcomings |
|---|---|---|---|---|
| Soutar et al. [85] | Fingerprint | Studied key binding algorithm in an optical correlation-based fingerprint authentication system | Pioneer work in this domain | Pre-aligned images required, Rigorous security analysis missing |
| Juels and Sudan [40] | No evaluation | Fuzzy vault scheme is proposed | Security is proved in terms of information theoretic sense | Pre-aligned images required, Not able to handle biometric variance. |
| Davida et al. [17] | Iris | Canonical IrisCode is generated from multiple iris scans and bounded distance decoding error-correcting code is constructed. | Privacy protection is high | Rigorous security analysis missing, Error-correcting bits is stored in the database and thus prone to attacks |
| Manrose et al. [59] | Keystroke, Voice | Keystroke biometrics are secured via passwords | Extensive experimentation | Complex algorithm |
| Clancy et al. [15] | Fingerprint | Improvement over Juels & Sudan | Ability to handle biometric variance | Assumed prealignment of fingerprints. |
| Uludag and Jain [95] | Fingerprint | Rotational and translational representations of invariant minutiae has been proposed based on orientation field | Automatic alignment of query with respect to template using helper data | System is developed for scenario where subject is expected to be cooperative, quite unrealistic in real cases |
| Rathgeb et al. [75] | Iris | Error-correcting codes are generated by employing Reed–Solomon and Hadamard error codes | Generic framework for building iris-based biometric cryptosystems | Evaluated on a single iris dataset |
| Li and Hu [51] | Fingerprint | Alignment free fuzzy vault system. Here minutiae structures are encoded and transformed that enables better security and de-correlation. | Robust against non-linear distortions, revocable and non-linkability | Difficult to implement in terms of computational complexity |
| Liu and Zhao [52] | Fingerprint | $l_1$ minimisation-based error correction code (ECC) is used for matching minutia cylinder code (MCC) in encrypted domain | Non-linkability | Computationally expensive |

## 2.4.3 CANCELABLE BIOMETRICS

In the past ten years, lot of work has been carried out in the cancelable domain owing to the increase in online biometric-based authentication. Recently, a taxonomy of CB techniques [56] is proposed, which divides it into six major categories, as shown in Figure 2.3. The concept of CB was first coined by Ratha et al. [71] in the year 2001. Later he suggested [73] three varied transformation functions (cartesian, polar, and surface folding) on fingerprint images. The sole aim of these transformation functions was to distort original feature vectors such that it is computationally infeasible and difficult to retrieve original feature vectors. Later studies [24], however, demonstrate potential security threats in this scheme, but it opens up new avenues for researchers in the cancelable domain. Since then, several works have been carried out in this field.

Table 2.4 demonstrates cancelable key approaches on fingerprint trait. Apart from fingerprints, iris is one of the most used and recognised biometric modalities. Thus, securing iris is also as much important as fingerprints. In Table 2.5, promising cancelable iris techniques along with its pros and cons are discussed. Apart from that, few works have been carried out on studying cancelability on multimodal biometrics. Like Chin et al. [14] proposed a template protection technique by fusing fingerprints and palmprint features on the basis of the user-specific key. In another notable work, Barrero et al. [30] proposed a bloom filter-based approach for protected face, finger-vein, and iris features. Very recently, a random distance method-based template protection technique [42] is proposed for protecting multiple templates that include the face, palmprint, palm-vein, and finger-vein. Recently, deep features extracted from finger knuckle modality have been secured via BioHashing technique, but the proposed approach [83] is not able to maintain the inexorable security-performance trade-off. Apart from traditional biometric traits, electrocardiogram (ECG) is emerging as a promising biometric trait in many authentication and verification applications. In Ref. [19], the authors investigated cancelable ECG biometrics using BioHash. In Ref. [11], the authors obtained excellent identification performance on highly compressed ECG data using Hadamard transform, but this could not achieve non-invertibility. Also, the applications of the compressive sensing theory for ECG have been investigated for compression [16]. So far, there have been very few prior works on cancelable ECG biometrics that deal with the issue of performance deterioration induced due to cancelable schemes and validation for CB criteria.

### 2.4.3.1    Deep Learning-Based Cancelable Techniques

With recent advances in AI and deep learning, an array of biometric-based authentication systems demonstrate outrageous performance and present unique security and privacy concerns. One of the pioneering works in this domain is performed by Talreja et al. [90]. In their work, they have proposed a secure multi-biometric system that uses a deep neural network and error-correcting codes. They have proposed two architectures: (i) fully connected architecture and (ii) bi-linear architecture for generating cancelable templates. In another notable work [37], highly discriminative facial features are learned via deep learning-based frameworks, which are further hashed using SHA-3, a well-known cryptographic technique. In Ref. [82], the authors proposed a novel CNN Network (FDFNet) for the extraction of the discriminative finger dorsal features. Then, BioHashing was used to hash the features extracted from each finger dorsal. In Ref. [3], a cancelable multi-biometric face recognition method was presented in which multiple CNNs extracted deep features from the face, eyes, nose, and mouth regions. In Ref. [67], the authors incorporated a classic

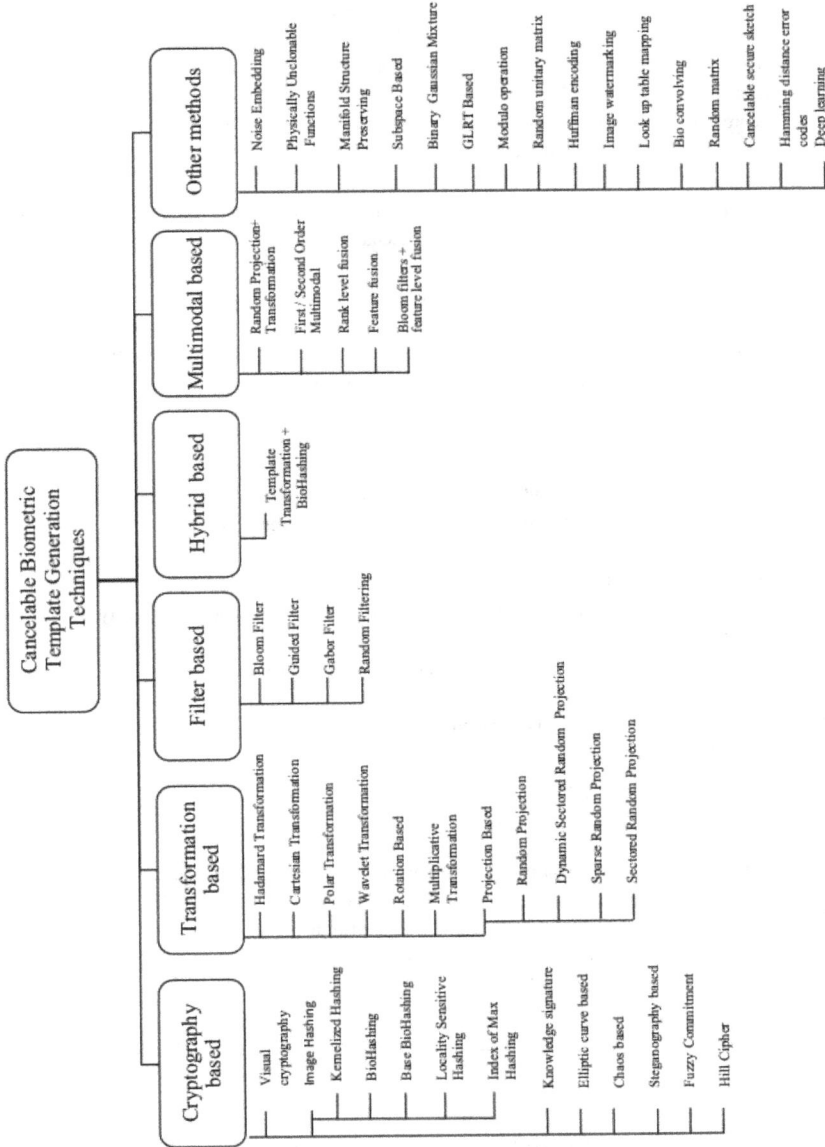

**FIGURE 2.3**    Taxonomy of cancelable biometric techniques. (Image is taken from [56].)

**TABLE 2.4**

**Key Cancelable Approaches on Fingerprints**

| Author | Description | Advantages | Shortcomings |
|---|---|---|---|
| Ratha et al. [73] | Pioneer work in this domain | Easy to implement | Pre-aligned images required |
| Jin et al. [33] | Random projection-based technique named BioHashing. It projects biometric feature to random space. By taking inner product of tokenised random vector with fingerprint features. | High performance | Performance degradation in stolen token scenario, prone to similarity based attack [21] |
| Lee et al. [49] | Rotational and translational invariant features are extracted from each minutiae | First alignment free cancelable template | Only theoretical justification of non-invertibility and revocability. No experimental validation. Unlinkability is not studied. |
| Ahn et al. [5] | Geometrical properties are explored to extract distinguish features from minutiae templates | Non-invertibility without loss of discriminative power, alignment free, low time complexity | Security analysis missing |
| Yang et al. [103] | Both local (distance, angle) and global (orientation, frequency) features of minutiae are explored to form non-invertible template | Non-invertible and unlinkable templates | Templates are unrevocable |
| Zhang et al. [106] | MCC was used for generating cancelable templates | Non-invertible, revocable, and unlinkable templates | Rigorous security analysis missing |
| Ferrara et al. [25] | KL transformation on MCC for generating cancelable templates named as P-MCC | | Non-revocable |
| Sandya and Prasad [80] | feature level fusion of fingerprint structures | ensures non-invertibility and revocability | Cross database attacks not studied |
| Arjona et al. [6] | A two factor fingerprint matching scheme that combines fingerprint identifier, i.e., protected MCC with device identifier, i.e., physically unclonable function generated from static random access memories | ensures discriminability, non-invertibility, revocability, and unlinkability | Only one dataset considered, i.e., FVC2002 |

deep learning approach into a BioCapsule-based facial authentication system to enhance recognition accuracy. In Ref. [105], a novel privacy-preserving finger-vein recognition system is developed based on binary decision diagram and multi-layer extreme learning machine paradigm. The proposed system ensures the safety of original finger-vein templates by ensuring non-invertibility and revocability.

## TABLE 2.5
## Key Cancelable Approaches on Iris

| Author | Description | Advantages | Shortcomings |
|---|---|---|---|
| Chin et al. [13] | Secure iris features coined as S-Iris encoding is proposed by iterating inner product between pseudo-random number and 1-D log Gabor iris features | Pioneer work | Security analysis missing |
| Zuo et al. [108] | Two salting-based approaches proposed named as GRAY SALT and BIN SALT | Irreversible | Deciding strength of noise pattern added to original iris template is quite challenging |
| Pillai et al. [68] | Cancelable iris template generation based on sectored random projections | sectored random projections was used for mitigating the performance degradation due to eyelids and eyelashes | Performance degradation in case of stolen token scenario [12] |
| Ouda et al. [66] | Bioencoding a template protection technique is proposed that extracts consistent bits from IrisCodes and further encoded by using randomly generated binary Codewords | Simple implementation and can be integrated with existing systems | Non-invertability is compromised when encoding factor is stolen [46]. |
| Rathgeb et al. [74] | Bloom filter-based cancelable iris template | High System Performance | Prone towards cross matching-based attacks [31] |
| Lai et al. [48] | Cancelable iris templates based on indexing first one hashing technique. The proposed framework is based on Min-Hashing and further strengthened by using modulo threshold function and P-order Hadamard product | Rigorous security analysis | testing on single dataset CASIA-V3 iris dataset |
| Umer et al. [96] | To improve the security of existing BioHashing technique two different tokens were used (i) User dependent (ii) User independent for generating cancelable templates | Evaluated on extensive dataset | Unlinkability is not studied |
| Sadhya and Raman [78] | Locality Sensitive Hashing is used for generating cancelable iris codes coined as Locality Sampled Codes | Extensive experimentation and security analysis | Genetic algorithm-based similarity attack not evaluated [21] |

## 2.4.3.2 Deep Learning versus Non-deep Learning Cancelable Techniques

With the usage of deep learning techniques in the cancelable domain, biometric feature extraction becomes less time consuming as compared to traditional feature extraction methods. Naive feature extraction methods often require pre-processing and parameter tuning according to the dataset in consideration, while a generalised trained deep model works quite well with different datasets (that belongs to the same biometric modality) without much tuning. On the other hand, non-deep learning-based techniques can even work on small dataset in the resource-constrained environment, but for training deep networks, large computational capability and the large dataset are required. It should be noted that usage of deep learning techniques in generating cancelable templates is in the infancy state as not much work has been carried out so far in this domain. Thus, deciding the supremacy between the two techniques at this point is not fair without harnessing the full advantages of deep learning-based techniques in generating cancelable templates.

## 2.5   PERFORMANCE MEASURES AND DATASETS IN CANCELABLE BIOMETRICS

In the case of CB, a PBI is generated by employing some transformation function and user-specific key. Here, matching is always performed in the transformed domain. As we know, for designing an effective CB scheme, four fundamental requirements, i.e., non-invertibility, revocability, unlinkability, and system usability, need to be addressed simultaneously. Thus, for measuring the performance of CB-based techniques, the four above-mentioned requirements need to be quantitatively assessed. It should be noted that in the literature, several CB-based techniques have been proposed, but not much work has been carried out in proposing metrics for quantitative assessment.

### 2.5.1   PERFORMANCE MEASURES FOR NON-INVERTIBILITY ANALYSIS

For the quantitative assessment of non-invertibility conditional Shannon, entropy can be used. But in the case of CB due to the generation of PBI, it is difficult to quantify Shannon entropy directly, thus to measure non-invertibility, the authors in Ref. [76] have proposed to study several attacks like zero effort attack, exhaustive attack, stolen biometric attack, stolen token attack, and worst-case attack. The decision results for any CB-based system can be defined as follows:

$$R_x = P(D_T(A_x, A_{xx})) \leq \epsilon \tag{2.1}$$

where $D_T$ stands for distance function in the transformed domain. $A_x$ represents the CB template of user $x$ [combination of feature vector say $f_x$ and $k_x$ (secret key or transformation)] generated during enrollment, while $A_{xx}$ (combination of query feature vector say $f_x$ and $k_x$) represents the CB template of user $x$ generated during authentication. Here, $R_x$ represents the decision result, and $E$ is a decision threshold chosen by the user. Zero effort attack is quantified as follows:

$$R_x = P(D_T(A_x, A_y)) \leq \epsilon \tag{2.2}$$

In this case, during authentication, imposter makes no effort and presents his own biometrics ($A_y$) to the system. While in case of brute force attack, imposter tries different random values of his own biometrics ($A_y$) with an intention that somehow $A_y$

matches with $A_x$. Other forms of attacks, i.e., stolen biometric, stolen token attack, and worst-case attack, are more serious forms of attack. Here, in these cases, imposter somehow got access either to the genuine user feature vector ($f_x$) or either to the genuine user transformation parameters ($k_x$) or in extreme case, i.e., the worst-case attack can get access to both $f_x$ as well as $k_x$.

### 2.5.2 Performance Measures for Unlinkability Analysis

Linkage across different databases can disclose different pieces of information about an individual and thus can allow an adversary attack by consolidating information. Thus, it is necessary to ensure unlinkability across biometric templates stored in different databases. Recently, two measures, a local and a global [29], have been proposed to quantify unlinkability.

a. **[local]** $D \leftrightarrow (s) \in [0,1]$: This metric depends upon the likelihood ratio between the mated (probe and gallery that belong to the same subject but transformed using different keys) and non-mated (probe and gallery that belong to different subjects and transformed using different keys) score distributions to evaluate the local linkability of a system at each score. In this measure, value of $D \leftrightarrow (s) = 0$ signifies "high" unlinkability, while $D \leftrightarrow (s) = 1$ signifies "low" unlinkability at score s.

b. **[global]** $D^{sys} \leftrightarrow \in [0,1]$: This metric is independent of the individual score, and it measures the global linkability of the entire system. In this measure, value of $D^{sys} \leftrightarrow = 0$ indicates "high" unlinkability, while $D^{sys} \leftrightarrow = 1$ indicates "low" unlinkability and defined as follows:

$$D^{sys} \leftrightarrow = \int p(s/H_m)D \leftrightarrow (s)ds \qquad (2.3)$$

where $p(s/H_m)$ indicates the score generated from distribution of mated samples.

### 2.5.3 Performance Measures for System Usability Analysis

Ensuring the usability of the system is a functional attribute. Measures for quantifying this attribute are the same as used for quantifying the performance of any traditional biometric system. These measures are mainly classified into two subcategories: (i) performance measures for verification and (ii) performance measures for identification. Mainly FAR, FRR) equal error rate (EER), and decidability index (DI) are used as metrics in the cancelable verification domain. These terms are described below.

**FAR**: It specifies how many unauthorised persons get access to the system. It is defined as follows:

$$FAR = \frac{\text{Number of unauthorized access granted by system}}{\text{Total number of identification attempts}} \qquad (2.4)$$

**FRR**: It specifies how many authorised persons are denied access by the system. It is defined as follows:

$$FAR = \frac{\text{Number of authorized identities denied by system}}{\text{Total number of identification attempts}} \quad (2.5)$$

**EER**: It is a point at which the FRR value is equal to FAR. Lower value of EER depicts superiority of the biometric system.

**DI**: This measure gives the separability between imposter and genuine score distributions, respectively. It is defined as follows:

$$DI = \frac{\mu_g - \mu_i}{\sqrt{((\sigma_g^2 - \sigma_i^2))^2}} \quad (2.6)$$

where $\mu_g$, $\mu_{im}$, $\sigma_g^2$, $\sigma_{im}^2$ are the mean and variances of genuine and imposter distributions. Apart from these regular verification metrics for measuring the system usability once, the biometric template is transformed into a metric, as recently proposed [76], which is defined as follows:

$$A_1 = \frac{AUC(FAR_T, FRR_T)}{AUC(FAR_O, FRR_O)} \quad (2.7)$$

where $FAR_T$ and $FRR_T$ in the numerator term represent the FAR and FRR of the transformed template, while $FAR_O$ and $FRR_O$ in the denominator term denote the FAR and FRR of original biometric templates. This metric measures the ratio of the receiver operating characteristic curve. Here, $A_1 = 1$ indicates ideal (perfect) scenario, while negative value of $A_1$ indicates deteriorating performance.

**Correct recognition rate (CRR)** is another a commonly used metric for assessing the CB identification performance. It measures the percentage of the match rate, and it is defined as follows:

$$CRR = \frac{\text{Number of correctly matched images}}{\text{Total number of images in the database}} \quad (2.8)$$

### 2.5.4  PERFORMANCE MEASURES FOR REVOCABILITY ANALYSIS

For ensuring the revocability, as suggested in Ref. [23], a distribution curve between the imposter and pseudo imposter distribution is drawn. The claim of revocability is preserved when the $\mu_{pseudo}$ (mean) and $var_{pseudo}$ (variance) of pseudo imposter is close to $\mu_{im}$ and $var_{im}$ of imposter and far from $\mu_g$ and $var_g$ of genuine distribution.

### 2.5.5  DATABASES USED IN CANCELABLE BIOMETRICS

Most of the work in the cancelable domain is mainly concentrated over three popular biometric traits, i.e., face, iris, and fingerprints. It is worth mentioning that in the cancelable domain, there is no standard protocol defined for training and

**TABLE 2.6**
**Key Databases Used in Cancelable Biometrics**

| Biometric Trait | Database | No of Subjects | Remarks |
|---|---|---|---|
| Face | CMU-PIE | 68 | |
| | FERET | 1199 | Largest and the most challenging dataset collected over 15 sessions |
| | AR | 126 | |
| | FEI | 200 | |
| Iris | CASIA IRS V3 | 396 | Commonly used cancelable iris dataset |
| | MMU1 DATASET | 100 | |
| | IITD | 224 | |
| | ND IRIS 0405 | 356 | |
| Fingerprint | FVC2002 DB-1,2,3,4 | 110 | Small dataset and not much challenging |

testing images. As a result, a different number of training and testing images are used by researchers in various works [48,96]. Table 2.6 illustrates key cancelable databases in the literature along with their advantages and limitations. It should be noted that most of the work in the cancelable domain has been conducted on small datasets despite the availability of large datasets, particularly in the face domain like MS-Celeb and FaceNet. Developing CB techniques for voluminous challenging datasets that can represent real-world population is the current need of the hour.

## 2.6 COMPARATIVE PERFORMANCE ANALYSIS: CANCELABLE BIOMETRICS

In this section, a comparative performance between well-known state-of-the-art studies has been made. For this, the ideal fingerprint and iris biometric traits under the same testing protocols and evaluation parameters have been selected.

**Fingerprint Cancelable Biometrics**: As a common thumb rule in CB, the matching of biometric traits takes place in the transformed domain so that original information can be protected. Thus, we consider those fingerprint studies only which performed matching in the transformed domain and used FVC datasets. Each of FVC dataset contains 800 fingerprint images with eight images per finger. To make a better understanding, four state-of-the-art fingerprint approaches [35,36,43,100] are considered. The results given in Table 2.7 reveal that ranking-based hashing approaches are a better choice in the cancelable domain for fingerprint images.

**Iris Cancelable Biometrics**: In this case, efforts are made to make a comparison between CNN-based approaches and traditional cancelable approaches. The results were evaluated on standard datasets like MMU, IITD, CASIA, and UBIRIS iris datasets, and the results are tabulated in Table 2.8. It can be observed that deep learning-based cancelable approaches are a better choice in the cancelable domain

**TABLE 2.7**
**Comparative Analysis of Cancelable Biometrics over Fingerprint Datasets**

| Approaches | Database | Accuracy (CRR %) | Accuracy (EER %) |
|---|---|---|---|
| RGHE [36] | FVC2000DB1 | 99.22 | 1 |
| RGHE [36] | FVC2000DB2 | 100 | 0.5 |
| PHT [100] | FVC2000DB1 | - | 1 |
| PHT [100] | FVC2000DB2 | - | 2 |
| PHT [100] | FVC2004DB2 | - | 13.2 |
| PR-NNLS [43] | FVC2000DB1 | 98.34 | 2.48 |
| PR-NNLS [43] | FVC2000DB2 | 97.01 | 1.51 |
| PR-NNLS [43] | FVC2004DB2 | 96.34 | 7.44 |
| URP-IOM [35] | FVC2000DB1 | - | 0.20 |
| URP-IOM [35] | FVC2000DB2 | - | 0.88 |
| URP-IOM [35] | FVC2004DB2 | - | 3.08 |

**TABLE 2.8**
**Comparative Analysis of Cancelable Biometrics over Iris Datasets**

| Approaches | Database | Accuracy (CRR %) | Accuracy (EER %) |
|---|---|---|---|
| LSC [78] | IITD | - | 1.4 |
| $f_{iris}$ [96] | IITD | 100 | 0.008 |
| $f_{iris}$ [96] | MMU | 100 | 0.006 |
| CNN-RP [87] | IITD | 98.66 | 0.12 |
| CNN-RP [87] | MMU | 95.57 | 0.15 |
| RD [77] | CASIAv3-I | - | 0.42 |
| RD [77] | CASIAv4-T | - | 2.07 |
| Morton filter [69] | IITD | - | $\approx 0$ |
| Morton filter [69] | CASIA-V4 | - | $\approx 0$ |

for fingerprint images. However, it is important to note that deep learning architecture in the above-mentioned studies was only utilised for iris feature extraction. There is no end-to-end deep learning architectures that are retrieved from the literature on cancelable iris biometrics.

## 2.7 CONCLUSIONS AND FUTURE PROSPECTIVE OF DEEP LEARNING IN BIOMETRICS

In order to leverage true benefits of deep learning in the biometric domain, voluminous challenging datasets are needed that can represent real-world scenarios. Currently available biometric datasets – although some of them contain a large number of images like MS-Celeb, FaceNet, and WebFaces – are far from representing the true world population. Another point of concern is that biometric-based models should be designed in such a manner that can be implemented in real-world situations at a reasonable cost with high computational speed. As we know, high computational cost is associated with deep neural networks. Thus, selecting the right kind of deep learning-based architecture with high accuracy and low computational cost is a challenging issue in the biometric domain. A possible solution is to design a network with comparable efficiency but with a less computational cost like the once designed by Felix et al. [39]. Here, an effective surrogate convolutional layer based on the domain knowledge is designed that affords sufficient parameter saving as compared to the traditional standard convolutional layer and thus can be deployed in a resource-constrained environment also.

With large-scale deployments, unimodal biometric systems often suffer from challenging issues like intra-class variations, non-universality, and many more. In such scenarios, multimodal biometric systems serve as a rescue measure that integrates multiple biometric modalities and thus helps in improving the recognition rate. It should be noted that some work has been carried out in biometric fusion using machine learning techniques, but it suffers from challenging issues, as depicted in Ref. [7]. Currently, a deep learning-based architecture that can amalgamate feature representation and aggregation from multiple biometric traits simultaneously is needed. Moreover, in biometrics, security is a prime concern in order to gain public trust and confidence in it. Mainly biometric samples need to be protected from adversarial attacks, template attacks, and presentation attacks.

## REFERENCES

1. European council, regulation of the European parliament and of the council on the protection of individuals. https://eur-lex.europa.eu/legal-content/EN/TXT/?uri=CELEX%3A52012PC0011.
2. Privaris Inc. http://www.privaris.com/.
3. E. Abdellatef, N. A. Ismail, S. E. S. A. Elrahman, K. N. Ismail, M. Rihan, and F. E. A. El-Samie. Cancelable multi-biometric recognition system based on deep learning. *The Visual Computer*, 36:1097–1109, 2020.
4. S. Ahmad and B. Fuller. Thirdeye: Triplet based iris recognition without normalization. *CoRR*, abs/1907.06147, 2019.
5. D. Ahn, S. G. Kong, Y.-S. Chung, and K. Y. Moon. Matching with secure fingerprint templates using non-invertible transform. In *Proceedings of the Congress on Image and Signal Processing (CISP)*, Sanya, China, pp. 29–33, 2008.
6. R. Arjona, M. Á. Prada-Delgado, I. Baturone, and A. Ross. Securing minutia cylinder codes for fingerprints through physically unclonable functions: An exploratory study. In *International Conference on Biometrics, ICB*, Gold Coast, Australia, pp. 54–60, 2018.
7. T. Baltrusaitis, C. Ahuja, and L. Morency. Multimodal machine learning: A survey and taxonomy. *IEEE Transactions on Pattern Analysis and Machine Intelligence*, 41(2):423–443, 2019.
8. N. Bansal. *Enhanced Rsa Key Generation Modeling Using Fingerprint Biometric*. PhD thesis, NIT Jamshedpur, 2018.
9. M. Baqar, A. Ghani, A. Aftab, S. Arbab, and S. Yasin. Deep belief networks for iris recognition based on contour detection. In *International Conference on Open Source Systems and Technologies (ICOSST)*, IEEE, pp. 72–77, 2016.
10. V. E. Brindha and A. M. Natarajan. Multi-modal biometric template security: Fingerprint and palmprint based fuzzy vault. *Biometrics and Biostatistics International Journal*, 3(150):1–6, 2012.

11. C. Camara, P. Peris-Lopez, and J. E. Tapiador. Human identification using compressed ECG signals. *Journal of Medical Systems*, 39(11):148, 2015.

12. K. H. Cheung, A. W. Kong, J. You, and D. Zhang. An analysis on invertibility of cancelable biometrics based on biohashing. In *Proceedings of the International Conference on Imaging Science, Systems, and Technology: Computer Graphics, CISST*, Las Vegas, NV, pp. 40–45, 2005.

13. C. S. Chin, A. T. B. Jin, and D. N. C. Ling. High security iris verification system based on random secret integration. *Computer Vision and Image Understanding*, 102(2): 169–177, 2006.

14. Y. J. Chin, T. S. Ong, A. B. J. Teoh, and M. O. M. Goh. Integrated biometrics template protection technique based on fingerprint and palmprint feature-level fusion. *Information Fusion*, 18:161–174, 2014.

15. T. C. Clancy, N. Kiyavash, and D. J. Lin. Secure smartcard-based fingerprint authentication. In *Proceedings of ACM Multimedia, Biometrics Methods and Applications Workshop*, pp. 45–52, 2003.

16. D. Craven, B. McGinley, L. Kilmartin, M. Glavin, and E. Jones. Adaptive dictionary reconstruction for compressed sensing of ECG signals. *IEEE Journal of Biomedical and Health Informatics*, 21(3):645–654, 2016.

17. G. I. Davida, Y. Frankel, and B. J. Matt. On enabling secure applications through off-line biometric identification. In *Security and Privacy Symposium on Security and Privacy*, Oakland, CA, pp. 148–157, 1998.

18. J. Deng, J. Guo, N. Xue, and S. Zafeiriou. Arcface: Additive angular margin loss for deep face recognition. In *IEEE Conference on Computer Vision and Pattern Recognition, CVPR*, Long beach, CA, pp. 4690–4699, 2019.

19. M. Dey, N. Dey, S. K. Mahata, S. Chakraborty, S. Acharjee, and A. Das. Electrocardiogram feature based inter-human biometric authentication system. In *2014 International Conference on Electronic Systems, Signal Processing and Computing Technologies*, IEEE, pp. 300–304, 2014.

20. Y. Dodis, L. Ostrovsky, L. Reyzin, and A. D. Smith. Fuzzy extractors: How to generate strong keys from biometrics and other noisy data. *SIAM Journal on Computing*, 38(1):97–139.

21. X. Dong, Z. Jin, and A. T. B. Jin. A genetic algorithm enabled similarity-based attack on cancellable biometrics. *CoRR*, abs/1905.03021:1–7, 2019.

22. Y. Duan, J. Lu, and J. Zhou. Uniformface: Learning deep equidistributed representation for face recognition. In *IEEE Conference on Computer Vision and Pattern Recognition, CVPR*, Long beach, CA, pp. 3415–3424, 2019.

23. R. Dwivedi, S. Dey, R. Singh, and A. Prasad. A privacy-preserving cancelable iris template generation scheme using decimal encoding and look-up table mapping. *Computers and Security*, 65:373–386, 2017.

24. Q. Feng, F. Su, A. Cai, and F. Zhao. Cracking cancelable fingerprint template of ratha. In *International Symposium on Computer Science and Computational Technology, ISCSCT*, Shanghai, China, pp. 572–575, 2008.

25. M. Ferrara, D. Maltoni, and R. Cappelli. Noninvertible minutia cylinder-code representation. *IEEE Transactions on Information Forensics and Security*, 7(6):1727–1737, 2012.

26. J. Galbally Herrero, J. Fierrez, and J. Ortega-Garcia. Vulnerabilities inbiometric systems: Attacks and recent advances in liveness detection. *Database*, 1(3):1–82, 2007.

27. A. K. Gangwar and A. Joshi. DeepIrisNet: Deep iris representation with applications in iris recognition and cross-sensor iris recognition. In *IEEE International Conference on Image Processing, ICIP*, pp. 2301–2305. IEEE, 2016.

28. A. Goh and D. N. C. Ling. Computation of cryptographic keys from face biometrics. In *Communications and Multimedia Security Advanced Techniques for Network and Data Protection, IFIP*, pp. 1–13, 2003.

29. M. Gomez-Barrero, J. Galbally, C. Rathgeb, and C. Busch. General framework to evaluate unlinkability in biometric template protection systems. *IEEE Transactions on Information Forensics and Security*, 13(6):1406–1420, 2018.

30. M. Gomez-Barrero, C. Rathgeb, G. Li, R. Raghavendra, J. Galbally, and C. Busch. Multi-biometric template protection based on bloom filters. *Information Fusion*, 42: 37–50, 2018.

31. J. Hermans, B. Mennink, and R. Peeters. When a bloom filter is a doom filter: Security assessment of a novel iris biometric template protection system. In *BIOSIG Proceedings of the 13th International Conference of the Biometrics Special Interest Group*, Darmstadt, Germany, pp. 63–74, 2014.

32. A. K. Jain, A. Ross, and U. Uludag. Biometric template security: Challenges and solutions. In *13th European Signal Processing Conference, EUSIPCO*, Antalya, Turkey, pp. 1–4, 2005.

33. A. T. B. Jin, D. N. C. Ling, and A. Goh. Biohashing: Two factor authentication featuring fingerprint data and tokenised random number. *Pattern Recognition*, 37(11):2245–2255, 2004.

34. A. T. B. Jin, D. N. C. Ling, and A. Goh. Personalised cryptographic key generation based on facehashing. *Computers & Security*, 23(7):606–614, 2004.

35. Z. Jin, J. Y. Hwang, Y.-L. Lai, S. Kim, and A. B. J. Teoh. Ranking-based locality sensitive hashing-enabled cancelable biometrics: Index-of-max hashing. *IEEE Transactions on Information Forensics and Security*, 13(2):393–407, 2017.

36. Z. Jin, M.-H. Lim, A. B. J. Teoh, and B.-M. Goi. A non-invertible randomized graph-based hamming embedding for generating cancelable fingerprint template. *Pattern Recognition Letters*, 42:137–147, 2014.

37. A. K. Jindal, S. Chalamala, and S. K. Jami. Face template protection using deep convolutional neural network. In *Proceedings of the IEEE Conference on Computer Vision and Pattern Recognition Workshops*, Salt Lake, UT, pp. 462–470, 2018.

38. M. Joshi, B. Mazumdar, and S. Dey. Security vulnerabilities against fingerprint biometric system. *CoRR*, abs/1805.07116, 2018.

39. F. Juefei-Xu, V. N. Boddeti, and M. Savvides. Local binary convolutional neural networks. In *IEEE Conference on Computer Vision and Pattern Recognition, CVPR*, Honolulu, HI, pp. 4284–4293, 2017.

40. A. Juels and M. Sudan. A fuzzy vault scheme. *Designs Codes Cryptography*, 38(2): 237–257, 2006.

41. A. Juels and M. Wattenberg. A fuzzy commitment scheme. In *Proceedings of the 6th ACM Conference on Computer and Communications Security*, Singapore, pp. 28–36, 1999.

42. H. Kaur and P. Khanna. Random distance method for generating unimodal and multimodal cancelable biometric features. *IEEE Transactions on Information Forensics and Security*, 14(3):709–719, 2019.

43. J. B. Kho, J. Kim, I.-J. Kim, and A. B. Teoh. Cancelable fingerprint template design with randomized non-negative least squares. *Pattern Recognition*, 91:245–260, 2019.

44. A. Kholmatov and B. A. Yanikoglu. Biometric cryptosystem using online signatures. In *Computer and Information Sciences ISCIS, 21th International Symposium*, Istanbul, Turkey, pp. 981–990, 2006.

45. S. Kim, B. Park, B. S. Song, and S. Yang. Deep belief network based statistical feature learning for fingerprint liveness detection. *Pattern Recognition Letters*, 77:58–65, 2016.

46. P. Lacharme. Analysis of the iriscode bioencoding scheme. *International Journal of Computer Science and Security (IJCSS)*, 6(5):315–321, 2012.

47. Y. Lai, J. Y. Hwang, Z. Jin, S. Kim, S. Cho, and A. B. J. Teoh. Symmetric keyring encryption scheme for biometric cryptosystem. *Information Sciences*, 502:492–509, 2019.

48. Y. Lai, Z. Jin, A. B. J. Teoh, B. Goi, W. Yap, T. Chai, and C. Rathgeb. Cancellable iris template generation based on indexing-first-One hashing. *Pattern Recognition*, 64: 105–117, 2017.

49. C. Lee, J. Choi, K. Toh, and S. Lee. Alignment-free cancelable fingerprint templates based on local minutiae information. *IEEE Transactions on systems, Man and Cybernetics*, 37(4):980–992, 2007.

50. M. B. Lee, Y. H. Kim, and K. R. Park. Conditional generative adversarial network-based data augmentation for enhancement of iris recognition accuracy. *IEEE Access*, 7:122134–122152, 2019.

51. C. Li and J. Hu. A security-enhanced alignment-free fuzzy vault-based fingerprint cryptosystem using pair-polar minutiae structures. *IEEE Transactions on Information Forensics and Security*, 11(3):543–555, 2016.

52. E. Liu and Q. Zhao. Encrypted domain matching of fingerprint minutia cylinder-code (MCC) with l1 minimization. *Neurocomputing*, 259:3–13, 2017.

53. H. Liu, X. Zhu, Z. Lei, and S. Z. Li. Adaptiveface: Adaptive margin and sampling for face recognition. In *IEEE Conference on Computer Vision and Pattern Recognition, CVPR*, Long beach, CA, pp. 11947–11956, 2019.

54. W. Liu, Y. Wen, Z. Yu, M. Li, B. Raj, and L. Song. Sphereface: Deep hypersphere embedding for face recognition. In *IEEE Conference on Computer Vision and Pattern Recognition, CVPR*, Honolulu, HI, pp. 6738–6746, 2017.

55. W. Liu, Y. Wen, Z. Yu, and M. Yang. Large-margin softmax loss for convolutional neural networks. In *Proceedings of the 33nd International Conference on Machine Learning, ICML*, New York City, NY, pp. 507–516, 2016.

56. Manisha and N. Kumar. Cancelable biometrics: A comprehensive survey. *Artificial Intelligence Review*, 53:1–19, 2019.

57. S. Minaee and A. Abdolrashidi. Deepiris: Iris recognition using a deep learning approach. *CoRR*, abs/1907.09380, 2019.

58. S. Minaee, A. Abdolrashidi, and Y. Wang. An experimental study of deep convolutional features for iris recognition. *CoRR*, abs/1702.01334, 2017.

59. F. Monrose, M. K. Reiter, and S. Wetzel. Password hardening based on keystroke dynamics. *International Journal of Information Security*, 1(2):69–83, 2002.

60. E. Mordini and S. Massari. Body, biometrics and identity. *Bioethics*, 22(9):488–498, 2008.

61. A. Nagar. *Biometric Template Security*. PhD thesis, Michigan State University, 2012.

62. A. Nagar, K. Nandakumar, and A. K. Jain. Multibiometric cryptosystems based on feature-level fusion. *IEEE Transactions on Information Forensics and Security*, 7(1):255–268, 2012.

63. T. A. T. Nguyen and T. K. Dang. Combining fuzzy extractor in biometric-kerberos based authentication protocol. In *International Conference on Advanced Computing and Applications, ACOMP*, Ho Chi Minh City, Vietnam, pp. 1–6, 2015.

64. T. A. T. Nguyen, T. K. Dang, Q. C. Truong, and D. T. Nguyen. Secure biometric-based remote authentication protocol using Chebyshev polynomials and fuzzy extractor. *CoRR*, abs/1904.04710, 2019.

65. R. F. Nogueira, R. de Alencar Lotufo, and R. C. Machado. Fingerprint liveness detection using convolutional neural networks. *IEEE Transactions on Information Forensics and Security*, 11(6):1206–1213, 2016.

66. O. Ouda, N. Tsumura, and T. Nakaguchi. Tokenless cancelable biometrics scheme for protecting iris codes. In *International Conference on Pattern Recognition, ICPR*, Istanbul, Turkey, pp. 882–885, 2010.

67. T. Phillips, X. Zou, F. Li, and N. Li. Enhancing biometric-capsule-based authentication and facial recognition via deep learning. In *Proceedings of the 24th ACM Symposium on Access Control Models and Technologies*, ACM, Toronto, Canada, pp. 141–146, 2019.

68. J. K. Pillai, V. M. Patel, R. Chellappa, and N. K. Ratha. Sectored random projections for cancelable iris biometrics. In *Proceedings of the IEEE International Conference on Acoustics, Speech, and Signal Processing, ICASSP*, Dallas, TX, pp. 1838–1841, 2010.

69. K. B. Raja, R. Raghavendra, S. Venkatesh, and C. Busch. Morton filters for superior template protection for iris recognition. *CoRR*, abs/2001.05290, 2020.

70. K. B. Raja, R. Raghavendra, S. Venkatesh, M. Gomez-Barrero, C. Rathgeb, and C. Busch. A study of hand-crafted and naturally learned features for fingerprint presentation attack detection. In *Handbook of Biometric Anti-Spoofing – Presentation Attack Detection*, Second Edition, Part of the Advances in Computer Vision and Pattern Recognition book series (ACVPR), Springer, pp. 33–48, 2019.

71. N. K. Ratha, S. Chikkerur, J. H. Connell, and R. M. Bolle. Generating cancelable fingerprint templates. *IEEE Transactions on Pattern Analysis and Machine Intelligence*, 29(4):561–572, 2007.

72. N. K. Ratha, J. H. Connell, and R. M. Bolle. An analysis of minutiae matching strength. In *Audio- and Video-Based Biometric Person Authentication, Third International Conference, AVBPA*, Halmstad, Sweden, pp. 223–228, 2001.

73. N. K. Ratha, J. H. Connell, and R. M. Bolle. Enhancing security and privacy in biometrics-based authentication systems. *IBM Systems Journal*, 40(3):614–634, 2001.

74. C. Rathgeb, F. Breitinger, and C. Busch. Alignment-free cancelable iris biometric templates based on adaptive bloom filters. In *International Conference on Biometrics, ICB*, Madrid, Spain, pp. 1–8, 2013.

75. C. Rathgeb and A. Uhl. Systematic construction of iris-based fuzzy commitment schemes. In *Advances in Biometrics, Third International Conference on Biometrics, ICB*, Alghero, Italy, pp. 940–949, 2009.

76. C. Rosenberger. Evaluation of biometric template protection schemes based on a transformation. In *Proceedings of the 4th International Conference on Information Systems Security and Privacy, ICISSP*, Funchal, pp. 216–224, 2018.

77. D. Sadhya, K. De, B. Raman, and P. P. Roy. Efficient extraction of consistent bit locations from binarized iris features. *Expert Systems with Applications*, 140:112884, 2020.

78. D. Sadhya and B. Raman. Generation of cancelable iris templates via randomized bit sampling. *IEEE Transactions on Information Forensics and Security*, 14(11): 2972–2986, 2019.

79. A. Sahai and B. Waters. Fuzzy identity based encryption. *In Advances in Cryptology - Eurocrypt, Springer*, 3494:457–473, 2005.

80. M. Sandhya and M. V. N. K. Prasad. Securing fingerprint templates using fused structures. *IET Biometrics*, 6(3):173–182, 2017.

81. H. Shao and D. Zhong. Few-shot palmprint recognition via graph neural networks. *Electronics Letters*, 55(16):890–892, 2019.

82. A. Singh, A. Arora, G. Jaswal, and A. Nigam. Comprehensive survey on cancelable biometrics with novel case study on finger dorsal template protection. *Journal of Banking and Financial Technology*, pp. 1–27, 2020.

83. A. Singh, A. Arora, S. H. Patel, G. Jaswal, and A. Nigam. FDFnet: A secure cancelable deep finger dorsal template generation network secured via. bio-hashing. In *5th IEEE International Conference on Identity, Security, and Behavior Analysis, ISBA*, Hyderabad, India, pp. 1–9, 2019.

84. A. Singh and A. Nigam. Effect of identity mapping, transfer learning and domain knowledge on the robustness and generalization ability of a network: A biometric based case study. *Ambient Intelligence and Humanized Computing, Springer*, 1868–5145: 1–18, 2019.

85. C. Soutar, D. Roberge, A. Stoianov, R. Gilroy, and V. K. Bhagavatula. Biometric encryption using image processing. *Proceedings of SPIE, Optical Security and Counterfeit Deterrence Techniques II*, 3314:178–188, 1998.

86. B. Stojanović, O. Marques, A. Nešković, and S. Puzović. Fingerprint ROI segmentation based on deep learning. In *2016 24th Telecommunications Forum (TELFOR)*, Sava Center, Serbia, pp. 1–4, 2016.

87. T. Sudhakar and M. Gavrilova. Multi-instance cancelable biometric system using convolutional neural network. In *International Conference on Cyberworlds (CW)*, IEEE, pp. 287–294, 2019.

88. V. Sujitha and D. Chitra. A novel technique for multi biometric cryptosystem using fuzzy vault. *Journal of Medical Systems*, 43(5):1–9, 2019.

89. N. H. Sultan, F. A. Barbhuiya, and N. Sarma. Pairvoting: A secure online voting scheme using pairing-based cryptography and fuzzy extractor. In *IEEE International Conference on Advanced Networks and Telecommunications Systems, ANTS*, Kolkata, India, pp. 1–6, 2015.

90. V. Talreja, M. C. Valenti, and N. M. Nasrabadi. Multibiometric secure system based on deep learning. In *IEEE Global Conference on Signal and Information Processing, GlobalSIP*, IEEE, Montreal, Canada, pp. 298–302, 2017.

91. Y. Tang, F. Gao, J. Feng, and Y. Liu. Fingernet: An unified deep network for fingerprint minutiae extraction. In *IEEE International Joint Conference on Biometrics, IJCB*, Denver, CO, pp. 108–116, 2017.

92. K. N. Thanh, C. Fookes, A. Ross, and S. Sridharan. Iris recognition with off-the-shelf CNN features: A deep learning perspective. *IEEE Access*, 6:18848–18855, 2018.

93. R. Tolosana, M. Gomez-Barrero, C. Busch, and J. Ortega-Garcia. Biometric presentation attack detection: Beyond the visible spectrum. *IEEE Transactions on Information Forensics and Security*, 15:1261–1275, 2020.

94. G. J. Tomko, C. Soutar, and G. J. Schmidt. Fingerprint controlled public key cryptographic system, U.S. Patent 5541994, 1996.

95. U. Uludag and A. K. Jain. Securing fingerprint template: Fuzzy vault with helper data. In *IEEE Conference on Computer Vision and Pattern Recognition, CVPR Workshops*, New York, NY, pp. 163–168, 2006.

96. S. Umer, B. C. Dhara, and B. Chanda. A novel cancelable iris recognition system based on feature learning techniques. *Information Sciences*, 406:102–118, 2017.

97. C. Vielhauer and R. Steinmetz. Handwriting: Feature correlation analysis for biometric hashes. *EURASIP Journal on Advances in Signal Processing*, 2004(4):542–558, 2004.

98. H. Wang, Y. Wang, Z. Zhou, X. Ji, D. Gong, J. Zhou, Z. Li, and W. Liu. Cosface: Large margin cosine loss for deep face recognition. In *IEEE Conference on Computer Vision and Pattern Recognition, CVPR*, Salt Lake City, UT, pp. 5265–5274, 2018.

99. M. Wang and W. Deng. Deep face recognition: A survey. *CoRR*, abs/1804.06655, 2018.

100. S. Wang, G. Deng, and J. Hu. A partial hadamard transform approach to the design of cancelable fingerprint templates containing binary biometric representations. *Pattern Recognition*, 61:447–458, 2017.

101. X. Wu, N. Qi, K. Wang, and D. Zhang. A novel cryptosystem based on iris key generation. In *Fourth International Conference on Natural Computation*, Washington, DC, pp. 53–56, 2008.

102. K. Xi, J. Hu, and F. Han. An alignment free fingerprint fuzzy extractor using near-equivalent dual layer structure check (NeDLSC) algorithm. In *6th IEEE Conference on Industrial Electronics and Applications*, Beijing, China, pp. 1040–1045, 2011.

103. H. Yang, X. Jiang, and A. C. Kot. Generating secure cancelable fingerprint templates using local and global features. In *Proceedings of the 2nd IEEE International Conference on Computer Science and Information Technology (ICCSIT)*, Beijing, China, pp. 645–649, 2009.

104. W. Yang, J. Hu, and S. Wang. A delaunay triangle-based fuzzy extractor for fingerprint authentication. In *IEEE International Conference on Trust, Security and Privacy in Computing and Communications*, Liverpool, UK, pp. 66–70, 2012.

105. W. Yang, S. Wang, J. Hu, G. Zheng, J. Yang, and C. Valli. Securing deep learning based edge finger-vein biometrics with binary decision diagram. *IEEE Transactions on Industrial Informatics*, 15(7):4244–4253, 2019.

106. N. Zhang, X. Yang, Y. Zang, X. Jia, and J. Tian. Generating registration-free cancel-able fingerprint templates based on minutia cylinder-code representation. In *IEEE Sixth International Conference on Biometrics: Theory, Applications and Systems, BTAS*, Arlington, VA, pp. 1–6, 2013.

107. Z. Zhao and A. Kumar. A deep learning based unified framework to detect, segment and recognize irises using spatially corresponding features. *Pattern Recognition*, 93:546–557, 2019.

108. J. Zuo, N. K. Ratha, and J. H. Connell. Cancelable iris biometric. In *19th International Conference on Pattern Recognition (ICPR)*, Tampa, FL, pp. 1–4, 2008.

# 3 On Training Generative Adversarial Network for Enhancement of Latent Fingerprints

*Indu Joshi, Adithya Anand,*
*Sumantra Dutta Roy, and Prem Kumar Kalra*
IIT Delhi

## CONTENTS

3.1    Introduction ........................................................................... 51
3.2    Related Work .......................................................................... 53
3.3    Proposed Algorithm ............................................................... 57
    3.3.1    Problem Formulation and Objective Function ................... 57
    3.3.2    Training Data Preparation ................................................ 59
    3.3.3    Network Architecture and Training Details ...................... 61
3.4    Performance Evaluation ......................................................... 63
    3.4.1    Databases and Tools Used ................................................ 63
    3.4.2    Evaluation Criteria ........................................................... 64
3.5    Results and Analysis ............................................................... 65
3.6    Challenges Observed .............................................................. 74
3.7    Conclusions ............................................................................ 78
Acknowledgements ........................................................................ 78
References ...................................................................................... 78

## 3.1    INTRODUCTION

Latent fingerprints are the impressions of the ridges on the fingertips which are unintentionally deposited on the surface of an object when the subject touches it. These fingerprints are lifted by forensic experts using specialised techniques like dusting or chemical processing. Latent fingerprints have unclear ridge structure, partial ridge information, and uneven contrast between ridges and valleys. They also possess structured noise due to overlapping text, lines, stains, and sometimes overlapping fingerprints in the background. Figure 3.1a showcases sample latent fingerprint images from IIITD-MSLF database [1].

Latent fingerprints picked up from the crime scene are matched with fingerprints in the law agency's fingerprint database, to find crime suspects. Standard fingerprint matching systems are designed for good quality fingerprints. However, due to

51

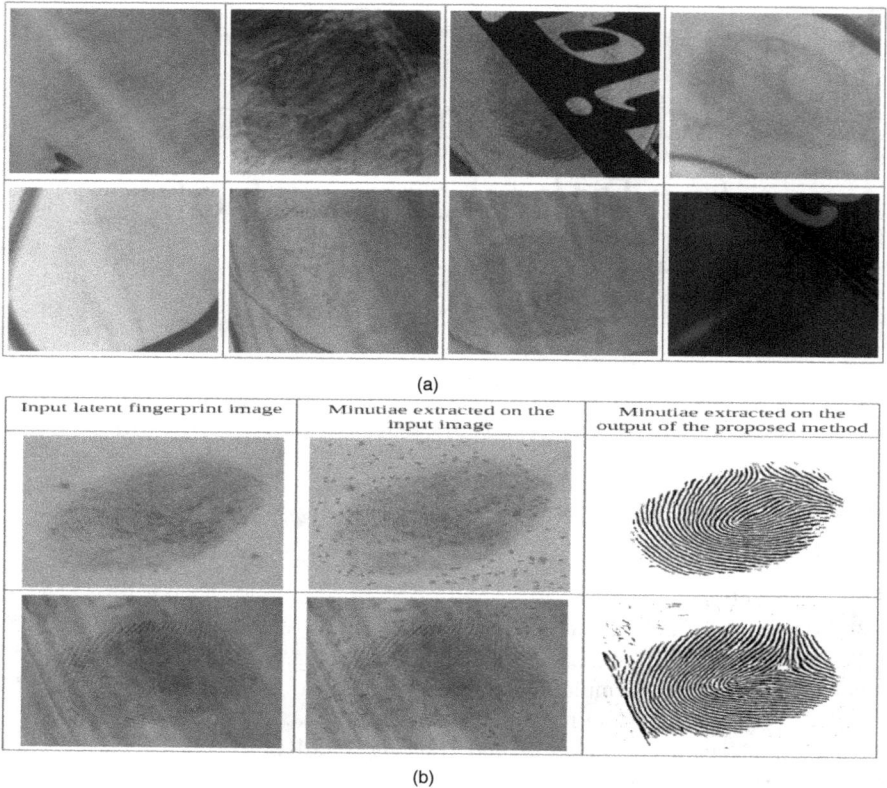

**FIGURE 3.1** (a) Sample latent fingerprints from IIITD-MSLF database depicting background noise, degraded fingerprint ridges, background with textures and multiple fingerprints overlapping with each other. (b) Fingerprints exhibiting the improvement of minutiae detection on enhanced images generated by proposed algorithm. Left column exhibits the original fingerprints, middle column showcases the minutiae detected (shown by blue dots) on original fingerprints using the NBIS tool [2]. Right column shows improved minutiae detection post enhancement.

the poor quality of latents, standard fingerprint feature (minutiae) extractors which perform well on plain and rolled fingerprints often fail on latent fingerprints [3]. Figure 3.1b showcases that many times, true minutiae are missed due to smudged and blurred ridges and many spurious minutiae are extracted due to background noise. As a result, the matching accuracy achieved by the standard fingerprint matchers on latent fingerprints is far from satisfactory to be used for latent fingerprint matching.

Due to this, latent fingerprints are manually matched by the latent fingerprint examiners which pose a huge burden on them. Furthermore, studies have reported inconsistency across evaluations of latent fingerprint examiners [4,5]. This poses a serious need to automate the process of latent fingerprint matching which can facilitate fast and accurate matching performance over the whole fingerprint database and not just a small subset of suspects. One of the key techniques to improve the

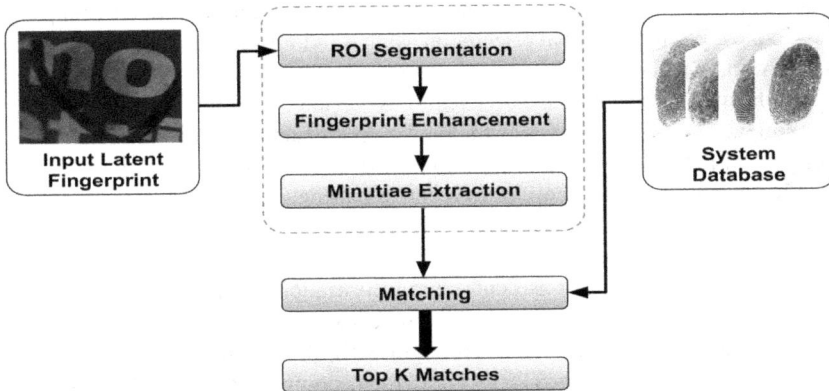

**FIGURE 3.2**  Latent fingerprint matching framework.

latent fingerprint matching performance is an enhancement module. An enhancement algorithm improves the contrast between ridges and valleys, removes background noise, and predicts the missing ridge information and thus facilitates correct minutiae extraction, in turn improving the matching performance. Figure 3.2 depicts the overall framework of latent fingerprint matching.

## 3.2  RELATED WORK

The early literature on latent fingerprint enhancement focuses on accurate estimation of orientation field of ridges in latent fingerprints. The estimated orientations are then fed to the Gabor filter to enhance latent fingerprints. Given below are the approaches of latent fingerprint enhancement which approximate the orientation field and utilise it to enhance latent fingerprints:

Yoon et al. [6] propose an orientation estimation algorithm that requires manually marked ROI (Region of Interest) and singular points. At first, the orientation skeleton image is derived from Verifinger [7] (the state-of-the-art commercial fingerprint matching tool). From these orientations, reliable and unreliable blocks are found out. Reliable blocks have orientations coherent with the neighbouring blocks. For the unreliable blocks, the re-estimation of orientations is performed by interpolations of orientations from the reliable blocks. Using the interpolated orientations, fingerprint rotation and skin distortion model are estimated. Furthermore, computation of orientations from singular points is carried out using zero-pole technique. Finally, orientation is estimated using orientation obtained through the zero-pole method and estimated distortion model. Gabor filtering is applied on the estimated orientation to obtain the enhanced image.

Yoon et al. [8] perform orientation field estimation assuming that the manually marked ROI and singular points are available for the input latent fingerprint image. The initial orientation field is computed by the Short Time Fourier Transform (STFT) enhancement algorithm. However, the performance of STFT can be easily affected by the unstructured background noise. They employ a two-level approach in which first they merge compatible orientation elements in a neighbourhood into an orientation group. Next, they generate top-ten best global orientation using Randomised Random sample consensus (R-RANSAC). Gabor filters with all the ten orientations are employed to obtain ten enhanced latent fingerprint images. Matching is carried out with all the ten images, and the maximum match score serves as the final output match score of the latent.

Feng et al. [9] argue that the orientation estimation is analogous to spelling correction in a sentence. They propose to create a dictionary of orientation patches estimated from good quality fingerprint patches. Creating a dictionary helps to eliminate non-word errors, i.e., predicting such orientations which cannot exist in real-life. They further discuss that just as contextual information can help in spelling correction, similarly orientation of neighbouring patches should be utilised for the estimation of orientation of a given patch. To begin with, they compute an initial estimate of the orientation field using STFT. They, then, compare the initial estimate with each dictionary element and identify potential candidates. They use compatibility between neighbouring patches to find the optimal candidate. Orientation information of all orientation patches is then summarised to obtain the final orientation field.

Yang et al. [10] utilise spatial locality information present in fingerprints to improve the quality of the estimate. Authors claim that only specific orientations occur at a given location, e.g., the orientations at the middle of fingerprints will be different than the orientations at the top of fingerprints. In order to exploit this information, they introduce localised dictionaries, i.e., create a dictionary for every location in a fingerprint. Due to this, each dictionary contains only a limited number of orientations leading to faster dictionary look-ups. Moreover, this technique leads to even fewer non-word errors.

Chen et al. [11] observe that the average size of noise is not the same in all latent fingerprints. Rather, it varies across different qualities of latent fingerprints. For a poor quality image, one can obtain better results by using a dictionary with bigger patch size and vice versa. So, a dictionary created for only a particular size of orientation patches will not work for all latent fingerprints. The authors solve this problem by creating multi-scale dictionaries, i.e., dictionaries of different patch sizes. They use compatibility between neighbours across different scales to find the optimal orientation patch for a given estimate.

Cao and Jain [12] discuss the limitations of dictionary-based methods. They further argue that there is a need for methods which can learn the orientation field from poor quality latent fingerprints. They formulate estimation of orientation field from a fingerprint image as a classification problem. They address this problem using a convolutional neural network (CNN)-based classification model. The real challenge in using a deep architecture is to have a large amount of latent fingerprints for training the network. For this purpose, they propose a model to simulate texture noise as present in latent fingerprints. Several structured and unstructured noise patterns are injected into good quality fingerprints for synthesising latent fingerprints. K-means clustering is performed on orientation patches of good quality images to select 128 representative orientation patch classes. They extract 1,000 orientation patches for each orientation class and train the network with the corresponding simulated latent. After training the model for each patch in input latent fingerprint, an orientation class is predicted by the model.

Liu et al. [13] pose the estimation of orientations as a denoising problem and propose sparse coding for denoising of orientation patches. Authors create multi-scale dictionaries from good quality fingerprints. After computing the initial estimate, they then reconstruct the orientation using a dictionary of smallest size with sparse coding. The quality of an orientation patch is then estimated based on compatibility with neighbours. If the quality is below a certain threshold, then the orientation patch is reconstructed using a dictionary of bigger patches. This process is continued until the quality of the reconstructed orientation patch is satisfactory.

Chaidee et al. [14] propose sparse coded dictionary learning in the frequency domain which fuses responses from Gabor and curved filters. In the offline stage, a dictionary is constructed from the frequency response. In the online stage, spectral response is computed which is then encoded by the spectral encoder. The sparse representation of the spectral code is computed and then decoded by the spectral decoder to reconstruct the Fourier spectrum. A weighted sum of the reconstructed image obtained from both the filters is computed to obtain the final enhanced image. Recently, the attention has been shifted to straight away generate enhanced fingerprint without explicitly approximating orientation field. We now describe such latent fingerprint enhancement algorithms:

Qu et al. [15] propose a deep regression neural network which outputs orientation angle values. The input latent fingerprint image is first pre-processed using total variation decomposition and Log-Gabor filtering. The pre-processed latent is then given as an input to the network, and orientation is estimated. Boosting is performed to further improve the prediction accuracy.

Li et al. [16] propose a multi-task learning-based enhancement algorithm which works on the patch level. An input latent fingerprint image is pre-processed using Total Variation Decomposition, and the texture component is used as an input for the proposed model. Proposed solution is based on encoder–decoder architecture trained with a multi-task learning loss. One branch enhances the latent fingerprint and the other branch predicts orientation for the input image. This algorithm requires orientation field information as a part of training data to train the network to generate the enhanced fingerprint image. Thus, this algorithm is beyond the scope of this chapter.

Svoboda et al. [17] suggest an end-to-end convolutional autoencoder architecture which implicitly minimises orientation and gradient loss between the target-enhanced fingerprint and the fingerprint produced by their model. The objective function is designed such that it only minimises $l2$-loss and it cannot address perceptual information. A brief summary of limitation of the state-of-the-art is provided in Table 3.1.

To summarise, the traditional state-of-the-art latent fingerprint enhancement algorithms focus on accurate orientation estimation for latent fingerprints and exploit only Gabor filters to enhance latent fingerprints. Recent state-of-the-art techniques, on the other hand, propose learning-based end-to-end latent fingerprint enhancement models which directly generate enhanced fingerprints without only relying on Gabor filters. The weights of the kernels in CNNs are rather learnt for the problem in hand. However, none of the above-mentioned latent fingerprint enhancement models exploit the perceptual information in the fingerprints.

Generative adversarial networks (GANs) generate sharper images compared to autoencoders which generate blurred images. As a result, GANs are better suited for generating fingerprint images as they can generate sharp images with clear ridge

**TABLE 3.1**

**Table Summarising the Literature on Latent Fingerprint Enhancement**

| Algorithm | Proposed Approach | Limitation | Reference |
|---|---|---|---|
| Classical Image Processing and hand-crafted models | Orientation estimation using zero-pole method and distortion model | Requires manually marked ROI and singular points | [6] |
| | R-RANSAC is used to find top-ten global orientations. All the ten enhanced images are used for matching | Requires manually marked ROI and singular points. Matching with ten enhanced images is an overhead | [8] |
| Dictionary Learning | Dictionary learning-based orientation estimation | Incorrect estimation around singular points, high computation time | [9] |
| | Localised dictionary learning-based orientation estimation | Algorithm first performs pose estimation and then orientation estimation leading to high computational complexity | [10] |
| | Multi-scale dictionary learning-based orientation estimation | Global multi-scale dictionaries are used due to which local a priori fingerprint information is not utilised | [11] |
| | Spectral dictionary | Requires manually marked core points | [14] |
| | Sparse coded dictionary learning-based orientation estimation | Global multi-scale sparse coded dictionaries are used due to which local a priori fingerprint information is not utilised | [13] |
| Deep Learning | Convolutional neural network-based classification for orientation estimation | Number of orientation patch classes is very limited, due to which the orientation estimation may not be accurate | [12] |
| | Deep regression neural network for orientation estimation | Requires pre-processing before orientation estimation. Moreover, algorithm is not evaluated on any of the publicly available latent fingerprint databases | [15] |
| | Multi-task learning-based autoencoder | The autoencoder is designed for pre-processed latent fingerprints | [16] |
| | Convolutional autoencoder that minimises orientation and gradient loss | Fails to preserve minutiae in case of poor quality input images | [17] |

structure and good ridge-valley contrast. This in turn facilitates improved minutiae extraction and matching performance.

The information on training GAN for latent fingerprint enhancement provided in this chapter is based on the latent fingerprint enhancement algorithm proposed by Joshi et al. [18]. The enhancement model proposed by the authors is trained not only with the reconstruction loss to preserve the ridge structure, but it also limits spurious pattern generation by employing a classification network trained with an adversarial loss to classify the reconstructed image as real or fake. Furthermore, the proposed GAN model is trained on synthetic latent fingerprint images due to which the training is not affected by the limited availability of publicly available latent fingerprint images.

## 3.3   PROPOSED ALGORITHM

### 3.3.1   PROBLEM FORMULATION AND OBJECTIVE FUNCTION

We propose a latent fingerprint enhancement algorithm based on conditional GAN [19,20]. Given a latent fingerprint, the proposed algorithm generates a fingerprint image with clear ridge structure and removes structured and non-structured background noise present in a latent fingerprint. The motivation behind using a conditional GAN is that the generator has to not only generate a "real-looking" binarised fingerprint image but it should also generate a fingerprint which has a similar ridge structure as the input latent fingerprint image. Thus, we formulate latent fingerprint enhancement as a conditional GAN-based image-to-image translation problem [21].

The proposed model has two networks: a latent fingerprint enhancer network and an enhanced fingerprint discriminator (See Figure 3.3). For a given latent fingerprint image $x$, the enhancer network generates a binarised enhanced image $[\mathcal{E}nh_L(x)]$. The enhancer network learns the transformation from a latent fingerprint to a binarised enhanced image, while preserving the overall ridge structure and ridge features including minutiae, without compromising the identity information in the fingerprint. The discriminator network classifies a given enhanced image as real or fake. Figure 3.3 depicts the proposed model for latent fingerprint enhancement. The loss function optimised by the proposed model is described below:

1. **Adversarial Loss:**

$$L_{\text{adv}} = E_{(x,y)\sim p(s=x,y)}[\log(\text{Dis}_E(x,y))] + E_{x\sim p_x(x)}[\log(1 - \text{Dis}_E(x,\mathcal{E}nh_L(x)))]$$

The enhancer network is trained such that the adversarial loss is minimised. On the other hand, the discriminator network is trained to maximise the adversarial loss. A penalty is imposed on the enhancer network if the image generated by the enhancer network $[\mathcal{E}nh_L(x)]$ is deemed fake by the

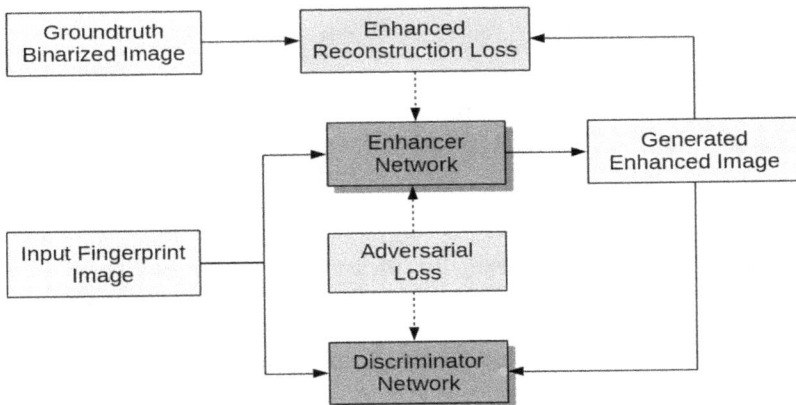

**FIGURE 3.3**   Proposed model for enhancement of latent fingerprints. The back propagation of losses while training enhancer network and discriminator network is shown by dotted lines.

discriminator. Due to this loss, the enhancer network learns the necessary transformation and associated features required to generate an enhanced fingerprint from a given latent fingerprint image.

The discriminator network is penalised if it misclassifies an enhanced fingerprint image generated by the enhancer network as a real fingerprint. As a result, discriminator learns the discriminating features for differentiating the enhanced images produced by the enhancer from the ground-truth binarised images.

Note that the discriminator is conditioned by the input latent fingerprint image so that the discriminator network doesn't just classify an enhanced image as real or fake, but the discriminator can also classify whether the enhanced fingerprint image has the ridge structure similar to the input latent fingerprint image.

**2. Enhanced Fingerprint Reconstruction Loss:**

$$L_{rec} = \| y - \mathcal{E}nh_L(x) \|$$

The task of generating a binarised enhanced image corresponding to an input latent fingerprint image is an ill-posed problem with only adversarial loss. We include fingerprint reconstruction loss into the objective function. This loss only penalises the enhancer network. It guides the enhancer network to generate enhanced fingerprint similar to the ground-truth binarised fingerprint image. The reconstruction loss facilitates the enhancer network to learn to preserve low-frequency details in the enhanced image. $l1$ norm is used in the loss function to encourage the enhancer to produce sharp images. $l2$ norm is not used as it generates blurred images.

**3. Overall Loss: The Final Objective Function is given as:**

$$\min_\alpha \max_\beta \; [E_{(x,y) \sim p(x,y)}[\log(\mathrm{Dis}_E(x,y)]$$

$$+ E_{x \sim p_x(x)}[log(1 - \mathrm{Dis}_E(x, \mathcal{E}nh_L(x))) + \lambda \| y - \mathcal{E}nh_L(x)) \|_l]]$$

where $\alpha$ and $\beta$ denote the parameters of enhancer and discriminator, respectively. $\lambda$ is the weight parameter for the reconstruction loss.

Reconstruction loss helps to preserve the low-frequency details in the fingerprint image. However, fingerprints are oriented textured patterns which have a lot of high-frequency details. To ensure that the proposed model is able to capture high-frequency details, we use a patch GAN-based model which classifies each $8 \times 8$ patch as real or fake. Furthermore, reconstruction loss is a pixel-based loss which assumes that each output pixel is independent of its neighbouring pixels. Patch GAN, on the other hand, considers the joint distribution of the pixels in a patch which introduces a texture loss which in turn forces the enhancer network to preserve fine ridge details including minutiae and thus helps to preserve the identity information in the fingerprint image.

## 3.3.2 Training Data Preparation

The proposed model is a supervised generative model which is trained to output an enhanced image given an input latent fingerprint image. Being a supervised model, it requires paired training data of latent fingerprints and their corresponding enhanced binarised images. However, there are no publicly available latent fingerprint datasets which have latent fingerprints and their corresponding enhanced images. Additionally, lack of large latent fingerprint database further complicates the training of a deep neural network-based latent fingerprint enhancement model. Thus, we need to generate synthetic latent fingerprints which have similar noise characteristics as observed in real latent fingerprints (see Figure 3.4) for training the proposed enhancement model.

The proposed model is trained on 9,042 synthetic latent fingerprint images and 2,423 fingerprint images from National Institute of Standards and Technology Special Database 4 (NIST SD4) and their corresponding binarised fingerprints. Due to training on synthetic latent fingerprints, the training of the proposed model is not affected by the limited availability of the latent fingerprint database. We now give details on preparing the training data for the proposed model.

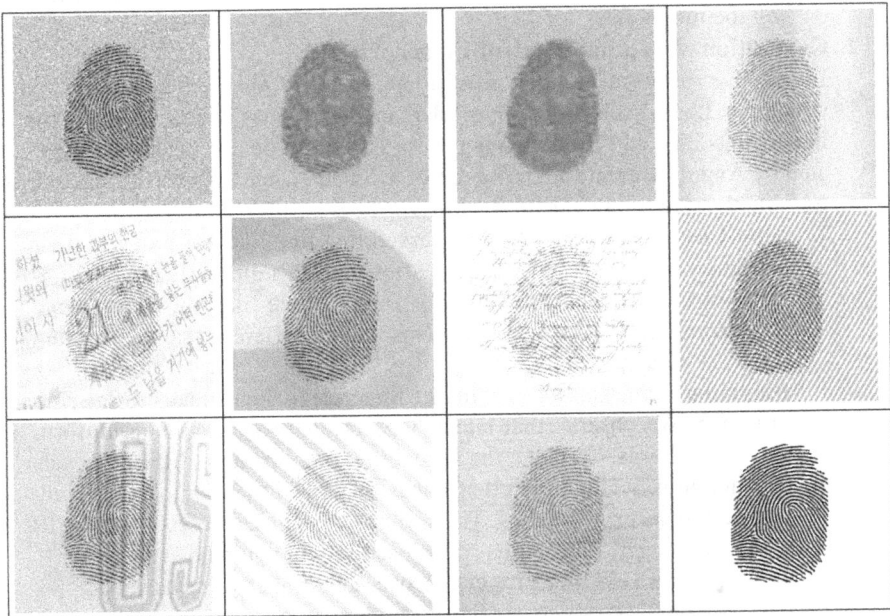

**FIGURE 3.4** Sample images showcasing the training dataset. The 11 fingerprints (from top-left) have the same binarised ground-truth image (bottom-right image). Varying textures and backgrounds are used for training the algorithm for simulating conditions of acquisition of latent fingerprint.

1. **Datasets for Preparing the Training Data:**
   i.  **Anguli**: Anguli [22] is an open-source implementation of the state-of-the-art synthetic fingerprint generator SFinGe [23], which simulates synthetic live fingerprints with similar features as real-live fingerprints. It can generate multiple impressions of a fingerprint with varying levels of noise.
   ii. **NIST SD4**: NIST SD4 [24] is a publicly available fingerprint database which has 2,000 rolled fingerprints. These are inked fingerprints with uniformly distributed fingerprint pattern type, namely left loop, right loop, arch, tented arch, and whorl. Due to the uniform distribution of the pattern type, the training dataset covers varieties of ridge patterns. Furthermore, as these fingerprints are inked prints, they have similar characteristics of non-uniform ink to latent fingerprints which have non-uniform powder content in many patches. We use NIST SD4 fingerprints with NIST Finger Image Quality 2 (NFIQ2) [25] quality score greater than or equal to 70. (NFIQ2 is an open-source state-of-the-art fingerprint quality assessment algorithm which gives a quality score in the range 1–100 to each fingerprint image where 1 denotes the worst quality and 100 denotes the best quality.) Although it is helpful to include poor quality inked prints as the training data, however, the ground-truth binarisation achieved through NBIS on poor quality fingerprints is poor which can adversely affect the performance of the model. So, we only use good quality NIST SD4 fingerprints for training the model.

2. **Generation of Synthetic Latent Fingerprints:**
   Latent fingerprints due to their acquisition conditions are often blurred and have structured noise such as lines, overlapping text, and sometimes overlapping fingerprints. We add the following noise into good quality fingerprints generated by Anguli to create a representative synthetic latent fingerprints dataset for training the proposed model:
   i.  **Line-Like Noise**: It has been observed that line-like noise due to their similarity with fingerprint ridges often lead to failure of standard fingerprint matching algorithms. To simulate line-like noise, we blend fingerprint images with straight lines having different orientations and different widths.
   ii. **Blurring**: Sometimes smudging of fingerprint ridges leads to missing minutiae. We observe that latent fingerprints often have non-uniform smudge patterns. To make the model invariant towards different levels of smudging, we add different levels of Gaussian noise on randomly selected fingerprint patches. The different patch sizes used are $10 \times 10$ and $40 \times 40$. The blur radius $= 2$ is used for Gaussian noise.
   iii. **Overlapping Text and Fingerprints**: Latent fingerprints have complex background noise which can have overlapping text and sometimes overlapping fingerprints. To simulate those scenarios, we blend fingerprint images with text images of varying fonts and styles. We also blend fingerprint images with partial fingerprint images to address challenges of overlapping fingerprints.

iv. **Different Surfaces**: Latent fingerprints can be collected from different surfaces. Surfaces can be plane/curved, porous/non-porous, shiny or can have uniform background. It has been reported that the surfaces which have high reflectance generate occluded ridge patterns [1]. Furthermore, the area and the quality of latent fingerprint left on a surface vary depending on the pressure exerted by the finger, surface characteristics, and adherence of the finger's natural secretions on that surface. Some surfaces have poor adherence property due to which the latent fingerprint deposited on such surfaces is often partial. To train the proposed model to be invariant towards various intra-class variations introduced due to various surfaces, we blend fingerprint image with varying textures such as wood surface, cardboard surface, plastic, and glass surface.

3. **Ground-Truth Binarisation:**
Ground-truth binarised image to train the proposed model is obtained using NIST Biometric Image Software (NBIS). A fingerprint image is binarised by NBIS based on the ridge flow direction. The image is divided into $7 \times 9$ grids, if there is a ridge pattern in a grid, the grid is rotated so that the grid is parallel to the ridge flow direction. For the pixel of interest, the neighbourhood grey values which also lie in the rotated grid are analysed to label a pixel as black or white.

### 3.3.3 NETWORK ARCHITECTURE AND TRAINING DETAILS

1. **Enhancer Network**: Enhancer network has an encoder–decoder (autoencoder) architecture. Convolutional layers (Conv1, Conv2, and Conv3) in the network extract features at different scales from the input latent fingerprint image capturing coarse to fine level details (See Figure 3.5). ResNet blocks help to circumvent the problem of vanishing gradient while training a deep network. Decoder layers (Deconv1, Deconv2, and Conv4) transform the features extracted from the latent fingerprint to an enhanced binarised fingerprint image.

2. **Discriminator Network**: The input latent fingerprint and binarised image are concatenated along the input channel dimension so that the discriminator can classify whether the binarised image corresponds to the input latent fingerprint image. Discriminator has a typical architecture as used in image classification. The convolutional layers in the discriminator (Conv5, Conv6, Conv7, Conv8, and Conv9) extract features at different scales capturing at different levels which helps the discriminator to classify an input fingerprint image as real or fake.

The details of the network architecture are given in Table 3.2. Adam optimiser is used to optimise the objective function. The following hyper-parameters are used: learning rate = 0.02, $\beta_1 = 0.5$, $\beta_2 = 0.999$, $\lambda = 10$ and batch size = 2. The model is trained on two GPUs each with 12 GB RAM.

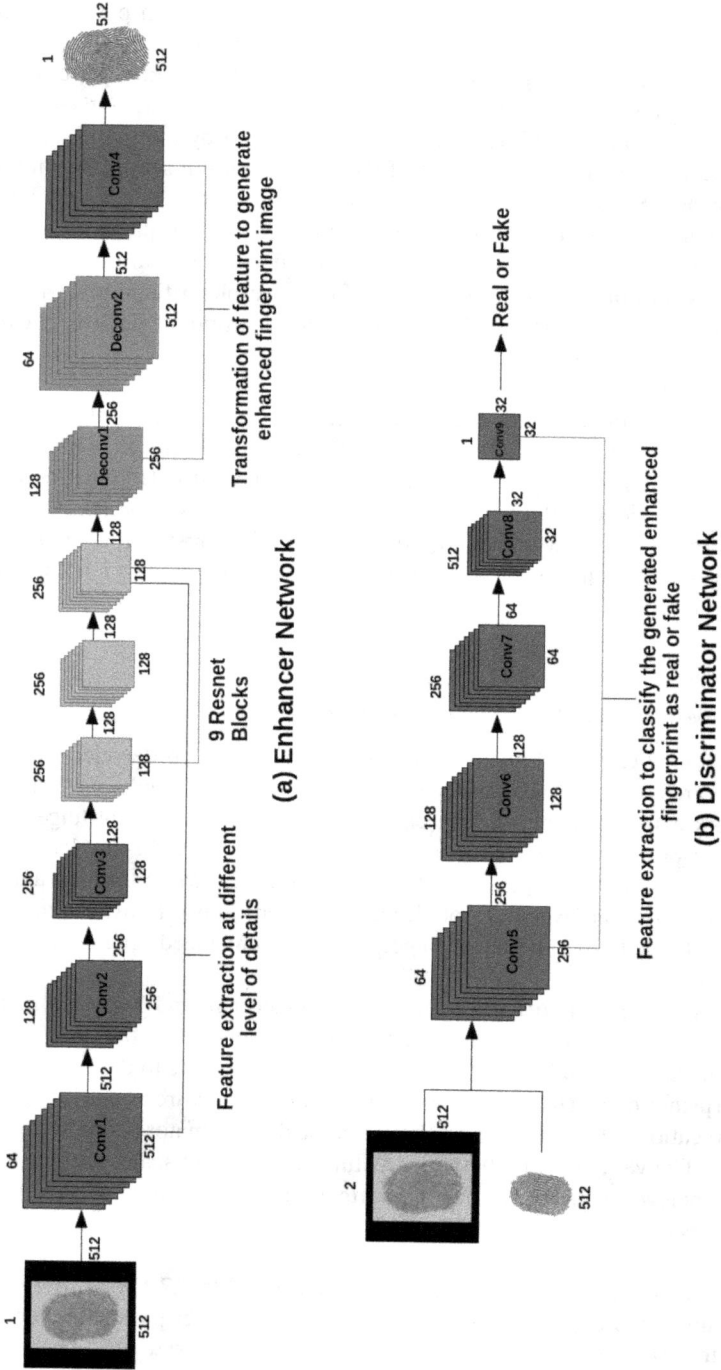

**FIGURE 3.5** Architecture of enhancer ($\mathcal{E}nh_L$) and discriminator ($\text{Dis}_E$).

**TABLE 3.2**

**Architecture of $\mathcal{E}nh_L$ and $\text{Dis}_E$**

| Block | Layers | Kernels | Size | Stride | Padding |
|---|---|---|---|---|---|
| Conv1 | Convolutional Layer + Batch Normalisation + ReLu | 64 | 7 | 1 | 3 |
| Conv2 | Convolutional Layer + Batch Normalisation + ReLu + Convolutional Layer + Batch Table 3.2: Architecture of $\mathcal{E}nh_L$ and $\text{Dis}_E$ | 128 | 3 | 2 | 1 |
| Conv3 | Convolutional Layer + Batch Normalisation + ReLu + Convolutional Layer + Batch Normalisation | 256 | 3 | 2 | 1 |
| ResNet Block | Convolutional Layer + Batch Normalisation + ReLu + Conv Layer + Batch Normalisation | 256 | 3 | 2 | 1 |
| Deconv1 | Convolutional Layer + Batch Normalisation Layer + ReLu + Convolutional Layer + Batch Normalisation | 128 | 3 | 2 | 1 |
| Deconv2 | Convolutional Layer + Batch Normalisation + ReLu + Conv Layer + Batch Normalisation | 64 | 3 | 2 | 1 |
| Conv4 | Convolutional Layer + Tanh | 1 | 7 | 1 | 3 |
| Conv5 | Convolutional Layer + LeakyReLu | 64 | 4 | 2 | 1 |
| Conv6 | Convolutional Layer + Batch Normalisation + LeakyReLu | 128 | 4 | 2 | 1 |
| Conv7 | Convolutional Layer + Batch Normalisation + LeakyReLu | 256 | 4 | 2 | 1 |
| Conv8 | Convolutional Layer + Batch Normalisation + LeakyReLu | 512 | 4 | 1 | 1 |
| Conv9 | Convolutional Layer | 1 | 4 | 1 | 1 |

## 3.4 PERFORMANCE EVALUATION

### 3.4.1 DATABASES AND TOOLS USED

The proposed model is evaluated on two publicly available latent fingerprint databases:

1. **IIITD-MOLF Database [26]:** Indraprastha Institute of Information Technology Delhi Multi-sensor Optical and Latent Fingerprint (IIITD-MOLF) is the biggest latent fingerprint database which is available in the public domain. It has latent fingerprints and live fingerprints acquired through different optical sensors. These fingerprints are collected from 100 subjects. This database has 4,400 latent fingerprints and 4,000 live fingerprints corresponding to each sensor.

2. **IIITD-MSLF Database [1]:** Indraprastha Institute of Information Technology Delhi Multi-surface Latent Fingerprint IIITD-MSLF database has latent fingerprints extracted from eight different surfaces like transparent glass, compact disc, ceramic mug, hardbound cover, etc. It has 551 latent fingerprints of 51 subjects.

The standard latent fingerprint database provided by NIST, NIST-SD27 has now been removed from the public domain due to which we cannot evaluate the proposed model on NIST-SD27 database. The proposed model is designed for the standard sized 500 dpi fingerprint image whose spatial dimensions are 512×512 pixels. The latent fingerprints are pre-processed and zero-padded to have a fixed size of 512×512. Table 3.3 provides the list of publicly available tools used in this work.

### 3.4.2  EVALUATION CRITERIA

Every fingerprint enhancement algorithm is designed to increase the clarity of ridges and valleys while preserving the ridge details to improve minutiae extraction and thereby improving fingerprint matching performance. We evaluate the proposed enhancement algorithm using the metrics given below:

1. **Fingerprint Quality Analysis**: Quality of a fingerprint image is determined as the ability of a fingerprint matcher to correctly match the image. Poor quality fingerprints often result in poor matching performance. We evaluate the fingerprint quality of latent fingerprints before and after enhancement using NIST Finger Image Quality (NFIQ) module of NBIS. NFIQ calculates quality of a fingerprint image using features such as: clarity of ridges and valleys, number of minutiae, size of the fingerprint image, etc. NFIQ scores a fingerprint image between 1 and 5 where 1 signifies the best fingerprint image quality and 5 means the worst quality. We compare the histogram of quality scores obtained by NFIQ before and after enhancement. Another publicly available tool to evaluate the quality of fingerprint images is NFIQ2 [25], which returns a score between 1 and 100. NFIQ2 is a more robust fingerprint quality assessment metric than NFIQ. However, NFIQ2 fails to process raw latent fingerprint images of

---

**TABLE 3.3**
**Table Summarising the Publicly Available Tools Used**

| Tool | Purpose | Usage |
|---|---|---|
| **MINDTCT** module of NBIS | Minutiae extraction | During testing, to extract minutiae from enhanced image and gallery images |
| **NFIQ** module of NBIS | Evaluates fingerprint image quality | During testing, to evaluate quality of enhanced fingerprints |
| **BOZORTH** module of NBIS | To match fingerprints | During testing, to perform fingerprint matching on minutiae extracted by MINDTCT |
| **MCC** fingerprint matcher | To match fingerprints | During testing, to perform fingerprint matching on minutiae extracted by MINDTCT |
| NFIQ2 | Evaluates fingerprint image quality | To evaluate quality of NIST SD4 images and keep good quality images for training the model |
| **Binarisation** module of NBIS | Binarise the fingerprint image | To generate the ground-truth binarisation of training images |

IIITD-MOLF database. As a result, we only compare fingerprint quality score obtained using NFIQ.

2. **Ridge Structure Preservation**: The most crucial factor for any fingerprint enhancement is that it should retain the ridge structure while improving clarity of ridges and valleys. To showcase ridge structure preservation (including minutiae) by the proposed model, we synthetically generate some test cases by adding noises and backgrounds on good quality fingerprints. We showcase the similarity between ground-truth binarisation and the enhanced fingerprint image generated by the proposed algorithm using the following two measures:

    i. We calculate **Structural Similarity Index Metric** (SSIM) [27] between the ground-truth binarised image and the enhanced fingerprint. SSIM is a metric which computes similarity between image $a$ and image $b$ based on the contrast, luminance, and structure:

    $$\text{SSIM}(a,b) = \frac{(2\mu_a\mu_b + C_1)(2\sigma_{ab} + C_2)}{(\mu_a^2 + \mu_b^2 + C_1)(\sigma_a^2 + \sigma_b^2 + C_2)}$$

    where $\mu_a$, $\mu_b$ are the mean, $\sigma_a$, $\sigma_b$ are the standard deviation, and $\sigma_{ab}$ is the covariance between image $a$ and image $b$.

    ii. We also calculate **match score (using Bozorth)** between ground-truth binarised image and the enhanced image generated by the proposed model. High match scores demonstrate that the proposed algorithm preserves minutiae while enhancing the input latent fingerprint image.

3. **Matching Performance**: The ultimate success of a fingerprint enhancement algorithm is when it is able to improve the fingerprint matching performance. We extract minutiae using the MINDTCT module of NBIS and use Bozorth and Minutia Cylinder Code (MCC) [28–30] fingerprint matchers to evaluate fingerprint matching performance. We compare matching performance before and after enhancement using Rank-50 accuracy. Rank-k accuracy is defined as:

Rank-k accuracy = no. of probe fingerprint for which the matching fingerprint in gallery achieved top-k scores × 100/total no. of probe fingerprints

We also plot cumulative matching curve (CMC), which is a Rank-k accuracy plot over varying values of k. CMC is a standard summarisation technique to quantify the matching performance of a closed-set identification system. We compare the CMC before and after enhancement in Figure 3.9.

## 3.4 RESULTS AND ANALYSIS

1. **Fingerprint Quality Analysis**: Figure 3.6 represents the histogram of NFIQ scores before and after enhancement. The average NFIQ score has improved from 4.96 to 1.91 after enhancement (smaller score means better quality) on IIITD-MOLF and 4.48 to 2.64 on IIIT-D MSLF database (see Table 3.4) which validates the improved clarity of ridges and valleys (thereby improving the quality score) in the enhanced fingerprints generated by the proposed model.

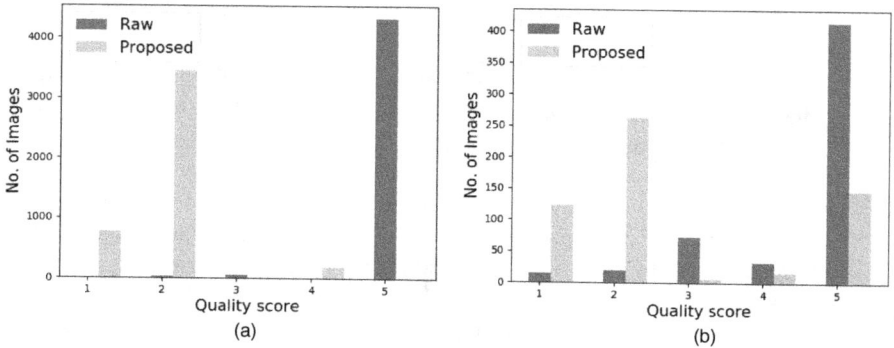

**FIGURE 3.6**  Evaluation of quality of fingerprint images using the NFIQ module of NBIS [2] for latent fingerprint images from (a) IIITD-MOLF and (b) IIITD-MSLF database.

**TABLE 3.4**

**Average NFIQ Scores Before and After Enhancement by the Proposed Model on IIITD-MOLF and IIITD-MSLF Databases**

| Dataset | Enhancement Algorithm | NFIQ Score |
|---|---|---|
| IIITD-MOLF | Raw Image | 4.96 |
| IIITD-MOLF | Raw Image | 1.91 |
| IIITD-MSLF | Proposed | 4.48 |
| IIITD-MSLF | Proposed | 2.64 |

2. **Ridge Structure Preservation:** In Figure 3.7, we present some sample test cases with their ground-truth binarisation and the output of the proposed algorithm. High match score and high SSIM value between the ground-truth binarised image and the output of the proposed method illustrate that the proposed algorithm preserves the ridge information of input latent fingerprint images including fingerprint class, orientation of ridges and minutiae, while enhancing them.

3. **Matching Latent Fingerprints to Multi-Sensor Fingerprints:** In Figure 3.8, we show CMCs for matching performance achieved by Bozorth and MCC matcher on the enhanced image generated by the proposed model across two different galleries. We also compare the Rank-50 accuracy of the proposed model with the recently proposed latent fingerprint algorithm [17] (see Table 3.5). The magnitude of improvement obtained over raw images using the proposed algorithm is much more than the previous work (Rank-50 accuracy of 34.43% on DB1 and 30.50% on DB2 gallery using the proposed algorithm compared to 22.36% on DB1 and 19.50% on DB2 gallery by the previous work [17]). This demonstrates that the proposed algorithm performs better than [17] in improving ridge-valley contrast, removing background noise while preserving ridge details

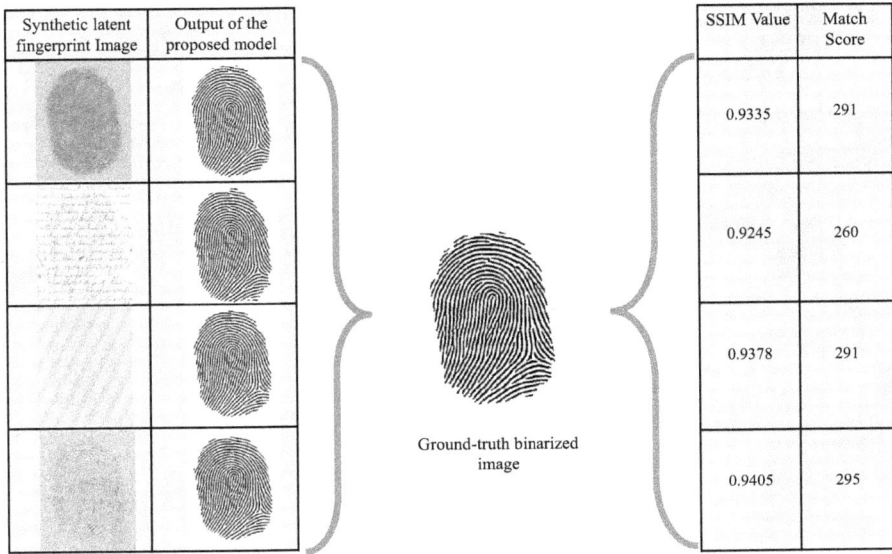

| Synthetic latent fingerprint Image | Output of the proposed model | | | SSIM Value | Match Score |
|---|---|---|---|---|---|
| | | | | 0.9335 | 291 |
| | | | | 0.9245 | 260 |
| | | | | 0.9378 | 291 |
| | | Ground-truth binarized image | | 0.9405 | 295 |

**FIGURE 3.7** Left side shows the enhanced fingerprint generated from the synthetic latent fingerprint, corresponding to the ground-truth binarised image (shown in the middle). Right side shows the SSIM value and the matching score (obtained using Bozorth) for each enhanced image corresponding to the ground-truth image.

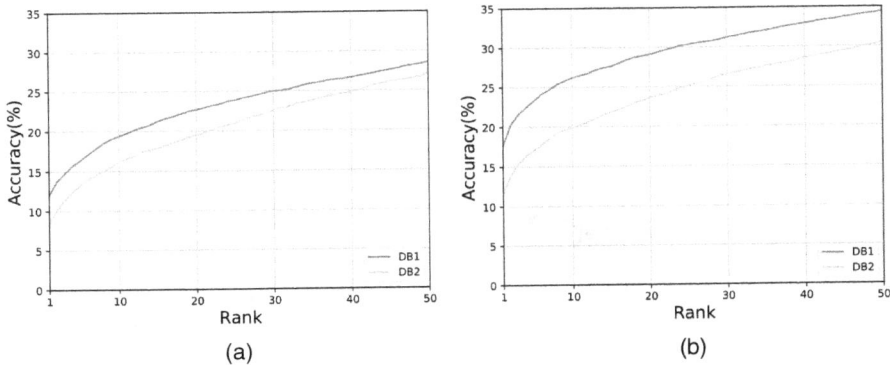

(a)　　　　　　　　　　　　　(b)

**FIGURE 3.8** CMC for proposed algorithm's matching performance for the IIITD-MOLF DB1 gallery and DB2 gallery at $\lambda = 5$, using (a) Bozorth (b) MCC.

due to which improved feature extraction and thereby, improved matching performance is obtained.

4. **Matching Multi-Surface Latent Fingerprints to Gallery of Live-Scan Fingerprints**: The Rank-50 accuracies before and after enhancement on IIITD-MSLF database using Bozorth are 11.43% and 12.80%, respectively. The CMC is shown in Figure 3.9. The accuracy obtained on IIITD-MSLF database is lesser compared to the accuracy achieved on IIITD-MOLF.

**TABLE 3.5**

**Rank-50 Obtained on IIITD-MSLF Database Before and After Enhancement by the Proposed Model**

| Enhancement Algorithm | Rank-50 Accuracy (DB1) | Rank-50 Accuracy (DB2) |
|---|---|---|
| Raw Image | 5.45 | 5.18 |
| Svoboda et al. [17] | 22.36 | 19.50 |
| Proposed | 34.43 | 30.50 |

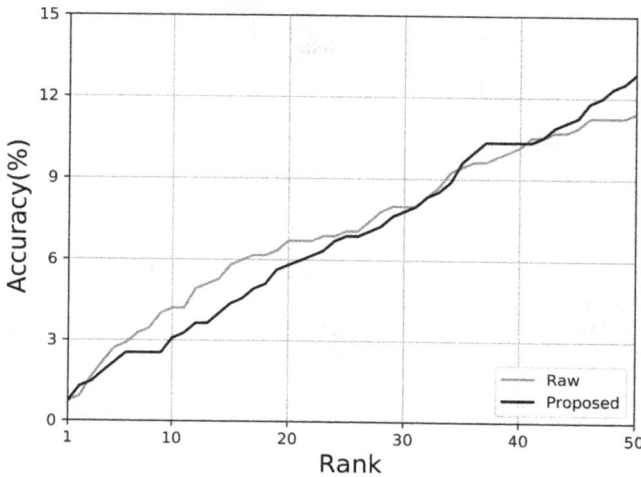

**FIGURE 3.9** CMC representing matching performance on IIITD-MSLF database, before and after enhancement by the proposed algorithm using Bozorth.

This is due to the complex background present in IIITD-MSLF database images. We observe that the intensity values of foreground and the background fingerprint regions have similar distribution in many images. This leads to spurious pattern generation by the proposed algorithm, which adversely affects the matching performance.

5. **Significance of Latent Fingerprint Reconstruction Loss**: To demonstrate the significance of the reconstruction loss in the objective function, we train the proposed model with only adversarial loss ($\lambda = 0$). We observe that the model becomes unstable and doesn't converge. In Figure 3.10, we show the sample results obtained with only reconstruction loss. Therefore, we conclude that the reconstruction loss is essential to stabilise the proposed model.

6. **Role of Hyper-Parameters**: During the various experiments conducted in this chapter, we find that the MCC is a better fingerprint matcher for latent fingerprints (see Tables 3.6–3.9). We conclude our observations based on the results obtained using the MCC matcher. By default, the hyper-parameters used are $\lambda = 10$, batch size = 2, and number of epochs = 200.

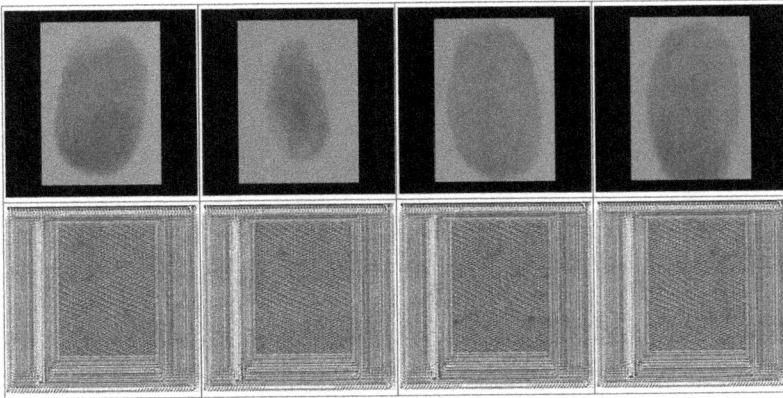

**FIGURE 3.10**  Sample-enhanced images obtained by the model when trained without latent fingerprint reconstruction loss.

**FIGURE 3.11**  Failure cases of NFIQ (lower score means better quality).

i.  **Weight Hyper-parameter** ($\lambda$): We observe that the Rank-50 achieved by the proposed model increases as the weight of enhanced fingerprint reconstruction loss is increased from 1 to 5 (as depicted in Table 3.9, Figures 3.12a and b and 3.13a and b). However, on increasing the weight further, the performance starts degrading. The best Rank-50 accuracies of 34.43% across DB1 and 30.50% across DB2 gallery are achieved for $\lambda = 5$. The quality, on the other hand, improves, while $\lambda$ is increased from 1 to 10. On increasing $\lambda$ further, the quality starts degrading (see Table 3.10 and Figure 3.14a). This suggests that the model is sensitive to the choice of weight parameters, and a careful combination of adversarial loss and reconstruction loss is required to efficiently train the proposed model.

**TABLE 3.6**

**Rank-50 Accuracy Obtained over Different Epochs on IIITD-MOLF Latent Fingerprints**

| Epoch | DB1(Bozorth) | DB2(Bozorth) | DB1(MCC) | DB2(MCC) |
|---|---|---|---|---|
| 30 | 24.80 | 23.02 | 28.16 | 25.98 |
| 60 | 28.11 | 25.05 | 33.61 | 29.36 |
| 90 | 28.61 | 25.05 | 33.14 | 29.43 |
| 120 | 28.63 | 26.75 | 33.55 | 30.14 |
| 150 | 24.66 | 24.05 | 29.23 | 26.93 |
| 180 | 28.77 | 26.70 | 33.34 | 29.93 |
| 200 | 27.25 | 25.64 | 32.02 | 29.32 |
| 210 | 25.93 | 24.50 | 30.50 | 28.30 |
| 240 | 25.34 | 23.84 | 30.16 | 27.05 |
| 270 | 24.75 | 23.84 | 29.34 | 26.59 |

**TABLE 3.7**

**Rank-50 Accuracy Obtained on IIITD-MOLF Latent Fingerprints with and without Adding NIST-SD4 Images in Training Data**

| Training Data | DB1(Bozorth) | DB2(Bozorth) | DB1(MCC) | DB2(MCC) |
|---|---|---|---|---|
| Without SD4 | 27.70 | 26.30 | 30.43 | 29.2045 |
| With SD4 | 27.25 | 25.64 | 32.02 | 29.32 |

**TABLE 3.8**

**Rank-50 Accuracy Obtained IIITD-MOLF Latent Fingerprints for Different Batch Size**

| Batch Size | DB1(Bozorth) | DB2(Bozorth) | DB1(MCC) | DB2(MCC) |
|---|---|---|---|---|
| 2 | 27.25 | 25.64 | 32.02 | 29.32 |
| 4 | 27.93 | 26.61 | 30.45 | 26.659 |
| 8 | 18.03 | 17.28 | 15.41 | 15.41 |

ii. **Number of Epochs**: As shown in Table 3.6 and Figures 3.12c and d and 3.13c and d, the Rank-50 accuracy initially improves with the number of epochs till 60 epochs, after which it fluctuates and is approximately the same till 200 epochs. The performance of the model starts degrading after 200 epochs due to over-fitting. NFIQ score, on the other hand, is improved initially till 150 epochs and then it fluctuates and no clear trend is found (see Table 3.11).

iii. **Batch Size**: We compare the Rank-50 accuracy achieved by the model at batch size = 2, 4, and 8 (see Table 3.8, Figures 3.12e and f and

**TABLE 3.9**

**Rank-50 Accuracy Obtained on IIITD-MOLF Latent Fingerprints over Different Values of $\lambda$**

| $\lambda$ | DB1(Bozorth) | DB2(Bozorth) | DB1(MCC) | DB2(MCC) |
|---|---|---|---|---|
| 1 | 24.00 | 22.0 | 28.09 | 23.70 |
| 3 | 28.11 | 24.89 | 31.70 | 28.11 |
| 5 | 28.52 | 27.11 | 34.43 | 30.50 |
| 10 | 27.25 | 25.64 | 32.02 | 29.32 |
| 15 | 26.60 | 25.27 | 31.89 | 25.39 |
| 20 | 25.66 | 23.43 | 29.34 | 27.55 |

3.13e and f). As the batch size is increased, the number of parameter updates per epoch reduces which leads to faster training of the model. Batch size = 2 turns out to be the best value of hyper-parameter batch size. The best performance at batch size = 2 is attributed to more parameter updates and thus better training.

Better quality score is obtained for batch size = 8 than batch size = 2 and batch size = 4, as can be seen in Table 3.12 and Figure 3.14c. This is a counter-intuitive result as the enhanced fingerprint generated at batch size = 8 has large missing regions compared to the fingerprints generated at batch size = 2. Many of the images generated at batch size = 2 have poor ridge smoothness compared to the image generated at batch size = 8 which have very small reconstructed fingerprint area but better smoothness (see Figure 3.11). Due to this, NFIQ gives a better score to images generated at batch size = 8 than the images generated at batch size = 2. Thus, we conclude that the anomaly in the quality score is due to the limitation of NFIQ. In reality, the quality of fingerprints generated at batch size = 2 and batch size = 4 is better than the quality of fingerprints generated at batch size=8 which is evident through the much higher Rank-50 accuracy achieved for batch size = 2 and batch size = 4 than batch size = 8.

7. **Effect of Training the Model with Real NIST SD4 Images**: The Rank-50 accuracy achieved by the images generated by the proposed model trained on both synthetic latent fingerprints and real fingerprints from NIST SD4 is better compared to the model trained on only synthetic latent fingerprints, as shown in Figures 3.15, 3.12g and h, and 3.13g and h). Similar trend is seen in the NFIQ quality scores of the enhanced fingerprints obtained using the proposed model (see Table 3.13 and Figure 3.14d). The real fingerprints have practical cases of non-linear distortion and non-uniform ridge width which are also observed in latent fingerprints. Thus, the real inked fingerprints like those of NIST SD4 database help the model to learn to be invariant to such distortions.

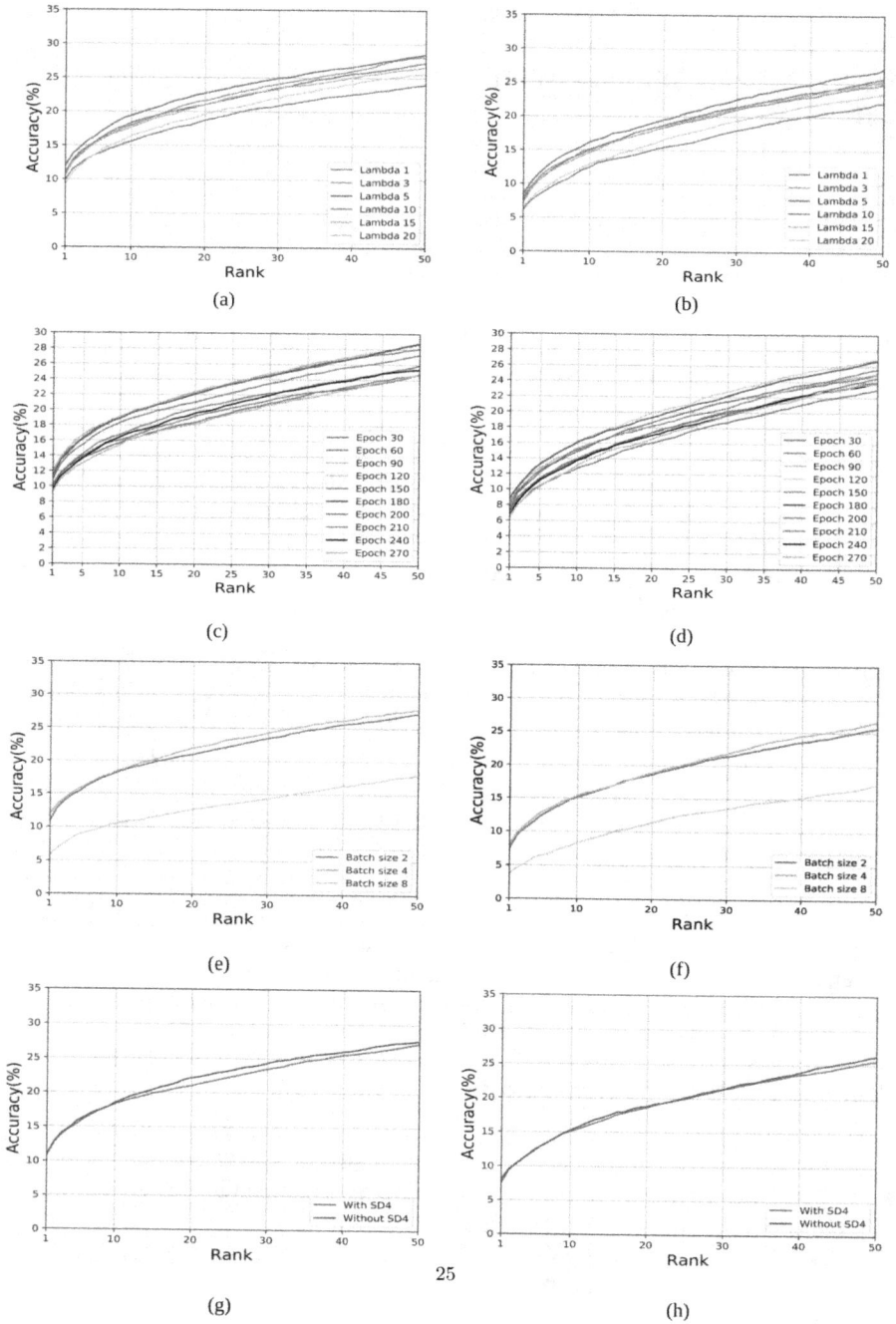

**FIGURE 3.12** CMC for matching the proposed algorithm using Bozorth on the IIITD-MOLF DB1 and DB2 galleries across different training settings.

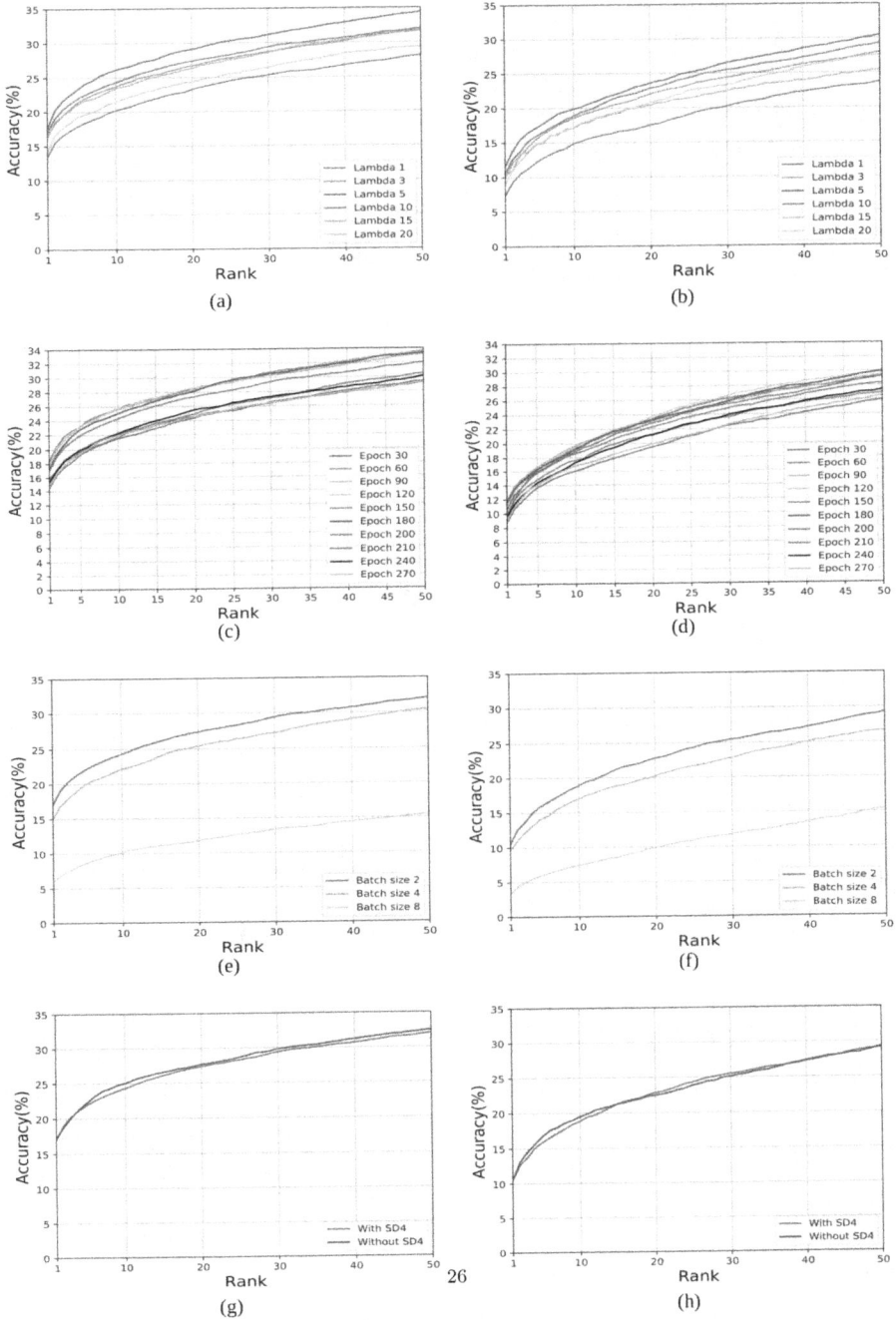

**FIGURE 3.13** CMC for matching the proposed algorithm using MCC on the IIITD-MOLF DB1 and DB2 galleries across different training settings.

**TABLE 3.10**

**Average NFIQ Scores of the Enhanced Fingerprints Obtained for IIITD-MOLF Database Using the Proposed Algorithm over Different Values of $\lambda$**

| $\lambda$ | NFIQ Score |
|---|---|
| 1 | 2.06 |
| 3 | 1.99 |
| 5 | 1.91 |
| 10 | 1.83 |
| 15 | 1.87 |
| 20 | 1.91 |

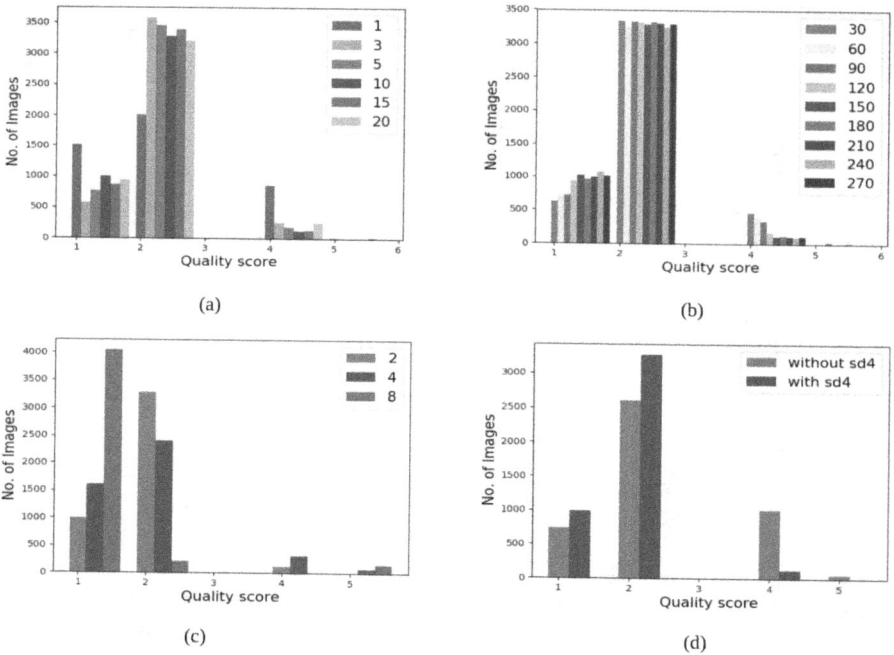

(a)

(b)

(c)

(d)

**FIGURE 3.14** NFIQ score distribution of the enhanced images produced by the proposed algorithm across different training settings.

## 3.6 CHALLENGES OBSERVED

While conducting different experiments, it has been found that the proposed algorithm improves matching performance (see Figure 3.16). However, we observe some cases where the proposed algorithm does not generate good results. Analysis of these cases is given in the following points:

**TABLE 3.11**
**Average NFIQ Scores of the Enhanced Fingerprints Obtained for IIITD-MOLF Database Using the Proposed Algorithm over Different Epochs**

| Epoch | NFIQ Score |
|---|---|
| 30 | 2.07 |
| 60 | 2.03 |
| 90 | 2.00 |
| 120 | 1.86 |
| 150 | 1.82 |
| 180 | 1.84 |
| 200 | 1.83 |
| 210 | 1.83 |
| 240 | 1.81 |
| 270 | 1.83 |

**TABLE 3.12**
**Average NFIQ Scores of the Enhanced Fingerprints Obtained for IIITD-MOLF Database Using the Proposed Algorithm over with and without Adding SD4 Images in Training Data**

| Batch Size | NFIQ Score |
|---|---|
| 2 | 1.83 |
| 4 | 1.83 |
| 8 | 1.18 |

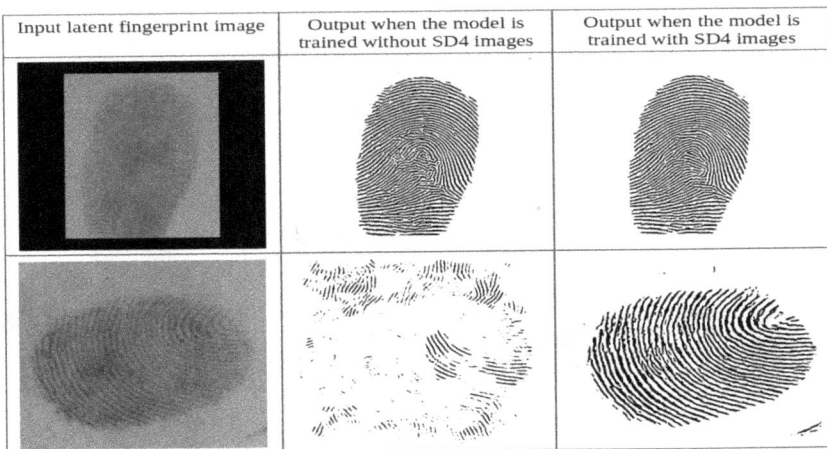

**FIGURE 3.15** Sample results obtained by the model when trained with and without NIST SD4 images in the training dataset.

**TABLE 3.13**

**Average NFIQ Scores of the Enhanced Fingerprints Obtained for IIITD-MOLF Database Using the Proposed Algorithm over Different Values of Batch Size**

| Training Data | NFIQ Score |
|---|---|
| Without SD4 | 2.33 |
| With SD4 | 1.83 |

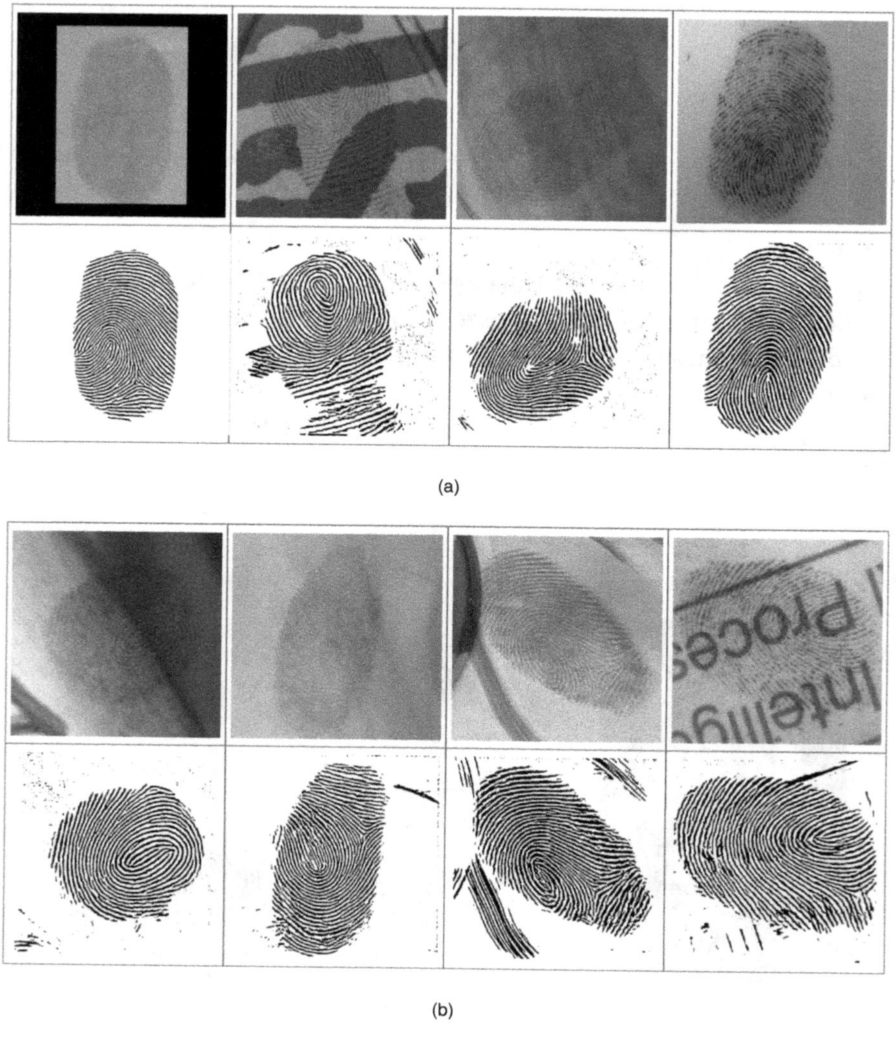

(a)

(b)

**FIGURE 3.16** Samples of successful enhancement of latent fingerprints by the proposed model.

1. We find that many of the input latent fingerprint images have low ridge information. However, even for such images, the proposed algorithm enhances those regions of the latent fingerprint image which have some ridge information (see the left-most column of Figure 3.17). We understand that it will be difficult for any enhancement algorithm to enhance such cases while preserving the minutiae details.

2. While matching latent fingerprint images, ROI is manually marked by forensic experts and the enhancement is performed only on ROI. However, the proposed algorithm automatically segments the foreground and background and then enhances the foreground fingerprint. Due to this, it sometimes misinterprets the background as foreground (see the last three columns from right in Figure 3.17) when the intensity distributions of background and foreground fingerprint are similar.

3. We found that the NFIQ is not a robust fingerprint quality assessment metric (see Figure 3.11). NFIQ2 is a more effective metric than NFIQ; however, it fails to process latent fingerprints. Thus, there is a need to introduce a more robust latent fingerprint quality assessment tool in the public domain to facilitate improved research in latent fingerprint matching.

4. The proposed model is observed to be highly sensitive to the choice of hyper-parameters and does not perform well if the training hyper-parameters are not carefully chosen.

5. The loss function is carefully designed for enhancement of latent fingerprints. Any change in the loss function can lead to unstable training of the model (as observed while training the model without enhanced reconstruction loss, as shown in Figure 3.10).

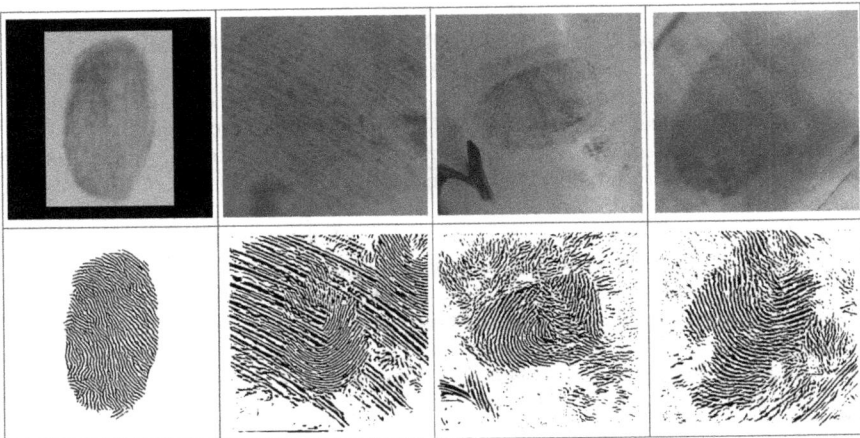

**FIGURE 3.17** Some challenging cases for the proposed model.

## 3.7  CONCLUSIONS

Motivated by the successful applications of GANs in various image processing applications, we formulate latent fingerprint enhancement like an image-to-image translation problem. The proposed model is trained using an enhancer and a discriminator network in an adversarial fashion. The model is trained using both synthetic and real fingerprints due to which it is robust to distortions observed in latent fingerprints. Moreover, the proposed model does not need a real latent fingerprint database to train the network. Two latent fingerprint databases available in the public domain are used for evaluating the proposed enhancement model. A detailed analysis of performance of model over hyper-parameters such as lambda, number of epochs, batch size is performed. We also gain insights on the role of real inked prints while training the model and the significance of reconstruction loss in the objective function.

We analyse the failure cases and some cases have been encountered when the ridge information is insufficient and the proposed algorithm generates spurious features. To address these limitations, the possibility of recoverability needs to be explored such that the algorithm can decide which portions of fingerprints can be reconstructed and which ones cannot. Training with a larger database with more variations in texture and background can help to achieve even better performance on IIITD-MSLF database. The proposed algorithm can also be utilised in challenging scenarios like latent to latent fingerprint matching.

## ACKNOWLEDGEMENTS

The authors thank IIT Delhi HPC facility for computational resources. The authors express their gratitude to Mayank Vatsa and Richa Singh from IIIT Delhi for their inputs. The authors also thank Himanshu Gandhi and Vijay Kumar for helpful discussions.

## REFERENCES

1. A. Sankaran, A. Agarwal, R. Keshari, S. Ghosh, A. Sharma, M. Vatsa, and R. Singh. Latent fingerprint from multiple surfaces: Database and quality analysis. In *IEEE 7th International Conference on Biometrics Theory, Applications and Systems*, 2015.
2. NBIS-NIST biometric image software. http://biometrics.idealtest.org/.
3. D. Nguyen, K. Cao, and A. K. Jain. Robust minutiae extractor: Integrating deep networks and fingerprint domain knowledge. In *2018 International Conference on Biometrics (ICB)*, 2018.
4. B. T. Ulery, R. A. Hicklin, J. Buscaglia, and M. A. Roberts. Repeatability and reproducibility of decisions by latent fingerprint examiners. *PloS one*, 7:e32800, 2012.
5. B. T. Ulery, R. A. Hicklin, J. Buscaglia, and M. A. Roberts. Accuracy and reliability of forensic latent fingerprint decisions. *Proceedings of the National Academy of Sciences*, 108, 2011.
6. S. Yoon, J. Feng, and A. K. Jain. On latent fingerprint enhancement. In *Biometric Technology for Human Identification VII*, 7667: 766707, 2010.
7. Neurotechnology inc., verifinger. http://www.neurotechnology.com.
8. S. Yoon, J. Feng, and A. K. Jain. Latent fingerprint enhancement via robust orientation field estimation. In *2011 International Joint Conference on Biometrics (IJCB)* pp. 1–8, 2011.

9. J. Feng, J. Zhou, and A. K. Jain. Orientation field estimation for latent fingerprint enhancement. *IEEE Transactions on Pattern Analysis and Machine Intelligence*, 35(4):925–940, 2013.

10. X. Yang, J. Feng, and J. Zhou. Localized dictionaries based orientation field estimation for latent fingerprints. *IEEE Transactions on Pattern Analysis and Machine Intelligence*, 36(5):955–969, 2014.

11. C. Chen, J. Feng, and J. Zhou. Multi-scale dictionaries based fingerprint orientation field estimation. In *International Conference on Biometrics* pp. 1–8, 2016.

12. K. Cao and A. K. Jain. Latent orientation field estimation via convolutional neural network. In *International Conference on Biometrics* pp. 349–356, 2015.

13. S. Liu, M. Liu, and Z. Yang. Sparse coding based orientation estimation for latent fingerprints. *Pattern Recognition*: 164–176, 2017.

14. W. Chaidee, K. Horapong, and V. Areekul. Filter design based on spectral dictionary for latent fingerprint pre-enhancement. In *2018 International Conference on Biometrics (ICB)* pp. 23–30, 2018.

15. Z. Qu, J. Liu, Y. Liu, Q. Guan, C. Yang, and Y. Zhang. Orienet: A regression system for latent fingerprint orientation field extraction. In *International Conference on Artificial Neural Networks*, pp. 436–446, 2018.

16. J. Li, J. Feng, and C.C. J. Kuo. Deep convolutional neural network for latent fingerprint enhancement. *Signal Processing: Image Communication*, 60:52–63, 2018.

17. J. Svoboda, F. Monti, and M. M. Bronstein. Generative convolutional networks for latent fingerprint reconstruction. In *IEEE International Joint Conference on Biometrics*, pp. 429–436, 2017.

18. I. Joshi, A. Anand, M. Vatsa, R. Singh, S. Dutta Roy, and P. K. Kalra. Latent fingerprint enhancement using generative adversarial networks. In *2019 IEEE Winter Conference on Applications of Computer Vision (WACV)*, pp. 895–903, 2019.

19. I. Goodfellow, J. Pouget-Abadie, M. Mirza, B. Xu, D. F. Warde, S. Ozair, A. Courville, and Y. Bengio. Generative adversarial nets. In *Advances in Neural Information Processing Systems 27*, pp. 2672–2680, 2014.

20. M. Mirza and S. Osindero. Conditional generative adversarial nets. *CoRR*, abs/1411.1784, 2014.

21. P. Isola, J. Zhu, T. Zhou, and A. A. Efros. Image-to-image translation with conditional adversarial networks. In *IEEE Conference on Computer Vision and Pattern Recognition*, 2017.

22. A. H. Ansari. *Generation and Storage of Large Synthetic Fingerprint Database*. Technical report, Indian Institute of Science, Bangalore, 2011.

23. R. Cappelli, D. Maio, and D. Maltoni. Sfinge: an approach to synthetic fingerprint generation. In *International Workshop on Biometric Technologies (BT2004)*, pp. 147–154, 2004.

24. NIST special database 4. http://www.nist.gov/srd/nistsd4.cfm.

25. E. Tabassi. NFIQ 2.0: NIST fingerprint image quality. *NIS-TIR* 8034, 2016.

26. A. Sankaran, M. Vatsa, and R. Singh. Multisensor optical and latent fingerprint database. *IEEE Access*, 3:653–665, 2015.

27. Z. Wang, A. C. Bovik, H. R. Sheikh, and E. P. Simoncelli. Image quality assessment: from error visibility to structural similarity. *IEEE Transactions on Image Processing*, 13, 2004.

28. R. Cappelli, M. Ferrara, and D. Maltoni. Minutia cylinder-code: A new representation and matching technique for fingerprint recognition. *IEEE Transactions on Pattern Analysis and Machine Intelligence*, 32(12): 2128–2141, 2010.

29. R. Cappelli, M. Ferrara, and D. Maltoni. Fingerprint indexing based on minutia cylinder-code. *IEEE Transactions on Pattern Analysis and Machine Intelligence*, 33(5): 1051–1057, 2011.

30. M. Ferrara, D. Maltoni, and R. Cappelli. Noninvertible minutia cylinder-code representation. *IEEE Transactions on Information Forensics and Security*, 7(6): 1727–1737, 2012.

# 4 DeepFake Face Video Detection Using Hybrid Deep Residual Networks and LSTM Architecture

*Semih Yavuzkiliç*
Fırat University

*Zahid Akhtar*
State University of New York Polytechnic Institute

*Abdulkadir Sengür*
Fırat University

*Kamran Siddique*
Xiamen University Malaysia

## CONTENTS

4.1   Introduction ............................................................................................82
4.2   Related Work ..........................................................................................84
    4.2.1   Categories of Face Manipulations .............................................84
        4.2.1.1   Face Synthesis.............................................................84
        4.2.1.2   Face Swap ...................................................................86
        4.2.1.3   Facial Attributes..........................................................86
        4.2.1.4   Face Expression ..........................................................87
    4.2.2   DeepFakes Detection..................................................................88
4.3   Proposed DeepFake Videos Detection Framework...............................91
    4.3.1   Convolutional Neural Networks (CNNs).....................................92
    4.3.2   Long Short-Term Memory (LSTM) .............................................93
    4.3.3   Residual Neural Network (ResNet) .............................................93
4.4   Experiments.............................................................................................94
    4.4.1   Datasets.......................................................................................94
        4.4.1.1   DeepFakeTIMIT Dataset.............................................94
        4.4.1.2   Celeb-DF Dataset........................................................94
    4.4.2   Figures of Merit .........................................................................94
    4.4.3   Experimental Protocol................................................................96
    4.4.4   Experimental Results..................................................................97

4.5    Challenges and Future Research Directions....................................................98
4.6    Conclusions..........................................................................................................99
Notes .............................................................................................................................100
References......................................................................................................................100

## 4.1    INTRODUCTION

Nowadays the fields of deep learning (DL) and artificial intelligence (AI) have expanded considerably. They are widely being used in many daily life applications and systems. Similarly, now the social networks play vital role in information and news dissemination. The advent of machine learning (ML)/AI together with social media has changed the perception of reality, especially in the digital world. For instance, the current technologies are being employed to create fake multimedia samples, which have become almost indistinguishable from the real ones. There exist readily available image and video manipulation software and apps (e.g., Face2Face, AgingBooth, FaceApp, PotraitPro Studio, and Adobe Photoshop), which do not need much technical knowledge. So that, the potential count of manipulated images and videos is very large. In fact, many of the videos posted either on the internet or on social media that become viral are fallacious and manipulated. Manipulated images/videos could be for benign reasons (e.g., images retouched for beautification) or antagonistic goals (e.g., fake news campaigns). In particular, fake face images/videos generated using DL techniques have recently gained a great public concern and attention. Such digitally manipulated facial samples are known as "DeepFakes", which are fake facial images/videos obtained by swapping the face of one individual by the face of other individual using AI/DL-based methods [1,2,3]. Representative examples of some of the most used publicly available apps are DeepFake, Face2Face, FaceApp, Face Swap Live. Such apps and software can be used to manipulate face age, facial hair, gender of the person, hair colour, facial expression, swapping two faces with each other, or generate synthetic facial samples of a person that does not exist in the real world, as also shown in Figure 4.1.

Although DeepFakes are mostly harmless and can be used for research or amusement purposes, the simple and easy to use software/apps can be utilised to produce audio and video imitations for theft, fraud, or revenge porn. DeepFake can influence the election results as fake videos can make people believe that a certain politician is saying things that he did not say or did. Likewise, fake evidences created with DeepFake

Age      Facial Hair      Gender      Hair Color      Expression      Face Swap      Synthetic Face

**FIGURE 4.1**    Examples of various face manipulations. First row: original face samples. Second row: manipulated face samples. Last column: a synthetically generated face.

techniques can be used against people in court, thereby innocent person can be charged with crimes they did not commit. On the other hand, guilty people can be released on the basis of false evidences. Also, people could alter their faces to appear younger or older to deceive age-based access controls. It has been shown that face ageing and face spoofing negatively affect the automated face recognition and identification systems [4–8]. In fact, DeepFakes not only can trick people but also degrade the accuracy of facial recognition system at the same time. For instance, Korshunov et al. [9] demonstrated that DeepFakes could escalate the error rates of VGG and FaceNet neural network-based face recognition approaches by 85.62% and 95.00%, respectively.

A typical countermeasure to DeepFakes is DeepFakes detection methods that target at distinguishing real face samples from manipulated faces [10–12]. For instance, The authors in Ref. [13] proposed a generalised metric-learning-based system that can detect Deepfakes from different datasets. Neubert et al. [14] designed and evaluated frequency and spatial domain feature spaces for face manipulation detection. Inspired by the recent success of DL frameworks in diverse set of applications such as object detection and autonomous car, researchers have studied and explored the efficacy of the DL techniques against face manipulation detection. Namely, in the last few years, DL schemes have successfully been used to detect DeepFakes. Especially, convolutional neural networks (CNNs) are employed to determine features from every frame for detection. Using part of the pre-trained CNN as the feature extractor is a proficient way to expand accuracy of face manipulation detection [15,16]. Also, a constrained convolutional layer [17], a statistical pooling layer [18], two-stream network [12], and two cascaded convolutional layers relied on the CNN [19] approaches were used for detection. Coherent survey of the prior DeepFakes methods demonstrated that the detection frameworks have progressed significantly and attained promising results but yet face difficulties in detecting sophisticated face manipulations [3]. Moreover, new and complicated face manipulations are hard to be noticed by existing forensics tools and human experts [13]. There is a huge demand to devise methods that attain impressive and improved accuracy.

In this chapter, we develop a hybrid framework method for DeepFake videos. The proposed method is composed of face detection, extraction of deep features, and long short-term memory (LSTM) classification. For a given video, first the face regions are detected in each frame. The detected face regions are fed to a pre-trained CNN model (i.e., FC1000 layer of the pre-trained deep residual network model) in order to extract feature. The extracted features are then used in seven layered LSTM model [i.e., input, two bidirectional LSTM (biLSTM), dropout, fully connected (FC), softmax, and output layers] for classification. Experimental analyses on the two public datasets (i.e., DeepFakeTIMIT and Celeb-DF) were performed using the false acceptance rate (FAR), the false rejection rate (FRR), and equal error rate (ERR) metrics. The proposed framework on DeepFakeTIMIT obtained 2.4217% EER and 0.0795% FRR@FAR10 (FRR percentage when FAR as 10% was used as the performance evaluation threshold). Similarly, on Celeb-DF dataset, it achieved 0.5014% EER and 0% FRR@FAR10. Moreover, the proposed framework outperformed the previously proposed DeepFake detection methods.

The remaining part of the chapter is structured as below. Section 4.2 outlines existing works on face manipulation. The developed method is detailed in Section 4.3.

Experimental database, figures of merit, experimental protocol, and empirical analyses are described in Section 4.4. Few future research directions and open issues are presented in Section 4.5. Section 4.6 outlines conclusions.

## 4.2  RELATED WORK

Advanced face manipulation can deceive both humans and automated face identification systems. Especially, DeepFakes is a significant issue as manipulated face/multimedia with false information could be briskly spread using messaging and social networking platforms. Such spread of fake information may create turbulent ramifications. For instance, an ex-lover can alter the content and person in a video (e.g., face swapping) to produce false revenge porn video, which may lead to the victim ending their life, especially if the victim is young person.

### 4.2.1  CATEGORIES OF FACE MANIPULATIONS

The manipulated faces are generally produced by altering facial features (e.g., gender and age), swapping two faces with each other (also known as face morphing), augmenting unnoticeable perturbations (also known as adversarial examples), artificially generating faces, or re-enacting/animating face expressions in the facial videos/images [2]. By analysing the existing face manipulations systematically, we can broadly group all manipulations into four categories: face synthesis, face swap, facial attributes, and face expression [3,20].

#### 4.2.1.1  Face Synthesis

This manipulation technique produces human faces that do not exist in real world, by generally using generative adversarial networks (GANs) [21]. These methods obtain astounding results, yielding face samples that are almost indistinguishable from real ones. For example, Karras et al. [22] proposed StyleGAN architecture, which is an enhanced version of ProGAN approach [23] and can generate entire nonexistent faces.

##### 4.2.1.1.1  Face Synthesis Generation Methods and Datasets
Algorithmic architectures of GANs can be described by using two neural networks that are named generator and discriminator. First, the generator creates fake face images of realistic quality, while the discriminator distinguishes face images among real and fake samples. When the discriminator cannot discriminate among real and fake images, the result is images that are not in reality but seemingly identical to reality. By taking advantage of the GAN, CycleGAN [24] is proposed that learns unsupervised image-to-image translation. Shen et al. [25] proposed FaceID-GAN that considers the facial identity classification as the third actor and contend with generator by discriminating the identifications of real and synthesised facials.

There exist some public face synthesis datasets for research. Figure 4.2 shows examples of three different datasets of synthesised faces. The common feature of these datasets is that none of them contain any real person's pictures. Therefore, researchers focusing on this kind of manipulation detection or recognition often use

**FIGURE 4.2**   Examples of different datasets, which are composed of synthesised faces.

real faces from popular public databases to train their systems. Following, we briefly describe the datasets:

- **100K-Faces**[1]: This dataset consists of 100,000 synthetic images created by using StyleGAN. In the dataset, the StyleGAN was trained with approximate 29,000 images of 69 different identities, and facial samples with a plain background were produced.
- **TPDNE**: This dataset contains 150,000 synthetic face images gathered on the website.[2] The synthetic face images are relied on the StyleGAN technique trained with the FFHQ[3] dataset.
- **DFFD**: Stehouwer et al. [20] presented a dataset, which is called Diverse Fake Face Dataset (DFFD). The authors used two pre-trained models for face synthesis manipulation. The 100,000 and 200,000 fake images were created using ProGAN and StyleGAN models, respectively.

#### 4.2.1.2   Face Swap

In this face alteration, a person's face is replaced by face of another one. Face swapping can be done by either traditional computer graphics-based schemes or new DL methods/techniques. There exist many popular mobile apps for this purpose, e.g., Snapchat. Moreover, for face swapping, recently many works have been published in the literature. For instance, Marcel et al. [9] created face DeepFakes dataset utilising a GAN-based face swapping scheme.[4]

##### 4.2.1.2.1   Face Swap Generation Methods and Datasets

Face Swap is one of the increasingly popular manipulation techniques. Below, we summarise some publicly available datasets.

- **UADFV [26]:** It comprises 49 real videos downloaded from YouTube and 49 manipulated videos obtained from these videos using the FakeApp application. Each video stands for a person with specifically $294 \times 500$ pixels resolution and an average of 11.14 seconds duration.
- **FaceForensics++ [1]:** This dataset contains 1,000 real videos selected on YouTube and 1,000 manipulated videos. Manipulations were generated using the faceswap[5] application.
- **DFDC[6]:** The DeepFake detection challenge (DFDC) dataset is presented by Facebook DFDC. It contains 1,131 real of 66 actors' videos and 4,119 forged videos at first. Forged videos were created using two different approaches; the details of these algorithms however are not disclosed. On December 11, 2019, the entire DFDC dataset was released and the competition started.

#### 4.2.1.3   Facial Attributes

Some face attributes (e.g., skin colour and gender) are altered in this category. Adobe Photoshop, AgeingBooth, and FaceApp are some of the popular apps for this type of manipulations. Also, He at al. [27] developed a scheme called attGAN, which can manipulate beard, young, age, hair colour, and mouth face traits while preserving identity of the person as well as other facial details.

##### 4.2.1.3.1   Face Attributes Generation Methods and Datasets

Since the code of most GAN techniques is publicly available, there are a few datasets known in the literature regarding face attributes exploiting such GAN techniques. Chang et al. [28] proposed a two-stage technique to generate face attribution: Texture Completion GAN (TC-GAN) and 3D Attribute GAN (3DA-GAN). The TC-GAN automatically removes the missing appearance from congestion and supplies a normalised UV texture. The 3DA-GAN operates on the UV texture area to create target attributes with the maximum protected identity of subject. Moreover, for complex picture alteration, Perarnau et al. [29] presented an approach, which is called IcGANs (Invertible Conditional GANs) based on a combination of an encoder utilised collectively accompanied by conditional GAN. This method gives certain outcomes for altering qualities, although it critically modifies one's facial identification.

Some facial attributes manipulation datasets have been made public, which can be utilised for research purposes. Some of them are detailed in the following.

- **CelebA [30]**: The Celeb-Faces Attributes (CelebA) dataset was obtained by tagging images chosen from a large-scale face feature dataset, CelebFaces [31]. The dataset consist of 10,177 identities, over than 202,000 facial images with five locations of landmark and 40 binary attribution for each images.
- **PubFig [28]**: This dataset consists of 58,797 images which belong to 200 people. Since it is obtained from the internet under uncontrolled conditions, it consists of remarkable variations in poses, expressions, etc. It labels 73 face attributes.
- **Attribute 25K [32]**: This dataset contains 24,943 people images, which are collected from Facebook. Not all features can be labelled for every image as images vary greatly in perspectives, poses, and occlusions. For example, if the person's head is not visible, it cannot be labelled with glasses.

### 4.2.1.4   Face Expression

This manipulation technique replaces one person's face expression by face expression of another one. Thies et al. [33] presented a technique that works on real-time videos for facial expression manipulation. The presented technique is called Face2Face.

#### 4.2.1.4.1   Face Expression Generation Methods and Datasets

One of the well-known databases that has focused on facial expression manipulation to date is FaceForensics ++ [1]. This dataset is an extension of FaceForensics [16]. At first, the FaceForensics dataset concentrated only on Face2Face, a computer graphics method that hands on the source identity expression to the target identity when preserving the identification of the target. It was accomplished by choosing of manual keyframe. Later, fake samples were created through transferring source expressions of every frame to the target video. Next, the same researchers introduced a new learning approach relied on NeuralTextures [34] in FaceForensics ++. The approach is rendering-based, which utilises real video data learning the neural appearance of target person, with the inclusion of a rendering network. The researchers rated it as a GAN-loss utilised in Pix2Pix [35], which is patch-based, in their applications. Only it was changed face expression corresponding to the mouth. The dataset contains 1,000 real videos downloaded from YouTube. It contains a total of 2,000 fake videos, 1,000 each, for each approach considered.

There are several apps available that can be utilised to manipulate facial expressions. For instance, Face2Face that is based on existing GAN algorithmic structures allowing it to easily change facial expressions. Similarly, with the StarGAN approach proposed in Ref. [36], the authors showed that a person's face image can be changed with different expressions. As far as we know the only database that is obtainable for research purposes is FaceForensics ++ [1].

## 4.2.2 DEEPFAKES DETECTION

DeepFakes detection could be considered like a two-class classification problem in which salient characteristics from the given face sample is extracted to be then fed to a classification scheme to attain the binary outcome: DeepFakes or Benign. DeepFakes detection techniques can be broadly categorised into three main classes: textural-, inherent attribute degradations-, and DL-based methods.

**Textural-Based Methods**: Approaches in this part analyse sample's textural properties to assign the actuality of the given sample. For example, Zahid et al. [37] studied the impact of ten local image feature descriptors for identifying DeepFakes, such as Frequency Decoded Local Binary Pattern (FDLBP), binarised statistical image features (BSIF), CENsus TRansform hISTogram (CENTRIST), and Binary Gabor Pattern (BGP). Similarly, authors in Ref. [38] employed local descriptors such as Local Binary Patterns (LBP), Histogram of Gradients (HOG), and Scale Invariant Feature Transform (SIFT) to detect manipulated faces.

**Inherent Attribute Degradations-Based Methods**: These algorithms analyse modifications in image/video's natural properties based on quality, noise, etc. For instance, the authors in Ref. [39] developed a scheme for manipulated faces that depends on analysis of the Fourier spectrum of sensor pattern noise. In Ref. [40], biological signals (head motion-based ballistocardiogram and photoplethysmography) were employed for determining DeepFakes.

**DL-Based Methods**: Frameworks in this group employ neural networks. For instance, in Ref. [12] the authors used two-stream CNNs with the underlying GoogleNet model [41] for face tampering detection. The first stream detects interfering artefacts on a facial, whereas the second one is a trained patch-based triple net. The framework was trained on the unpublished dataset, which was created by authors with the help of SwapMe and FaceSwap applications. Their method obtained 99.9% detection accuracy. In Ref. [11], the authors suggested a CNN-based detection method. The proposed method centred on mesoscopic properties of images by using a deep neural network with few layers. The technique was trained on the datasets that was collected by authors and were generated from the hyper realistic forged videos. Authors reported 98.4% and 95.3% accuracy for Deepfake and Face2Face datasets, respectively. In Ref. [10], the authors presented a temporal-aware pipeline for detection of DeepFake. Frame features were extracted using CNNs that were then utilised to train a recurrent neural network (RNN). The results of the experiment were reported on a dataset collected by the authors with 97.1% accuracy. Li et al. [42] suggested a new DL-based approach for distinguishing DeepFake videos. The proposed method detected the artefacts by matching the synthesised facial regions and their neighbourhood areas with a CNN model. In particular, they used residual network with 50 layers (ResNet-50) [43] model to detect the DeepFake videos to uncover the facial warping artefacts brought about by rescaling and interpolation processing in fundamental DeepFake builder algorithms. Experiments were performed on UADFV [26] and DeepfakeTIMIT LQ and DeepfakeTIMIT HQ [9] datasets, with 97.4%, 99.9%, and 93.2% accuracy rate, respectively.

McCloskey et al. [44] examined the GAN pipeline for detection of various artefacts among real and fake pictures. The authors introduced a detection method relied on colour characteristics and a linear support vector machine (SVM) to classify.

The method performed on NIST MFC2018 [45] dataset. The authors obtained 70.0% of area under the curve (AUC) for the proposed approach. In Ref. [46], the authors provided a detection approach originated in natural image statistics and steganalysis. The proposed technique was particularly relied on a compound of pixels co-occurrence matrices and CNNs. Their method was primarily tested on a dataset of different objects generated using the CycleGAN approach. In addition, the researchers made an attractive analysis to observe the validity of the suggested method to fake images generated through two GAN structures (CycleGAN and StarGAN) with well generalisation consequences. Later, Neves et al. [47] applied this detection method on the 100K-Faces dataset and obtained an EER result of 7.2%.

Tariq et al. [48] introduced CNNs to detect face attribute manipulations. They used various CNN approaches such as VGG16 [49], VGG19 [49], residual neural network (ResNet) [43], and XceptionNet [50]. The CelebA dataset is used for the real face images. Two different approaches were taken into account for the fake images: (i) machine approaches relied on GANs and (ii) Adobe Photoshop CS6-based manual approach with the inclusion of manipulations such as makeup, glasses. They achieved 99.99% and 74.9% AUC scores for the machine-created scenario and manual-created scenarios, respectively. In Ref. [20], the authors presented a detection method relied on CNN with attention techniques to operate and develop feature maps for the classifier model. The considered attention map could be applied comfortably and inserted into present backbone networks. In this study, the authors performed the detection method on DFFD dataset. The considered method was obtained 99.43% of AUC and 3.1% of EER particularly for face swap detection. In Ref. [51], the authors suggested that a DL method depend upon restricted Boltzmann machine for the detection of face images with digital retouching. The detection method input contains facial patches to learn discriminating features for classifying each picture as original or forged. The experiments were performed on two different fake face datasets, which are ND-IIITD dataset (collection B) and containing facial pictures from a number of celebrities downloaded on the Internet. They obtained 96.2% accuracy rate for the dataset and 87.1% accuracy for the ND-IIITD dataset.

Nguyen et al. [52] employed capsule structures [53] that rely on a VGG19 [49] network to detect forged face images and videos. The method achieved 97.05% performance accuracy on FaceForensics++ [1]. Matern et al. [54] presented a detection approach predicated on capture visual artefacts in the eyes, teeth, and face circumferences of the forged facials. Two different approaches were used which were multi-layer feedforward neural network and logistic regression (LogReg) model classifiers. A 0.866 AUC value was obtained on an unpublished dataset which Deep-Fake videos from YouTube, and images were cropped from CelebA dataset [30]. Dolhansky et al. [55] using the DFDC dataset have produced basic results using three simple detection approaches. To detect low-level image manipulations, they used CNN architecture. Then, an XceptionNet approach was trained using just facial frames. Finally, another XceptionNet approach was trained using the entire image. The detection technique relied on XceptionNet with just the facial images that obtained 93.0% and 8.4% scores precision and recall, respectively. In Ref. [56], the authors presented a method for detection of fake samples that relied on temporal data that exist in the stream. The perception behind the proposed method is to take advantage of temporal differences between the frames. Therefore, instead of using a pre-trained model, they used a recurrent convolutional network similar to [10] end-to-end trained. The authors performed experiments on FaceForensics++ dataset, used only low-quality videos, and obtained 96.3% and 96.9% of AUC scores for the FaceSwap and DeepFake, respectively.

The work in Ref. [57] introduced a detection approach that relied on head movements and facial expressions. To extract features, OpenFace2 toolkit [58] was taken into account. They obtained 18 different face actions with an intensity and occurrence. Moreover, for head movements, four features were considered. SVM was performed for the last classification. The researchers created their own datasets by downloading videos from YouTube. They obtained 96.3% of AUC for the best performance. Nguyen et al. [59] designed a multitask learning CNN model for simultaneously detecting manipulated videos/images and locating the manipulated regions in the given image/video. In total, 92.77%, 92.50%, 52.32%, and 83.71% classification accuracy rates were achieved for FaceForensics [16] and FaceForensics++ [49]. Amerini et al. [60] introduced a two-step method. First, they applied optical flow fields [61] for feasible inter-frame dissimilarity features. Then, these features were fed to CNN classifiers. After training on CNN, Resnet-50 [43] and VGG16 [49] were used for testing. The proposed model achieved 75.46% and 81.61% accuracy rate for Resnet-50 and VGG16, respectively, on FaceForensics++ dataset. Yang et al. [26] detected DeepFake videos by examining the incoherencies in the head poses with an SVM model. The areas under receiver operating characteristic curve based on videos and frames were 0.974 and 0.89, respectively, using UADFV dataset. A brief summary of the model architecture, method, used datasets, and publication year for several DeepFake manipulation detection methods is given Table 4.1. Moreover, all in all, the DeepFakes generation, detection, and recognition taxonomy, as also defined in Ref. [2], can be depicted as in Figure 4.3.

---

**TABLE 4.1**

**A Representative List of DeepFake Detection Methods**

| Methods | Techniques | Dataset | Year |
|---|---|---|---|
| Zahid et al. [37] | Local Image Features | DeepfakeTIMIT [9] | 2019 |
| Ciftci et al. [40] | Photoplethysmography + Power Spectrum + Statistical Features | FaceForensics [16] | 2019 |
| Zhou et al. [12] | GoogleNet model [41] | Private dataset | 2017 |
| Afchar et al. [11] | Designed CNN | Private dataset | 2018 |
| Güera and Delp [10] | RNN | Private dataset | 2018 |
| Li et al. [42] | ResNet-50 [43] | Private dataset | 2018 |
| Yang et al. [26] | SVM | UADFV [26] | 2018 |
| Nguyen et al. [52] | Capsule Network [53] | FaceForensics++ [1] | 2019 |
| Matern et al. [54] | CNN and Logistic Regression Model | Private dataset | 2019 |
| Nguyen et al. [59] | CNN | FaceForensics [16] and FaceForensics++ [1] | 2019 |
| Amerini et al. [60] | Resigned CNN | FaceForensics++ [1] | 2019 |
| Stehouwer et al. [20] | CNN + Attention Mechanism | DFFD [20] | 2019 |
| Dolhansky et al. [55] | CNN | DFDC[6] | 2019 |
| Sabir et al. [56] | CNN + RNN | FaceForensics++ [1] | 2019 |
| Nataraj et al. [46] | CNN | 100K-Faces[1] | 2019 |

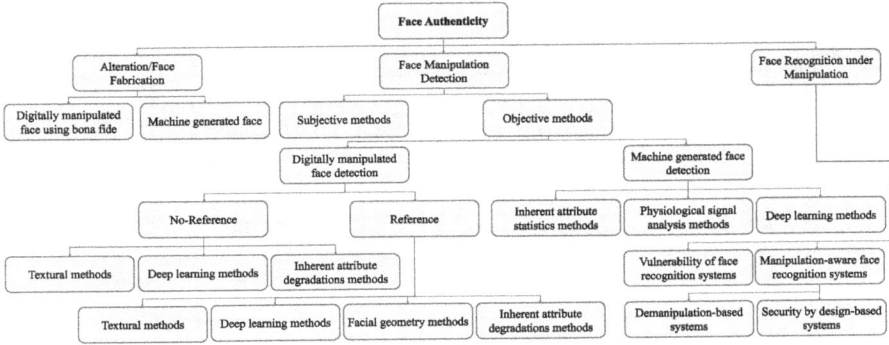

FIGURE 4.3 Taxonomy of DeepFakes detection, generation, and recognition [2].

## 4.3 PROPOSED DEEPFAKE VIDEOS DETECTION FRAMEWORK

Face DeepFakes detection could be discerned as a two-class classification system. In this system, the input video has to be flagged as either real or fake. Figure 4.4 represents the overall schematic of the developed DeepFake detection technique. The proposed technique is composed of three stages, namely, face detection in given frames, feature extraction on detected face regions, and feature classification. In this study, the cascade face detector is used to detect faces [62]. Specifically, the Viola–Jones algorithm is employed in the cascade face detector where the algorithm detects the faces, noses, eyes, and mouth [63]. The detected face regions are then cropped and resized to 224 × 224 in order to be compatible with the input of the ResNet-50 model. The ResNet-50 model is a deep CNN architecture that contains 50 layers. Most of these layers are convolutional layers, and few layers are pooling layers. The FC layer (fc1,000) of the ResNet-50 model is utilised to extract deep features for the cropped face regions. The extracted features are 1,000 dimensional, which are given to the sequential input layer of the LSTM classifier. The biLSTM layer comes after sequential input layer. FC, softmax, and classification layers follow each other to detect the fake faces.

In the following, we detail the CNNs, LSTM, and ResNet.

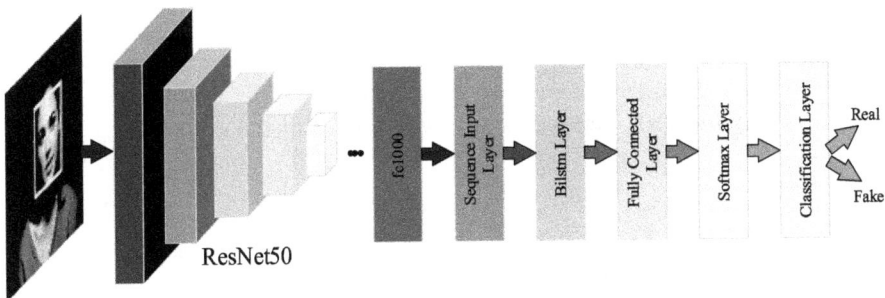

FIGURE 4.4 Proposed DeepFake video detection framework.

## 4.3.1 CONVOLUTIONAL NEURAL NETWORKS (CNNs)

A CNN architecture is usually made up of different layers, which are utilised consecutively to construct different architectures corresponding to different tasks. These layers may be convolution, pooling, normalisation, and FC layers [64]. The convolution layers are performed to produce features via input pictures. Let $X_i^{l-1}$ be the features extracted from the previous layers, $b_j^l$ be the training bias operated to avoid overfitting and $k$ be the learnable kernels [64]. The feature map output is evaluated as follows:

$$X_j^l = f\left( \sum_{i \in M_j} X_i^{l-1} \times k_{ij}^l \right) \tag{4.1}$$

where $M_j$ indicates the input map choice, and $f(.)$ is the activation operator (function). The pooling layer is performed to accomplish the feature maps down sampling, which is transmitted through the convolution layer. In the literature, different pooling techniques like mean and maximum pooling are utilised. The computational nodes are reduced via pooling layers, and pooling layers estop the overfitting problem in the CNN structure [65]. The pooling is identified as follows:

$$X_j^l = \text{down}\left( X_j^{l-1} \right) \tag{4.2}$$

where the down(.) function conducts the down sampling operation. It should be noted that down sampling provides a summary of topical features that are then used in the following layers. FC layers pass all connections with all activations in the foregoing layer. FC layers supply distinguishing properties to classify the input frame into different classes. The FC layers' activations are calculated using matrix multiplication followed by the bias [65]. The CNN's training is conducted by employing an optimisation scheme in (Equation 4.3). For neural networks, adaptive moment estimation (ADAM) and Stochastic gradient descent accompanied by momentum (SGDM) are two acknowledged training methods. The weights in the SGDM method are updated on a regular basis for each training set to achieve the target at the earliest point [66]:

$$V_t = \beta V_{t-1} + \alpha \nabla_w L(W, X, y) \tag{4.3}$$

where $W$, $\alpha$, and $L$ denote the weights, learning rate, and the loss function, respectively. Through the CNN training, new weights are computed as follows:

$$W = W - \alpha V_t \tag{4.4}$$

The optimiser of ADAM uses the mean of the second moments of slopes, updates the learning rate in each iteration, and adopts the learning rate parameter predicated on the mean of the first moment in the RMSProp method [67].

## 4.3.2 LONG SHORT-TERM MEMORY (LSTM)

The LSTM is an exclusive type of RNNs [68–70]. LSTM is usually considered far robust than feed forward neural networks and RNNs because of memory blocks and recurrent connections in the recurrent hidden layer [71]. The LSTM is very effective in classification and regression problems [69,70]. Memory blocks of the LSTM have self-connected memory cells, which at every time step store the transient states of the network. Information flow is supplied via an input to memory units/cells. Then, it passes from there to the other units by the gates. A forget unit (gate) is employed to scale internal condition/state of the cell before adding to the memory cell as an input. It is performed by repeating the memory cell itself and, if necessary, sets anew or omits memory of the cell. The forget gate is controlled by an activation function with a one-layer neural network identified as below:

$$f_t = \alpha\big(W\big[x_t,\, h_{t-1},\, C_{t-1}\big] + b_f\big) \qquad (4.5)$$

where $C_{(t-1)}, h_{(t-1)}, x_t,$ and $b_f$ are the previous LSTM block memory, the output of previous block, the input sequence, and the bias vector, respectively. The logistic sigmoid function and the weight vector assigned for each input are denoted as $\alpha$ and $W$, respectively. The activation operator is implemented to the foregoing memory structure/block. It determines the preceding memory structure/block effect on the ongoing LSTM with element-wise accumulation (multiplication). The value of activation set/vector output is checked and if it is almost zero, then preceding memory is forgotten.

In the input gate, a simple neural network produces a new memory by taking into account the impact of preceding memory block and the tanh activation function. The related process is as follows:

$$i_t = \alpha\big(W\big[x_t,\, h_{h-1},\, C_{t-1}\big] + b_i\big) \qquad (4.6)$$

$$C_t = f_t C_{t-1} + i_t \tanh\big(W\big[x_t,\, h_{h-1},\, C_{t-1}\big] + b_c\big) \qquad (4.7)$$

where $i_t$, $b_i$, and $W$ indicate outcome of the input gate, the bias vector, and weights, respectively. $h(t-1)$ shows the outcome of preceding block, $C(t-1)$ demonstrates the foregoing LSTM memory, and $\alpha$ parameter denotes the activation function [70–72]. In respect of the output (outcome) gate, it can be considered as a branch where outcome of the ongoing LSTM structure/block is generated by considering the following formulas:

$$o_t = \alpha\big(W\big[x_t,\, h_{h-1},\, C_{t-1}\big] + b_o\big) \qquad (4.8)$$

$$h_t = \tan h\big(C_t\big) \cdot o_t \qquad (4.9)$$

## 4.3.3 RESIDUAL NEURAL NETWORK (RESNET)

The ResNet was developed by He et al. with 152-layer-deep CNN architecture [43]. The ResNet attempts to address the vanishing gradient problem occurring during back-propagation of CNN. The ResNet architecture presented residual connections (skip connections) to prevent loss of information during deep network training. Skip connection technique enables to train very deep networks to improve the model performance. The residual blocks are the main building blocks of the ResNet architecture. The architecture of ResNet contains connections through residual blocks, while the consecutive hidden layers are connected to another one in shallow neural networks. The preservation of the gained knowledge throughout training session and increasing the network capacity resulting in speeding up the time of the training of the model are two of the most significant advantages of residual connections in the architecture of ResNet. In this study, we focused on the ResNet-50, which is the residual DL network with 50 layers.

## 4.4  EXPERIMENTS

This section presents an empirical evaluation of developed DeepFakes detection framework.

### 4.4.1  DATASETS

Two publicly available datasets, i.e., DeepFakeTIMIT and Celeb-DF, were used in this study.

#### 4.4.1.1  DeepFakeTIMIT Dataset

The dataset of DeepFakeTIMIT [9] consists of two equal sized subsets of low-quality (LQ) and high-quality (HQ) DeepFakes generated using the dataset of VidTIMIT [70]. Both LQ and HQ subsets include 320 videos with the pixels of $64 \times 64$ and $128 \times 128$, respectively. In this work, we used HQ subsets as it is more difficult subset and showing results on this subset shows the efficacy of the proposed framework.

#### 4.4.1.2  Celeb-DF Dataset

The Celeb-DF dataset [73] consists of 590 and 5,639 real and DeepFake videos, respectively. It corresponds to more than two million video frames. The real videos are taken from YouTube videos of 59 celebrities with different gender, age groups, and ethnic groups. The generation of the DeepFake videos is done by face swapping for each pair of the specified 59 subjects. Figure 4.5 shows some examples of real video frames from both datasets, whereas Figure 4.6 shows the corresponding detected face regions. Similarly, Figure 4.7 shows examples of fake video frames from both datasets, and Figure 4.8 shows the corresponding detected fake face regions.

### 4.4.2  FIGURES OF MERIT

A face DeepFakes detection framework is subject to two kinds of errors, i.e., false rejection rate (FRR) and FAR. The FRR is percentage of real samples classified as DeepFakes, while the FAR is percentage of DeepFake samples incorrectly classified as real samples. In this work, the efficacy of the developed framework was evaluated using FRR, FAR, and EER. EER metric describes the accuracy of the system when FAR is equal to FRR, i.e., FAR% = FRR%.

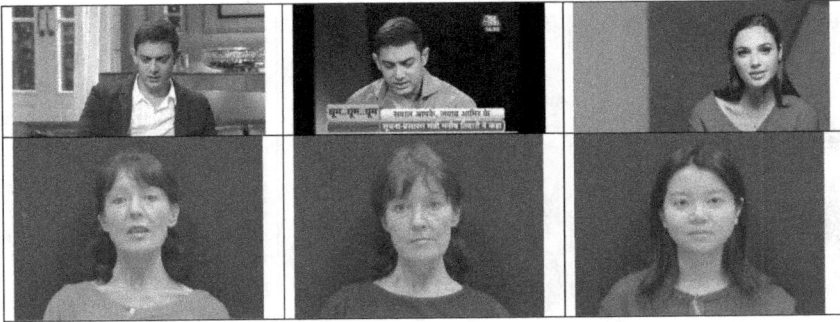

**FIGURE 4.5** Real video frames: the first row shows some frames from Celeb-DF dataset and the second row is some frames from DeepFakeTIMIT dataset [9].

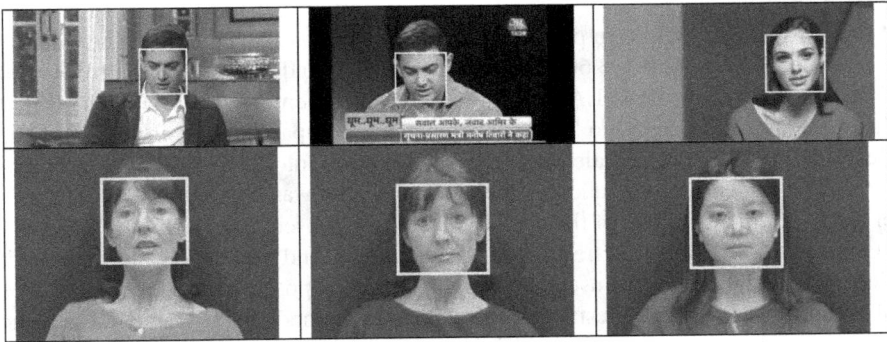

**FIGURE 4.6** Detected face regions on given frames. The first row shows some detected faces for Celeb-DF dataset and the second row shows some detected faces for DeepFakeTIMIT dataset [9].

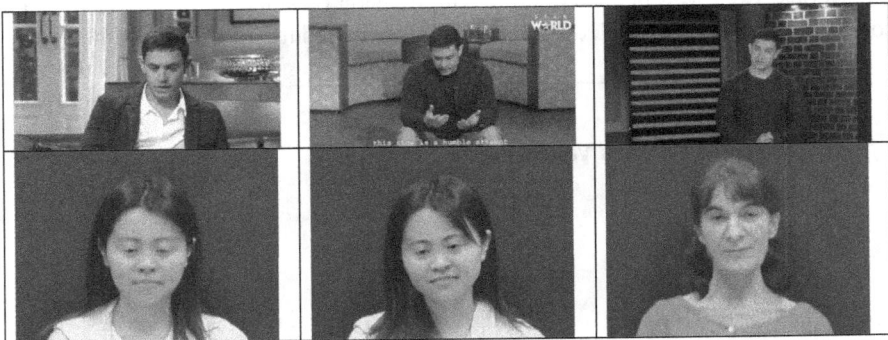

**FIGURE 4.7** Fake video frames. The first row shows some fake frames from Celeb-DF dataset and the second row shows some fake frames from DeepFakeTIMIT dataset [9].

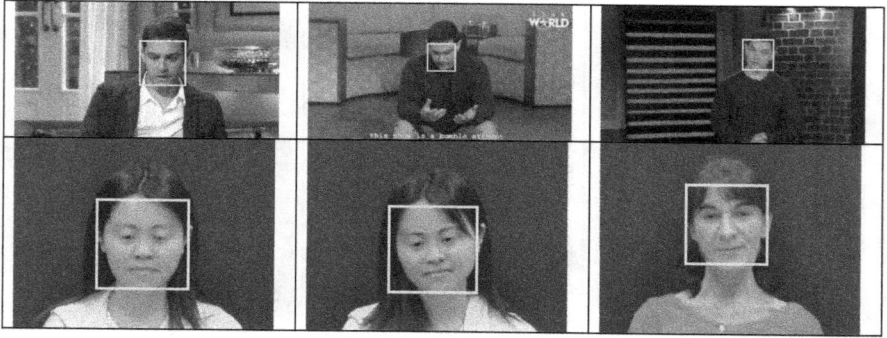

**FIGURE 4.8** Detected face regions on given fake frames. The first row shows some detected fake faces for Celeb-DF dataset and the second row shows some detected fake faces for DeepFakeTIMIT dataset [9].

### 4.4.3 Experimental Protocol

All empirical analyses were performed on a workstation that contained the Intel(R) Xeon(R) CPU E5-1650 @3.60 GHz 64 GB memory and NVIDIA Quadro M4000 GPU. We utilised the MATLAB (R2018b). As it was mentioned earlier, the detected face regions were resized to $224 \times 224$ for being compatible with the input of the ResNet-50 model. The rescaled face regions were fed into ResNet-50 to extract features. The dimension of sequence intake (input) layer was set to be 1,000. The size of hidden elements (units) of the biLSTM layer was chosen to be 100. The outputs of FC were two. In Table 4.2, training criterion, values, and parameters are presented. The 75% of datasets were used in training of the proposed method, and rest 25% of datasets were used for testing of the proposed method. Using Table 4.2 and the parameters described in there, the training procedure was conducted out 7,560 recurrences. The number 63 was set as the iteration number in every epoch. The zero-centre normalisation was employed to normalise the data before it was applied to the LSTM network. To this end, each feature-wise average and standard divergence of entire sequence calculation were performed. Then, for every training input/sample, the average value was deducted along with division by standard deviation. "Adam" solver, in this study, was selected as the training procedure for the LSTM network.

### TABLE 4.2
### Training Variables Values

| Methods | Techniques |
|---|---|
| Maximum epoch | 120 |
| Mini-batch size | 20 |
| Initial learning rate | 0.001 |
| Learn rate schedule | Piecewise |
| Learn drop period | 100 |
| Learn drop factor | 0.001 |
| Gradient threshold | 1 |

### 4.4.4  EXPERIMENTAL RESULTS

In Table 4.3, we report the evaluation of the devised DeepFake video detection frame-work, for both datasets, in terms of EER and FRR@FAR10% by employing the thresh-old when FAR = 10%. It can be seen in Table 4.3 that for both datasets, the proposed method was able to produce reasonable results. For example, 2.4217 EER(%) and the 0.0795 FRR@FAR10% (%) scores were obtained for DeepFakeTIMIT dataset, while 0.5014 EER(%) and the 0 FRR@FAR10% (%) scores were obtained for Celeb-DF data-set. These scores are quite low when compared with other methods in the literature.

We further compared the performance of the proposed DeepFake video detection method with some of the existing methods from the literature. Table 4.4 shows some results that were obtained by using various local descriptors adopted in Ref. [37] such as LBP, Pyramid of Histogram of Oriented Gradients (PHOG), SIFT, CENTRIST, BSIF, Local Phase Quantisation (LPQ), BGP, Quaternionic Local Ranking Binary Pattern (QLRBP), FDLBP, and Speeded Up Robust Features (SURF) [37].

**TABLE 4.3**
**Performance Evaluation of the Developed Method on Both Dataset in Terms of EER (%) and FRR@FAR10% (%)**

| Dataset | EER (%) | FRR@FAR10% (%) |
|---|---|---|
| DeepFakeTIMIT | 2.4217 | 0.0795 |
| Celeb-DF | 0.5014 | 0.00 |

**TABLE 4.4**
**Performance Comparison of the Proposed Method with Existing Methods on DeepFakeTIMIT Dataset in Terms of EER (%) and FRR@FAR10% (%)[a]**

| Method | EER (%) | FRR@FAR10% (%) |
|---|---|---|
| IQM + SVM [9] | 8.97 | 9.05 |
| LPB [37] | 17.16 | 43.02 |
| FDLBP [73] | 37.19 | 88.30 |
| QLRBP [73] | 27.70 | 59.49 |
| BGP [73] | 13.33 | 15.99 |
| LPQ [73] | 13.69 | 16.53 |
| BSIF [73] | 60.88 | 93.69 |
| CENTRIST [73] | 11.43 | 13.12 |
| PHOG [73] | 89.70 | 100 |
| SIFT [73] | 57.58 | 95.43 |
| SURF [73] | 67.26 | 98.17 |
| Proposed method | **2.42** | **0.080** |

[a]  Bold values are indicating the best results.

**TABLE 4.5**

**Accuracy Comparison of the Proposed Framework with Prior Techniques on Celeb-DF Dataset in Terms of EER (%)**

| Method | EER (%) |
|---|---|
| Face X-ray-Blended [74] | 31.16 |
| Xception-based method [16] | 59.64 |
| Face X-ray-Blended+FaceForensics [74] | 26.70 |
| Proposed method | 0.5014 |

In Table 4.4, it can be observed that among local image descriptors employed in Ref. [37], the CENTRIST descriptor attained the better classification accuracy of 11.43% EER(%) and 13.12% FRR@FAR10% (%). Moreover, the BGP and LPQ descriptors produced 13.33% EER(%) and 15.99% FRR@FAR10% (%) and 13.69% EER(%) and 16.53% FRR@FAR10% (%), respectively. The worst achievement was produced by the PHOG descriptors, where 89.70% EER(%) and 100% FRR@FAR10% (%) scores were obtained. However, the proposed method outperformed the other considered local descriptors such that 2.43% EER(%) and 0.080% FRR@FAR10% (%) scores were obtained by the proposed technique. Similarly, the proposed technique achieved better results than the image quality feature-based method developed in Ref. [9]. This high performance of the method presented in this study was obtained because of the deep features where both colour and texture features were combined in the deep CNN model.

In Table 4.5, we report a comparison of the proposed framework with the prior techniques on Celeb-DF dataset in terms of EER (%). In the table, we could see that the developed framework outperformed prior techniques in detecting DeepFake videos. For instance, the proposed method attained 0.5014% EER, while Face X-ray-Blended [74] and Xception-based method [16] achieved 31.16% and 59.64%, respectively. The authors in Ref. [74] proposed Face X-ray technique based on blending operation and CNNs with different training procedures such as blended forgeries/images (i.e., Face X-ray-Blended) and state-of-the-art manipulations (Face X-ray-Blended+FaceForensics). On the other hand, the authors in Ref. [16] employed pre-trained XceptionNet [50] with fine-tuning for DeepFake video detection. Compared to the proposed method, the frameworks developed in Refs. [74] and [16] require large training datasets in order to attain lower error rates, as also pointed out by the authors and other publications. However, the framework developed in this study comparatively demands a smaller number of training samples to obtain good performances.

## 4.5   CHALLENGES AND FUTURE RESEARCH DIRECTIONS

This section discusses few research directions and open issues for DeepFakes detection.

   i. **Generalised DeepFakes Detectors**: In spite of the advancement, most prior mechanisms are limited to their ability in detecting manipulated faces. Namely, the performance of existing methods drops significantly when they encounter DeepFakes with different manipulations or dataset sources that were not the part of the training. All in all, they have low generalisation capability. There is a huge demand for DeepFakes detection frameworks that have higher generalisation capabilities and attain lower error rates for new manipulations, tools, and datasets absent in the training phase. More research efforts must be focused on developing new generalised DeepFakes detection schemes.

ii. **Adversary-Aware Face Recognition Systems**: It has been shown that performance of the face recognition systems goes down under manipulated face samples. Moreover, it is easy to see in the literature that there are very limited works that attempted to address the issue of DeepFakes. Studies should be directed towards developing demanipulation-based systems (i.e., where first the faces are de-manipulated and then utilised for recognition/identification) and security by design-based systems (i.e., algorithms particularly developed to take into account the face manipulations).

iii. **Wearable/Mobile Manipulation Detection**: Majority of the DeepFakes detection frameworks are designed for personal computer, which are usually not usable on mobile/wearable platforms owing to high computational cost. To make DeepFakes detection more practical, scientists must address the issue of DeepFakes on mobile/wearable devices by designing novel compact and efficient DeepFakes detectors.

iv. **Large-Scale Database**: Very few sizeable DeepFakes datasets are publicly available. There is a need of large-scale benchmark datasets with several types of manipulations. Moreover, high-grade synthetic face generation techniques that can be utilised to produce datasets is an exigent problem. Such challenges have stymied advancement in the field of DeepFakes.

## 4.6 CONCLUSIONS

Daily many manipulated videos are being shared on social media. Manipulated face videos, known as DeepFakes, have attracted concerns as they can fool human as well as face recognition systems. There is need of efficient methods that can detect the manipulated videos before they cause any danger. Thus, in this chapter, a proficient framework is developed for discrimination of the fake and genuine face videos. The proposed approach is based on hybrid paradigm that uses the discriminative powers of the deep CNN features by combining CNN with LSTM architectures. In particular, the efficient pre-trained ResNet-50 model and the LSTM classifier were adopted. Experimental analysis using two public DeepFake videos datasets showed that the deep features and LSTM classifier have great potential in discriminating the fake faces videos from real ones. The proposed DeepFake detection framework outperformed the existing techniques. As deep features utilised both colour and texture, thereby quite efficient than a dozen of local descriptors and prior methods. In the feature, we are planning to extend our work on other face video manipulation types and techniques. Moreover, we will apply the proposed method on more challenging datasets. Also, other pre-trained deep models will be used for improving the performance.

## NOTES

1 https://generated.photos/
2 https://thispersondoesnotexist.com
3 https://github.com/NVlabs/ffhq-dataset
4 https://github.com/shaoanlu/faceswap-GAN
5 https://github.com/deepfakes/faceswap
6 https://deepfakedetectionchallenge.ai/

## REFERENCES

1. A. Rossler, D. Cozzolino, L. Verdoliva, C. Riess, J. Thies, and M. Nießner, "Faceforensics++: learning to detect manipulated facial images", *arXiv preprint arXiv*:1901.08971, pp. 1–14, 2019.
2. Z. Akhtar, D. Dasgupta, and B. Banerjee, "Face authenticity: an overview of face manipulation generation, detection and recognition", *International Conference on Communication and Information Processing (ICCIP)*, pp. 1–8, 2019.
3. R. Tolosana, R. Vera-Rodriguez, J. Fierrez, A. Morales, and J. Ortega-Garcia, "DeepFakes and beyond: a survey of face manipulation and fake detection", *arXiv preprint arXiv*:2001.00179, 2020.
4. Z. Akhtar, A. Rattani, A. Hadid, and M. Tistarelli, "Face recognition under ageing effect: a comparative analysis", *17th International Conference on Image Analysis and Processing (ICIAP)*, pp. 309–318, 2013.
5. Z. Akhtar, G. Fumera, G.L. Marcialis, and F. Roli, "Robustness evaluation of biometric systems under spoof attacks", *16th International Conference on Image Analysis and Processing (ICIAP)*, pp. 159–168, 2011.
6. Z. Akhtar, G. Fumera, G.L. Marcialis, and F. Roli, "Robustness analysis of likelihood ratio score fusion rule for multimodal biometric systems under spoof attacks", *45th IEEE Int'l Carnahan Conference on Security Technology (ICCST)*, pp. 237–244, 2011.
7. Z. Akhtar and A. Rattani, "A face in any form: new challenges and opportunities for face recognition technology", *IEEE Computer*, vol. 50, no. 4, pp. 80–90, 2017.
8. Z. Akhtar, A. Hadid, M. Nixon, M. Tistarelli, J.L. Dugelay, and S. Marcel, "Biometrics: in search of identity and security (Q & A)", *IEEE MultiMedia*, vol. 25, no. 3, pp. 22–35, 2018.
9. P. Korshunov and S. Marcel, "Deepfakes: a new threat to face recognition? assessment and detection", *arXiv preprint arXiv*:1812.08685, pp. 1–5, 2018.
10. D. Guera and E. J. Delp, "Deepfake video detection using recurrent neural networks", *IEEE International Conference on Advanced Video and Signalbased Surveillance (AVSS)*, pp. 1–6, 2018.
11. D. Afchar, V. Nozick, J. Yamagishi, and I. Echizen, "Mesonet: a compact facial video forgery detection network", *IEEE International Workshop on Information Forensics and Security (WIFS)*, pp. 1–7, 2018.
12. P. Zhou, X. Han, V. I. Morariu, and L. S. Davis, "Two-stream neural networks for tampered face detection", *IEEE Conference on Computer Vision and Pattern Recognition Workshops (CVPRW)*, pp. 1–9, 2017.
13. B. Bekci, Z. Akhtar, and H.K. Ekenel, "Cross-dataset face manipulation detection", *IEEE Conference on Signal Processing and Communications Applications (SIU)*, pp. 1–5, 2020.
14. T. Neubert, C. Kraetzer, and J. Dittmann, "A face morphing detection concept with a frequency and a spatial domain feature space for images on eMRTD", *Proceedings of ACM Workshop on Information Hiding and Multimedia Security*, pp. 95–100, 2019.

15. R. Raghavendra, K. B. Raja, S. Venkatesh, and C. Busch, "Transferable deep-CNN features for detecting digital and print-scanned morphed face images", *IEEE International Conference on Computer Vision Workshop (ICCV)*, pp. 1822–1830, 2017.

16. A. Rossler, D. Cozzolino, L. Verdoliva, C. Riess, J. Thies, and M. Nießner, "Faceforensics: a large-scale video dataset for forgery detection in human faces", *arXiv preprint arXiv*:1803.09179, pp. 1–21, 2018.

17. B. Bayar and M. C. Stamm, "A deep learning approach to universal image manipulation detection using a new convolutional layer", *ACM Workshop on Information Hiding and Multimedia Security (IH&MMSEC)*, pp. 1–6, 2016.

18. N. Rahmouni, V. Nozick, J. Yamagishi, and I. Echizen, "Distinguishing computer graphics from natural images using convolution neural networks", *IEEE International Workshop on Information Forensics and Security*, pp. 1–7, 2017.

19. W. Quan, K. Wang, D. M. Yan, and X. Zhang, "Distinguishing between natural and computer-generated images using convolutional neural networks", *IEEE Transactions on Information Forensics and Security*, pp. 2772–2787, 2018.

20. J. Stehouwer, H. Dang, F. Liu, X. Liu, and A. Jain, "On the detection of digital face manipulation", *arXiv preprint arXiv*:1910.01717, pp. 1–10, 2019.

21. I. Goodfellow, J. Pouget-Abadie, M. Mirza, B. Xu, D. Warde-Farley, S. Ozair, A. Courville, and Y. Bengio, "Generative adversarial nets", *Proceedings of Advances in Neural Information Processing Systems*, pp. 1–9, 2014.

22. T. Karras, S. Laine, and T. Aila, "A style-based generator architecture for generative adversarial networks", *Proceedings of Conference on Computer Vision and Pattern Recognition*, pp. 4401–4410, 2019.

23. T. Karras, T. Aila, S. Laine, and J. Lehtinen, "Progressive growing of GANs for improved quality, stability, and variation", *arXiv preprint arXiv*:1710.10196, 2018.

24. J.Y. Zhu, T. Park, P. Isola, and A.A. Efros, "Unpaired image-to-image translation using cycle-consistent adversarial networks", *arXiv preprint arXiv*:1703.10593, 2017.

25. Y. Shen, P. Luo, J. Yan, and X. Wang, and X. Tang, Faceid GAN: learning a symmetry three-player GAN for identitypreserving face synthesis. In *CVPR*, pp. 821–830, 2018.

26. X. Yang, Y. Li, and S. Lyu, "Exposing deep fakes using inconsistent head poses", *International Conference on Acoustics, Speech, and Signal Processing (ICASSP)*, pp. 1–4, 2019.

27. Z. He, W. Zuo, M. Kan, S. Shan, and X. Chen, "AttGAN: facial attribute editing by only changing what you want", *IEEE Transactions on Image Processing*, pp. 5464–5478, 2019.

28. F.J. Chang, X. Yu, R. Nevatia, and M. Chandraker, "Pose-variant 3D facial attribute generation", *arXiv preprint arXiv*:1907.10202, 2019.

29. G. Perarnau, J. V. D. Weijer, B. Raducanu, and J. Alvarez, "Invertible conditional GANs for image editing", arXiv preprint arXiv:1611.06355, 2016.

30. Z. Liu, P. Luo, X. Wang, and X. Tang, "Deep learning face attributes in the wild", *IEEE International Conference on Computer Vision Workshop (ICCV)*, pp. 1–11, 2015.

31. Y. Sun, Y. Chen, X. Wang, and X. Tang, "Deep learning face representation by joint identification-verification", *Advances in Neural Information Processing Systems (NIPS)*, pp. 1988–1996, 2014.

32. N. Zhang, M. Paluri, M. Ranzato, T. Darrell, and L. Bourdev, "Panda: pose aligned networks for deep attribute modeling", *Proceedings of the IEEE Conference on Computer Vision and Pattern Recognition (CVPR)*, pp. 1637–1644, IEEE, 2014.

33. J. Thies, M. Zollhofer, M. Stamminger, C. Theobalt, and M. Nießner, "Face2face: real-time face capture and reenactment of RGB videos", *Proceeding of Conference on Computer Vision and Pattern Recognition*, pp. 2387–2395, 2016.

34. J. Thies, M. Zollhofer, and M. Nießner, "Deferred neural rendering: image synthesis using neural textures", *ACM Transactions on Graphics*, vol. 38, no. 66, pp. 1–12, 2019.

35. P. Isola, J. Zhu, T. Zhou, and A. Efros, "Image-to-image translation with conditional adversarial networks", in *Proceedings of IEEE Conference on Computer Vision and Pattern Recognition*, pp. 1125–1134, 2017.

36. Y. Choi, M. Choi, M. Kim, J. Ha, S. Kim, and J. Choo, "StarGAN: unified generative adversarial networks for multi-domain imageto-image translation", in *Proceeding of IEEE Conference on Computer Vision and Pattern Recognition*, pp. 8789–8797, 2018.

37. Z. Akhtar and D. Dasgupta, "A comparative evaluation of local feature descriptors for DeepFakes detection", *IEEE International Symposium on Technologies for Homeland Security*, pp. 1–5, 2019.

38. U. Scherhag, C. Rathgeb, and C. Busch, "Performance variation of morphed face image detection algorithms across different datasets", *International Workshop on Biometrics and Forensics (IWBF)*, pp. 1–6, 2018.

39. L.B. Zhang, F. Peng, and M. Long, "Face morphing detection using Fourier spectrum of sensor pattern noise", *IEEE International Conference on Multimedia and Expo (ICME)*, pp. 1–6, 2018.

40. U.A. Ciftci, and I. Demir, "FakeCatcher: detection of synthetic portrait videos using biological signals", *arXiv*:1901.02212, pp. 1–5, 2019.

41. C. Szegedy, W. Liu, Y. Jia, P. Sermanet, S. Reed, D. Anguelov, D. Erhan, V. Vanhoucke, and A. Rabinovich, "Going deeper with convolutions", *IEEE Conference on Computer Vision and Pattern Recognition*, pp. 1–12, 2015.

42. Y. Li and S. Lyu, "Exposing DeepFake videos by detecting face warping artifacts", *IEEE Conference on Computer Vision and Pattern Recognition Workshops (CVPRW)*, pp. 1–7, 2019.

43. K. He, X. Zhang, S. Ren, and J. Sun, "Deep residual learning for image recognition", *IEEE Conference on Computer Vision and Pattern Recognition*, pp. 1–12, 2016.

44. S. McCloskey and M. Albright, "Detecting GAN-generated imagery using color cues", *arXiv preprint arXiv*:1812.08247, 2018.

45. H. Guan, M. Kozak, E. Robertson, Y. Lee, A. Yates, A. Delgado, D. Zhou, T. Kheyrkhah, J. Smith, and J. Fiscus, "MFC datasets: largescale benchmark datasets for media forensic challenge evaluation", in *Proceeding of IEEE Winter Applications of Computer Vision Workshops*, pp. 63–72, 2019.

46. L. Nataraj, T. Mohammed, B. Manjunath, S. Chandrasekaran, A. Flenner, J. Bappy, and A. Roy-Chowdhury, "Detecting GAN generated fake images using co-occurrence matrices", *arXiv preprint arXiv*:1903.06836, 2019.

47. J. Neves, R. Tolosana, R. Vera-Rodriguez, V. Lopes, and H. Proença, "Real or fake? spoofing state-of-the-art face synthesis detection systems", *arXiv preprint arXiv*:1911.05351, 2019.

48. S. Tariq, S. Lee, H. Kim, Y. Shin, and S. Woo, "Detecting both machine and human created fake face images in the wild", *Proceedings of International Workshop on Multimedia Privacy and Security*, pp. 81–87, 2018.

49. K. Simonyan and A. Zisserman, "Very deep convolutional networks for large-scale image recognition", *arXiv preprint arXiv*:1409.1556, pp. 1–14, 2014.

50. F. Chollet, "Xception: deep learning with depthwise separable convolutions", *IEEE Conference on Computer Vision and Pattern Recognition (CVPR)*, pp. 1800–1807, 2017.

51. A. Bharati, R. Singh, M. Vatsa, and K. Bowyer, "detecting facial retouching using supervised deep learning", *IEEE Transactions on Information Forensics and Security*, vol. 11, no. 9, pp. 1903–1913, 2016.

52. H. H. Nguyen, J. Yamagishi, and I. Echizen, "Use of a capsule network to detect fake images and videos", *arXiv preprint arXiv*:1910.12467, pp. 1–14, 2019.

53. S. Sabour, N. Frosst, and G. E. Hinton, "Dynamic routing between capsules", *Neural Information Processing Systems (NeurIPS)*, pp. 1–11, 2017.

54. F. Matern, C. Riess, and M. Stamminger, "Exploiting visual artifacts to expose DeepFakes and face manipulations", *IEEE Winter Applications of Computer Vision Workshops (WACVW)*, pp. 83–92, 2019.
55. B. Dolhansky, R. Howes, B. Pflaum, N. Baram, and C. Ferrer, "The deepfake detection challenge (DFDC) preview dataset", *arXiv preprint arXiv:1910.08854*, 2019.
56. E. Sabir, J. Cheng, A. Jaiswal, W. AbdAlmageed, I. Masi, and P. Natarajan, "Recurrent convolutional strategies for face manipulation detection in videos", in *Proceeding of Conference on Computer Vision and Pattern Recognition Workshops*, pp. 80–87, 2019.
57. S. Agarwal, H. Farid, Y. Gu, M. He, K. Nagano and H. Li, "Protecting world leaders against deep fakes", in *Proceedings of IEEE Conference on Computer Vision and Pattern Recognition Workshops*, pp. 38–45, 2019.
58. T. Baltrusaitis, A. Zadeh, Y. Lim, and L. Morency, "OpenFace 2.0: facial behavior analysis toolkit", *Proceedings of International Conference on Automatic Face & Gesture Recognition*, pp. 1–10, 2018.
59. H. H Nguyen, F. Fang, J. Yamagishi, and I. Echizen, "Multi-task learning for detecting and segmenting manipulated facial images and videos", *IEEE International Conference on Biometrics: Theory, Applications and Systems (BTAS)*, pp. 1–8, 2019.
60. I. Amerini, L. Galteri, R. Caldelli, and D. Bimbo, "DeepFake video detection through optical flow based CNN", *IEEE International Conference on Computer Vision (ICCV)*, pp. 1–3, 2019.
61. S. S. Beauchemin and J. L. Barron, "The computation of optical flow", *ACM Computing Surveys (CSUR)*, pp. 1–57, 1995.
62. R. Lienhart, A. Kuranov, and V. Pisarevsky, "Empirical analysis of detection cascades of boosted classifiers for rapid object detection", *25th DAGM Symposium on Pattern Recognition,* pp. 297–304, 2003.
63. P. Viola and M. Jones, "Rapid object detection using a boosted cascade of simple features", *IEEE Computer Society Conference on Computer Vision and Pattern Recognition*, pp. 511–518, 2001.
64. E. Basaran, Z. Comert, and Y. Çelik, "Convolutional neural network approach for automatic tympanic membrane detection and classification", *Biomedical Signal Processing Control*, pp. 1–10, 2020.
65. C. Xu, J. Yang, H. Lai, J. Gao, L. Shen, and S. Yan, "UP-CNN: un-pooling augmented convolutional neural network", *Pattern Recognition Letters*, vol. 119, pp. 34–40, 2019.
66. M. D. Zeiler, "ADADELTA: an adaptive learning rate method", *arXiv preprint arXiv:1212.5701*, pp. 1–4, 2012.
67. R. Shindjalova, K. Prodanova, and V. Svechtarov, "Modeling data for tilted implants in grafted with Bio-Oss maxillary sinuses using logistic regression", *AIP Conference Proceedings*, vol. 1631, pp. 58–62, 2014.
68. S. Hochreiter, and J. Schmidhuber, "Long short-term memory", *Neural Computing*, vol. 9, pp. 1735–80, 1997.
69. F. A. Gers, J. Schmidhuber, and F. Cummins, "Learning to forget: continual prediction with LSTM", *International Conference on Artificial Neural Networks (ICANN)*, vol. 2, pp. 850–855, 1999.
70. F. A. Gers, N.N. Schraudolph, and J. Schmidhuber, "Learning precise timing with LSTM recurrent networks", *Journal of Machine Learning Research*, vol. 3, pp. 115–143, 2002.
71. U. Budak, V. Bajaj, Y. Akbulut, O. Atilla, A. Sengur, "An effective hybrid model for EEG-based drowsiness detection", *IEEE Sensors Journal*, vol. 19, pp. 7624–7631, 2019.
72. C. Sanderson and B. C Lovell, "Multi-region probabilistic histograms for robust and scalable identity inference", *International Conference on Biometric*, pp. 199–208, 2009.

73. Y. Li, X. Yang, P. Sun, H. Qi, and S. Lyu, "Celeb-DF: a new dataset for DeepFake forensics", *arXiv preprint arXiv*:1909.12962, pp. 1–6, 2019.
74. L. Li, J. Bao, T. Zhang, H. Yang, D. Chen, F. Wen, and B. Guo, "Face X-ray for more general face forgery detection", *arXiv preprint arXiv*:1912.13458, pp. 1–10, 2019.

# 5 Multi-spectral Short-Wave Infrared Sensors and Convolutional Neural Networks for Biometric Presentation Attack Detection

*Marta Gomez-Barrero*
Hochschule Ansbach

*Ruben Tolosana*
Universidad Autonoma de Madrid

*Jascha Kolberg and Christoph Busch*
Hochschule Darmstadt

## CONTENTS

5.1 Introduction ...................................................................................... 106
5.2 Definitions......................................................................................... 108
5.3 Related Works.................................................................................... 109
5.4 Proposed PAD Method ..................................................................... 114
    5.4.1 Hardware: Multi-Spectral SWIR Sensor ............................... 115
    5.4.2 Software: Multi-Spectral Convolutional Neural Networks ........... 116
        5.4.2.1 Multi-Spectral Samples Pre-Processing ........................... 117
        5.4.2.2 CNN Models ..................................................................... 117
        5.4.2.3 Score Level Fusion........................................................... 120
5.5 Experimental Setup ........................................................................... 120
    5.5.1 Database............................................................................... 120
    5.5.2 Evaluation Metrics............................................................... 123
    5.5.3 Experimental Protocol ......................................................... 123
5.6 Experimental Evaluation ................................................................... 123
    5.6.1 Baseline: Handcrafted RGB Conversion .............................. 123
    5.6.2 Input Pre-Processing Optimisation ...................................... 125
    5.6.3 Final Fused System.............................................................. 126

5.7   Conclusions and Future Research.................................................................. 129
Acknowledgements............................................................................................ 130
References........................................................................................................... 130

## 5.1   INTRODUCTION

There is currently no doubt about the importance of subject authentication in a wide range of applications or about the numerous advantages offered by biometric recognition with respect to password or token-based systems (Jain 2007). We can safely say that biometrics overcomes problems like forgetting or losing a key, and that it provides a stronger link between the subject and the claimed identifier. Such and other advantages have led to an ever-growing deployment of biometric recognition systems in the market and in national-wide identification scenarios (Government of India 2012).

However, biometric systems are still vulnerable to external attacks. Among the possible attack points described by Ratha, Connell, and Bolle (2001), which include both inner modules of the system and communication channels, the biometric capture device is probably the most exposed one. The main difference with respect to any other attack lies on the knowledge required by the individual launching the attack: he does not need to know *anything* about the inner functioning of the system. Such attacks directed to the capture device are known in the literature as *presentation attacks* (PAs) and defined within the ISO/IEC 30107 standard on biometric presentation attack detection (PAD) as the "presentation to the biometric data capture subsystem with the goal of interfering with the operation of the biometric system" (ISO/IEC JTC1 SC37 Biometrics 2016). In other words, an attacker can present the capture device with a *presentation attack instrument* (PAI), such as a face mask, a gummy finger, or a fingerprint overlay, instead of his own bona fide biometric characteristic. His intentions may be to impersonate someone else (i.e., active impostor) or to avoid being recognised due to black-listing (i.e., identity concealer).

Given the serious threat posed by PAs, PAD methods have been developed in the last decade to automatically distinguish between bona fide (i.e., real or live) presentations and access attempts carried out by means of PAIs (Marcel et al. 2019). Research in this new area has been fostered by the organisation of international competitions such as the LivDet series (Ghiani et al. 2017; Orrú et al. 2019), and by several international projects, such as the European TABULA RASA (2010), BEAT (2012), and RESPECT (2019), or the US ODIN research program (ODNI and IARPA 2016). Such initiatives and funding programs have consequently led to the development of specific PAD methods for iris (Galbally and Gomez-Barrero 2017), fingerprint (Marasco and Ross 2015; Sousedik and Busch 2014), or face (Galbally, Marcel, and Fierrez 2014), among other biometric characteristics.

In general, PAD methods can be broadly divided into software- and hardware-based methods. Whereas the former, in the particular case of fingerprint, utilise the output of traditional optical and capacitive sensors, the latter introduce specific sensors to capture other properties of a bona fide fingerprint (Marasco and Ross 2015; Sousedik and Busch 2014). The LivDet competitions focus on software-based methods, since only conventional sensors are used to capture the fingerprint samples used in the benchmarks. For these datasets, very high detection rates, close to a 100% accuracy, have been achieved. However, it should be noted that only a limited number of different PAI species (i.e., 11) is included in those benchmarks. In a recent study (Kanich, Drahansky, and Mézl 2018), the authors analyse the vulnerabilities of commercial off-the-shelf (COTS) fingerprint sensors to PAIs fabricated with 21 different materials. Their results highlight the vulnerabilities to most of the materials used, which are in many cases not included in the LivDet benchmarks (e.g., wax). There is therefore a clear need to further analyse the detection capabilities of current and eventually new PAD techniques for larger databases, including a higher variability in terms of PAI species.

However, before developing new PAD techniques, we should remember that fingerprint sensors are designed to capture the ridge and valley patterns on the finger in order to achieve the best possible recognition accuracy. This may not be the best approach to discriminate between bona fide and attack presentations. On the contrary, the use of other technologies can help increasing the PA detection rates. In fact, it has been recently shown that images acquired within the short-wave infrared (SWIR) spectrum can yield very accurate PAD approaches both for face and fingerprint (Steiner et al. 2016; Tolosana et al. 2019). This is due to the fact that all skin types according to the Fitzpatrick scale (Fitzpatrick 1988) present very similar remission curves for these wavelengths, and at the same time quite different from other materials commonly utilised for the fabrication of PAIs (e.g., silicone or paper) (Steiner et al. 2016). Therefore, the task of discriminating skin (i.e., bona fide presentations) from other non-skin-materials (i.e., PAs) becomes easier in this part of the spectrum, in contrast to other wavelengths for which the skin types are very different among themselves and at the same time similar to, for instance, coloured silicone. We therefore analyse in this Chapter the use of SWIR finger images in combination with the latest deep learning algorithms to detect a large number of fingerprint PAI species: up to 41, fabricated with 35 different materials.

Among the different works recently carried out on fingerprint PAD for SWIR images (Tolosana et al. 2019; Hussein et al. 2018; Gomez-Barrero, Kolberg, and Busch 2018, 2019; Gomez-Barrero and Busch 2019), Tolosana et al. (2019) carried out a thorough study on the soundness of using deep convolutional neural networks (CNNs) in combination with SWIR images. In particular, the sensor utilised in that work captures four grayscale images of the finger at different SWIR wavelengths. Given that most pre-trained CNN models expect RGB images (i.e., three channels: red, green, and blue), the authors defined a handcrafted pre-processing of the samples to convert the four grayscale images into three channels. These RGB images were used as input to three different CNN models [i.e., VGG19 (Simonyan and Zisserman 2015), MobileNet (Howard et al. 2017), and a self-designed ResNet (Szegedy, Ioffe, and Vanhoucke 2016)]. On the experimental evaluation, tested on a large dataset, including 35 PAI species, remarkably low error rates were achieved. Given that this is also the work evaluated on the largest database in terms of PAI species so far, we build upon it for developing an improved PAD method.

In this chapter, we propose an automatic pre-processing of the four grayscale images via an additional convolutional layer, integrated with the CNN model and trained together (end-to-end approach). This way, the four grayscale images can be regarded as a single four-channel image, and the network can learn the most discriminant features for the subsequent layers to process, thereby enhancing the overall detection performance. In addition to the three networks analysed by Tolosana et al. (2019) (i.e., a ResNet trained from scratch and the pre-trained MobileNet and VGG19 models), we have studied (i) the newer MobileNetV2 model (Sandler et al. 2018), which includes residual connections in the form of inverted bottlenecks and (ii) the VGGFace network (Parkh, Vedaldi, and Zisserman 2015), pre-trained on facial images for recognition purposes. Since VGGFace has been trained on more skin data, this could be beneficial for the PAD task. Then, all PAD partial scores (i.e., one per CNN model) are combined with a weighted sum rule to achieve a more robust PAD scheme.

In addition to the aforementioned improvements on the software side of the PAD method, the capture device used to acquire the fingerprint data has also been improved. The main limitation of the sensor used in (Tolosana et al. 2019; Gomez-Barrero and Busch 2019) was the low resolution of the images (i.e., $64 \times 64$ px.), with the consequent loss on textural information. The capture device developed within the BATL project has been accordingly improved to capture $320 \times 245$ px. images with a better focus on the region of interest (ROI); i.e., the fingerprint. The performance of the proposed PAD approach is thus evaluated on a newly acquired database comprising 8,214 bona fide and 3,310 PA samples, stemming from 41 different PAI species. This new dataset hence includes a higher number of PA samples, stemming from more PAI species, which allows for a more realistic evaluation of the detection capabilities of the proposed method.

## 5.2   DEFINITIONS

In the following, we include the main definitions stated within the ISO/IEC 30107-3 standard on biometric PAD – part 3: testing and reporting (ISO/IEC JTC1 SC37 Biometrics 2017), which will be used throughout this chapter:

- **Bona Fide Presentation**: "interaction of the biometric capture subject and the biometric data capture subsystem in the fashion intended by the policy of the biometric system". That is, a normal or genuine presentation.
- **Presentation Attack (PA)**: "presentation to the biometric data capture subsystem with the goal of interfering with the operation of the biometric system". That is, an attack carried out on the capture device to either conceal your identity or impersonate someone else.
- **Presentation Attack Instrument (PAI)**: "biometric characteristic or object used in a presentation attack". For instance, a silicone 3D mask or an ecoflex fingerprint overlay.
- **PAI Species**: "class of presentation attack instruments created using a common production method and based on different biometric characteristics".

In order to evaluate the vulnerabilities of biometric systems to PAs, the following metrics should be used:

- **Attack Presentation Classification Error Rate (APCER)**: "proportion of attack presentations using the same PAI species incorrectly classified as bona fide presentations in a specific scenario".

- **Bona fide Presentation Classification Error Rate (BPCER)**: "proportion of bona fide presentations incorrectly classified as attack presentations in a specific scenario".

Derived from the aforementioned metrics, the detection equal error rate (D-ERR) is defined as the error rate at the operating point where APCER = BPCER. In addition, to evaluate the convenient operating point recommended by the IARPA Odin program, the APCER at a BPCER = 0.2% is denoted as $APCER_{0.2\%}$.

## 5.3 RELATED WORKS

Following the discussion in Section 5.1, we describe in this section the latest fingerprint hardware-based PAD methods, with a special focus on deep learning-based techniques, due to their superior detection performance in comparison with approaches based on handcrafted features. The most relevant works are summarised in Table 5.1. For more details on other PAD approaches, the reader is referred to the corresponding surveys on the topic (Marasco and Ross 2015; Sousedik and Busch 2014; Marcel et al. 2019).

### TABLE 5.1
### Summary of the Most Relevant Methodologies for Fingerprint PAD Based on Non-conventional Sensors

| Technology | Reference | Approach | Performance | # PAIs |
|---|---|---|---|---|
| OCT | Meissner, Breithaupt, and Koch (2013) | Sweat glands detection* | APCER = 16% BPCER = 7% | – |
| | Chug and Jain (2019) | Patch-wise CNNs | APCER = 0.17% BPCER = 0.2% | 8 |
| VIS multi-spectral | Rowe, Nixon, and Butler (2008) | Wavelet transform* | APCER = 0.9% BPCER = 0.5% | 49 |
| LSCI | Keilbach et al. (2018) | Texture descriptors and SVMs* | APCER = 10.97% BPCER = 0.84% | 32 |
| | Kolberg, Gomez-Barrero, and Busch (2019) | Texture descriptors and fusion of classifiers* | APCER = 9.05% BPCER = 0.05% | 35 |
| | Mirzaalian, Hussein, and Abd-Almageed (2019) | LSTM | APCER = 12.9% BPCER = 0.2% | 6 |
| SWIR | Tolosana et al. (2019) | Full image CNNs | APCER ≈ 7% BPCER = 0.2% | 35 |
| | Gomez-Barrero and Busch (2019) | Multi-spectral CNNs | APCER = 1.35% BPCER = 0.2% | 35 |
| | Proposed Approach | Multi-spectral CNNs | APCER = 1.16% BPCER = 0.2% | 41 |
| SWIR + LSCI | Hussein et al. (2018) | Patch-based CNNs | APCER = 0% BPCER = 0.2% | 17 |
| | Gomez-Barrero, Kolberg, and Busch (2019) | Texture descriptors + CNNs | APCER ≈ 2% BPCER = 0.2% | 35 |

The three approaches marked with * represent methods based on handcrafted features.

As a first alternative to conventional fingerprint sensors, multi-spectral capture devices have been designed for fingerprint recognition and PAD purposes. In particular, Rowe, Nixon, and Butler (2008) developed a pioneering multi-spectral fingerprint sensor a decade ago, which has now evolved into a COTS device. The Lumidigm sensor captures multi-spectral images in four different wavelengths: 430, 530, and 630 nm, as well as white light. In their article, the authors study not only the fingerprint recognition accuracy achieved with the multi-spectral images but also the feasibility of implementing PAD methods. To that end, absolute magnitudes of the responses of each image to dual-tree complex wavelets are computed. In a self-acquired database, including 49 PAI species, an APCER of 0.9% is reported for a BPCER of 0.5%. Even if these results are remarkable, the PAD methods used are not described in detail, and not much information about the acquired database or the experimental protocol are available. Therefore, it is difficult to establish a fair benchmark with similar works.

More recently, another set of approaches based on multi-spectral images captured within the SWIR spectrum has been developed (Gomez-Barrero, Kolberg, and Busch 2018; Tolosana et al. 2019; Gomez-Barrero and Busch 2019) within the BATL (2017) project, motivated by the initial works of Steiner et al. (2016) for facial images. In this case, samples are captured at four different wavelengths: 1200, 1300, 1450, and 1550 nm. As mentioned in Section 5.1, this area of the spectrum is especially relevant for performing a skin vs. non-skin classification, since all skin types present similar remission curves for the aforementioned wavelengths. In other words, the intra-class variability of the bona fide samples is minimised. In a preliminary evaluation on a small dataset of 60 SWIR samples, comprising 12 different PAI species, Gomez-Barrero, Kolberg, and Busch (2018) showed the feasibility of using pixel-level spectral signatures extracted from SWIR data. However, the detection performance of those handcrafted features was clearly outperformed by deep learning architectures, in particular a pre-trained VGG19 network: Tolosana et al. (2018) achieved perfect results over the same small database.

In a follow-up study, Tolosana et al. (2019) thoroughly analysed the use of deep learning architectures in combination with SWIR data for PAD purposes. In the first step, the four images, acquired at different wavelengths, were combined into a single RGB image with a linear operation in order to have the adequate input for the CNNs. Then, the authors tested both pre-trained models (MobileNet and VGG19) and a self-designed residual network trained from scratch, denoted as ResNet. Over a database comprising over 4,700 samples and 35 different PAI species, and using only 260 samples for training and 180 for validation (i.e., almost 4,300 samples for testing), the score level fusion of MobileNet and ResNet achieved the best performance: $APCER_{0.2\%} \approx 7\%$. More recently, Gomez-Barrero and Busch (2019) were able to improve those results by including an additional convolutional layer in the models which substitutes the handcrafted RGB conversion of the samples. In particular, the score level fusion of three networks yielded an $APCER_{0.2\%} = 1.35\%$.

In addition to those multi-spectral devices, fingerprint PAD methods have been proposed for two different technologies widely used for biomedical applications: optical coherence tomography (OCT) and laser speckle contrast imaging (LSCI). In both cases, the analysis of inner parts of the finger, below the surface, allows to extract a number of features which can help discriminating bona fide from attack presentations. On the one hand, OCT scanners acquire high-resolution, cross-sectional images of internal tissue microstructures by measuring their optical reflections (Huang et al. 1991). To that end, a beam of near infrared (NIR) light is split into a sample or object of interest and a reference mirror. When the difference between the distance travelled by the light for the sample and the reference paths is within the coherence length of the light source, an interference pattern representing the depth profile at a single point is produced. This is known as *A-scan*. A lateral combination of several A-scans yields a cross-sectional scan, referred to as *B-scan*. Furthermore, 3D volumetric representations can be created by stacking multiple B-scans. Such representations of the inner layers of the finger skin allow the analysis of eccrine glands and capillary blood flow. Following this line of thought, since 2006 different laboratories worldwide have carried out visual analysis of the aforementioned B-scans to discriminate between bona fide and presentations attacks (Cheng and Larin 2006, 2007; Bossen, Lehmann, and Meier 2010; Liu and Buma 2010; Moolla et al. 2019).

In spite of those promising studies, due to the large amounts of time necessary to capture the OCT data and the cost of the scanners, no systematic analysis had been carried out on large- or medium-size datasets – only up to 153 samples had been acquired by Bossen, Lehmann, and Meier (2010). To tackle this issue, Sousedik, Breithaupt, and Busch (2013) and Sousedik and Breithaupt (2017) proposed an enhanced pipeline to pre-process the massive raw OCT data into more manageable representations in a short time. In addition, an automatic gland detection approach was proposed, which the authors argued could be used for PAD. In fact, Meissner, Breithaupt, and Koch (2013) used helical eccrine gland ducts to distinguish bona fide from attack presentations over the largest database acquired so far, comprising almost 7,500 bona fide images and 3,000 PA samples. Even if not many details are provided on their algorithms, the authors report an APCER = 16% for a BPCER = 7%. In 2019, Liu, Liu, and Wang (2019) achieved a remarkable 0% APCER and BPCER only analysing the peaks of 1D depth signals to detect four different PAI species of different thicknesses over a rather small dataset comprising 90 samples.

In contrast to the previous OCT-based works, based on handcrafted features and mostly evaluated on rather limited datasets, Chug and Jain (2019) analysed a database comprising 3,413 bona fide samples and 357 PAs, stemming from eight different PAI species. In more details, the proposed method trained the Inception-v3 network (Szegedy et al. 2016) from scratch on local patches extracted from fingerprint depth profiles from cross-sectional B-scans. The local patches were selected in areas where at least 25% of the pixels have non-zero values in order to have enough depth information. On a five-fold cross-validation protocol over the aforementioned dataset, using approximately 3,000 samples for training and 760 for testing, almost perfect detection rates were reported: an $APCER_{0.2\%}$ of 0.17%. Even if in this case, the acquisition time remains below one second (i.e., it can be considered for real-time applications), the capture device costs over 80,000 USD, which is still the main drawback of this otherwise promising technology.

On the other hand, LSCI techniques are based on a different interference phenomenon of coherent light (i.e., a laser). When such a coherent light is reflected by a rough surface, a granular pattern of dark and bright spots appears as the light scatters on the surface and the waves either add up or cancel out. This is called a speckle pattern (Goodman 1975). Furthermore, since the laser light has a certain penetration depth, if moving scatterers are present (i.e., blood), the speckle pattern will change over time (Vaz et al. 2016). Therefore, speckle patterns can be used to detect blood flow and, eventually, PAs. To that end, within the biomedical applications, the raw LSCI sequences are pre-processed either in the temporal or in the spatial domain to compute the speckle contrast.

Based on that principle, Keilbach et al. (2018) analysed the PAD capabilities of LSCI sequences over a large database also captured within the BATL (2017) project, comprising 32 PAIs and more than 750 samples. In the first step, LSCI sequences were captured from three contiguous regions of the finger, and temporally pre-processed in order to obtain a single averaged LSCI image per region. Afterwards, several descriptors were extracted from the averaged LSCI images, including the well-known local binary patterns (LBPs), binarised statistical image features (BSIFs), and the histogram of oriented gradients (HOGs). The extracted features were subsequently classified using support vector machines (SVMs). A final cascaded score level fusion yielded an APCER = 10.97% for a BPCER = 0.84%. It should be noted that in this case, only 136 samples were used for training the SVMs, in contrast to the larger training sets required by most deep learning approaches.

In a subsequent work, Kolberg, Gomez-Barrero, and Busch (2019) reduced the captured regions with the LSCI sensor from three to two, since a deeper analysis of the database showed that the region under the fingernail presented undesired noise. Then, using the same descriptors as Keilbach et al. (2018), the authors established a benchmark, including nine different classifiers. They found that the best results with grey-scale histograms and LBP were achieved with random forests, with SVMs for BSIF, and with stochastic gradient descent for HOG. Therefore, a multi-algorithm fusion of the aforementioned features and classifiers led to an APCER = 9.01% for a BPCER = 0.05% over the extended database and protocol established by Tolosana et al. (2019).

Since the capture device used for the database acquisition in Keilbach et al. (2018) and Tolosana et al. (2019) can acquire both LSCI and SWIR data simultaneously, Gomez-Barrero, Kolberg, and Busch (2019) tested a score level fusion of the aforementioned handcrafted LSCI features (Keilbach et al. 2018) and the SWIR deep learning approach first presented by Tolosana et al. (2019). Evaluated over the same dataset, and following the same protocol as Tolosana et al. (2019), the $APCER_{0.2\%} \approx 7\%$ was reduced down to $APCER_{0.2\%} \approx 2\%$. The reason of this improvement lies on the fact that, whereas the SWIR images allow for an analysis of the surface of the finger or the PA, the LSCI technology enables an analysis of the inner side of the finger, as mentioned above. Therefore, both approaches focus on complementary information, which, when combined, lead to a more robust PAD method.

In contrast to that combination of handcrafted and learned feature-based approach, Hussein et al. (2018) proposed a full deep learning method to fuse SWIR and LSCI data. Whereas all previous works were based on a fixed ROI, in this case, a variable size ROI was used for training, depending on the PAI species. Then, 8 8 px. patches were extracted from the images, resulting in either 4-dimensional tensors for the SWIR data, or 100-dimensional vectors for the LSCI data (i.e., first 100 frames out of the total 1,000 LSCI frames acquired). Those patches were fed to a simplified version of AlexNet (Krizhevsky, Sutskever, and Geoffrey 2012), which produced a score per patch. The average score was used for the final decision. This approach was tested over a dataset comprising 551 bona fide and 227 PA samples, stemming from 17 PAI species. Evaluated on a five-fold protocol, with 552 samples for training, 86 for validation, and 140 samples for testing, the SWIR-based network achieved an APCER = 2.5% at BPCER = 0% and the LSCI-based approach achieved an APCER = 8.9% at BPCER 1.3%. Similar to Gomez-Barrero, Kolberg, and Busch (2019), the fusion of both technologies further reduced the error rates to APCER = BPCER = 0%.

In a subsequent work, Mirzaalian, Hussein, and Abd-Almageed (2019) conducted a study on different patch-wise DNN architectures for raw LSCI sequences, over a larger database than that considered by Hussein et al. (2018). In particular, they analysed the baseline architecture first tested by Hussein et al. (2018), a modification of the former, including residual connections, a shallower version of the GoogLeNet architecture (Szegedy et al. 2015), and a double-layer long short-term memory (LSTM) network. The latter has the advantage of being able to process temporal sequences, such as the acquired LSCI data, which, in this work, is not pre-processed but used in its raw form. The evaluation dataset consisted on 3,743 bona fide samples and 218 PA samples, including six different PAI species. Over a six-fold leave-one-attack-out partition of the database, the LSTM network achieved the best detection performance: an $APCER_{0.2\%}$ of 8.81%. It should be noted that in this case over 3,800 samples were used for training and validation and only around 160–180 for testing (depending on the fold).

In summary, we can extract the following take-away messages from the current literature on both deep learning and handcrafted-based PAD techniques:

- Deep learning approaches have clearly outperformed handcrafted feature-based algorithms for both SWIR images – $APCER_{0.2\%} \approx 45\%$ for handcrafted features vs. 7% for a CNN fusion (Tolosana et al. 2019) – and for OCT samples – APCER = 16% for BPCER = 7% for handcrafted features (Meissner, Breithaupt, and Koch 2013) vs. APCER = 0.17% for BPCER = 0.2% with CNNs in (Chug and Jain 2019).
- In the case of LSCI sequences, the detection performance improvement is not clear yet. This is probably due to the nature of the raw data, which resembles random images, and needs to be manually pre-processed in order to be further utilised for PAD purposes. Some further research on how to pre-process these data with CNNs or any other deep learning-based technique still needs to be carried out.
- The main drawback of deep learning is its requirement for big amounts of training data. This can be, however, tackled using pre-trained models or small networks designed ad hoc for the problem at hand (Tolosana et al. 2019), or by using data augmentation, based for instance on a patch-wise approach (Hussein et al. 2018; Chug and Jain 2019).
- Regarding computational resources, training CNNs requires considerably more time than handcrafted-based approaches, also leading to a higher computational load. It should be noted that the CNN models need to be trained only once. Once the system has been deployed, testing whether a

sample is a bona fide or an attack presentation is fast. Therefore, the impact of using deep learning approaches on practical scenarios is minimised. The only remaining issue is the memory: CNNs tend to comprise a high number of parameters [from 319,937 to 20,155,969 in (Tolosana et al. 2019)], which need to be stored in the device memory. This is usually not the case compared to other traditional classifiers such as SVMs.

- Finally, two major challenges in the field are related to the detection of unknown PAs and cross-sensor scenarios. Most studies so far have shown a detection performance degradation. To alleviate this, different approaches consider data augmentation techniques through a synthetic PA sample generator (Chugh and Jain 2019), or handcrafted features embedding (Gonzalez-Soler et al. 2019) have been proposed for traditional fingerprint sensors. Their applicability to other technologies, such as SWIR, LSCI, or OCT, still needs to be explored.

## 5.4  PROPOSED PAD METHOD

The PAD methodology presented in this Chapter is summarised in Figure 5.1. First, a dedicated capture device (Section 5.4.1) acquires images of the finger at four different wavelengths within the SWIR spectrum. Then, those images are processed by deep learning algorithms (Section 5.4.2). In particular, the four channel images are fed to five different CNN models, which include first an additional pre-processing layer at the beginning of the model (Section 5.4.2.1). The models and the differences between them are described in Section 5.4.2.2 and Figure 5.4. Finally, the output of different models can be fused at score level, as presented in Section 5.4.2.3, in order to achieve a more robust PAD module.

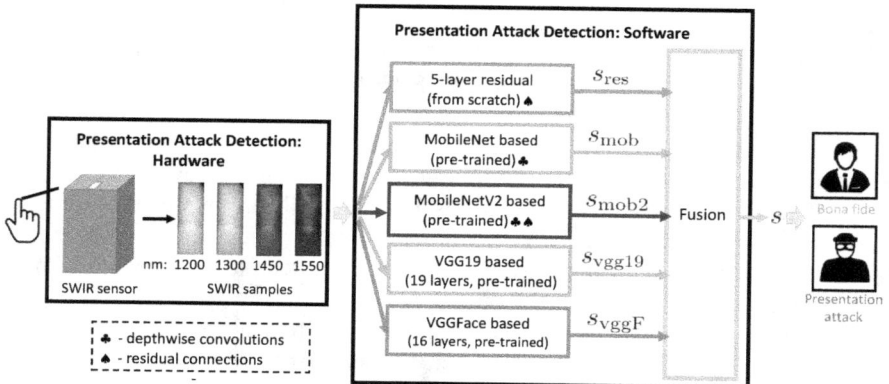

FIGURE 5.1  General diagram of the proposed PAD method. First, images in four different SWIR wavelengths are acquired from the finger. These are later used to train five different CNN models (a five-layer residual network, reduced versions of MobileNet and MobileNetV2, VGG19, and VGGFace, see Figure 5.4 for details). An initial pre-processing module is included to convert the four wavelengths into three channel images (see Figure 5.3). Finally, a score level fusion is carried out.

## 5.4.1 HARDWARE: MULTI-SPECTRAL SWIR SENSOR

The SWIR finger capture device used in the present work was developed by our partners at USC within the BATL (2017) project. In essence, the SWIR sensor is embedded in a closed box with a slot on the top for the finger (see Figure 5.1, left), with the camera and lens placed inside the box. When the finger is placed over the slot, all ambient light is blocked and therefore only the desired wavelengths are considered during the acquisition. Furthermore, in order to avoid any interference, two images are captured at each wavelength: one with the LEDs on, and another one, "dark image", with the illumination off. By subtracting both images, undesired light noise can be suppressed.

In contrast to the capture device described in Tolosana et al. (2019), where $64 \times 64$ px. images were captured at 1000 fps, in this work the sensor used (a Xenics Bobcat 320) is able to capture $320 \times 256$ px. images at 100 fps, with a 35 mm focal length lens. This way, higher resolution images, including more textural details and thereby more suited for deep learning studies, are acquired. As in Tolosana et al. (2019) and Steiner et al. (2016), images are captured at four different SWIR wavelengths, namely: 1,200, 1,300, 1,450, and 1,550. The differences between the images acquired by the new and the previous sensor can be observed in Figure 5.2: not only have the new images (top) at a higher resolution, but the focus has also been improved to eliminate some of the blur existing in the images acquired with the previous sensor (bottom).

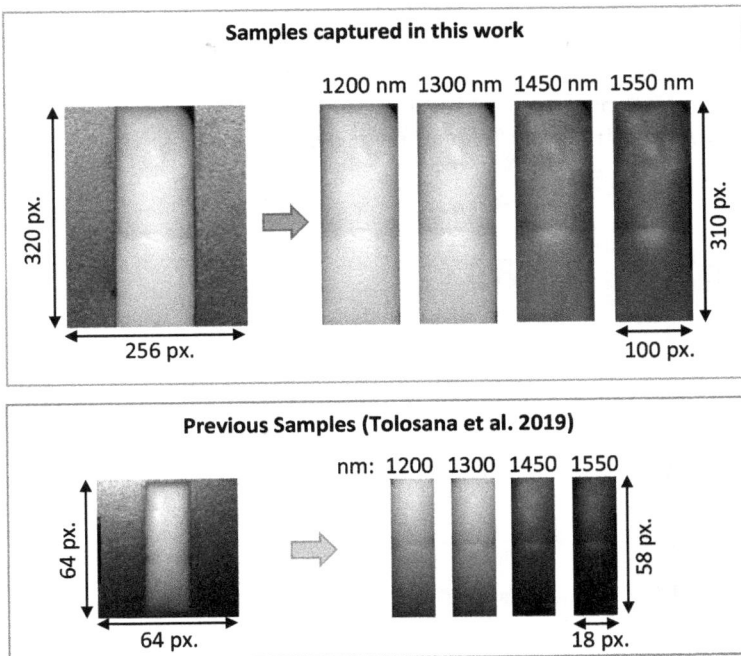

**FIGURE 5.2** Samples comparison. Top: samples captured by a new capture device. Bottom: samples captured with the previous device. In both cases, the complete sample at 1,200 nm is shown on the left, and the ROIs at all wavelengths on the right.

It should also be noted that the camera captures the finger slot and the surrounding area of the box. Since the finger is always placed over the fixed open slot, and the camera does not move, the ROI can be extracted using a simple fixed size cropping: The final ROI has a size of 310 × 100 px. The four ROIs for a bona fide, from now on referred to simply as images or samples, are depicted in Figure 5.2 (top right).

Finally, even though the main aim of this work is the development of PAD techniques, it is important not to forget about the fingerprint recognition task. A single capture device needs to acquire samples which can be processed for fingerprint recognition and for PAD purposes in a single acquisition attempt. Otherwise, a potential attacker would provide his own bona fide finger for PAD testing and subsequently a PAI for recognition. Therefore, the multi-modal capture device utilised also contains a second 1.3 MP camera with a 35 mm VIS-NIR lens in order to capture finger photographs from which contactless fingerprint recognition can be carried out. Kolberg et al. (2019) showed how COTS can extract minutiae correctly from these samples, in order to allow compatibility with conventional fingerprint sensors.

### 5.4.2 Software: Multi-Spectral Convolutional Neural Networks

The software PAD approach proposed in this Chapter is summarised in Figure 5.3 and compared to the workflow described in Tolosana et al. (2019). As mentioned in Section 5.1, most CNN pre-trained models have been trained on the ImageNet (Krizhevsky, Sutskever, and Geoffrey 2012) or VGGFace (Parkh, Vedaldi, and Zisserman 2015) databases, and thus expect RGB images. However, the SWIR sensor described in Section 5.4.1 outputs four different grey-scale images acquired at four different wavelengths. Therefore, in order to be able to use pre-trained models and benefit from transfer learning techniques, some kind of pre-processing

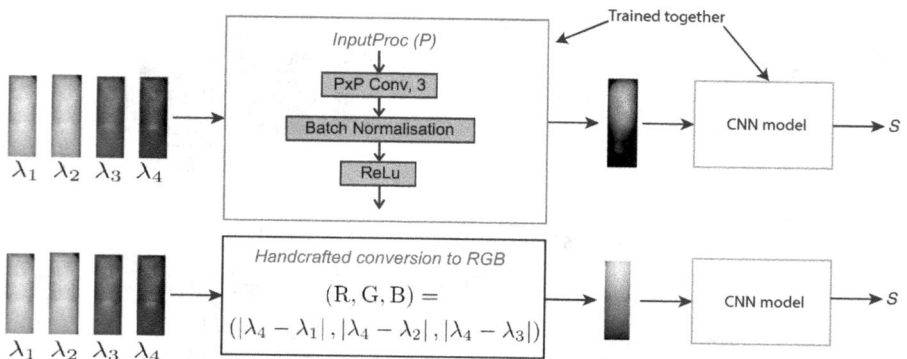

**FIGURE 5.3** PAD software diagram. Top: As proposed in this chapter, the four SWIR images are automatically processed by the corresponding CNN model using a single convolutional layer with three filters of size $P \times P$ and a stride of 1. In addition, batch normalisation and a ReLu activation are used to facilitate convergence. The result is a three channel image. Bottom: the handcrafted RGB conversion proposed by Tolosana et al. (2019). After the pre-processing step, the corresponding three channel image is processed by the CNN model at hand, which outputs the PAD score $s$.

needs to be added to convert the four channel samples to three channel images (see Section 5.4.2.1). After that, regular pre-trained CNN models can be applied (see Section 5.4.2.2).

### 5.4.2.1 Multi-Spectral Samples Pre-Processing

The PAD method proposed in Tolosana et al. (2019) included a handcrafted conversion of the four SWIR samples to an RGB image, as depicted in Figure 5.3 (bottom). In spite of the low error rates reported in that work, such a manual pre-processing presents several drawbacks. On the one hand, the linear transformation from three to four channels was optimised in terms of the average pixel intensity variance for the bona fides (i.e., intra-class variability, to be minimised) and the corresponding differences between bona fide and attack presentations (i.e., inter-class variability, to be maximised). Such a line of thought optimises the problem from a human vision perspective and leads to a single input which will be further processed by different CNN models, which will in principle learn different features starting from the input data. In addition, it should be noted that a single transformation is applied to the whole image, even if the finger may have a non-uniform illumination, as shown in Figure 5.2, and some PAI species may only cover part of the finger.

In contrast to the aforementioned handcrafted conversion, we propose to let the network itself convert the four grey-scale input channels into RGB images (i.e., tensors) comprising three channels. This way, the network can apply different linear and non-linear combinations to each region of the image and learn the most suitable features for the following layers. To that end, we include at the beginning of each CNN model the pre-processing module shown in Figure 5.3 (top). This new convolutional layer has a four-dimensional tensor as input, a stride of one in order to preserve the image size, and a filter of size $P \times P$ px. The value of $P$ needs to be optimised ad hoc for each model, since different CNN models may learn features at different scales during training. In addition, to facilitate convergence, batch normalisation and a ReLu activation function are added to the convolutional layer. The corresponding parameters will be trained together (i.e., end-to-end) with the last layers of the pre-trained models, or the full residual network trained from scratch, so that the updates can propagate through the whole network in each training epoch.

### 5.4.2.2 CNN Models

We consider five different CNN models, whose architectures are shown in Figure 5.4. First, we analyse the three models studied in Tolosana et al. (2019), namely: (i) the 5-layer ResNet, (ii) a reduced version of MobileNet (Howard et al. 2017), and (iii) the pre-trained VGG19 (Simonyan and Zisserman 2015). In addition, we also study two further CNN architectures: (iv) a reduced version of MobileNetV2 (Sandler et al. 2018), which is an improved version of MobileNet, and (v) the pre-trained VGGFace (VGG16) (Parkh, Vedaldi, and Zisserman 2015), which has been trained on facial images, thus containing skin, instead of training it on the more general ImageNet database as the remaining pre-trained models. All strategies have been implemented under the Keras framework using Tensorflow as back-end, with a NVIDIA GeForce GTX 1080 GPU. An Adam optimiser is considered with a learning rate value of 0.0001 and a loss function based on binary cross-entropy.

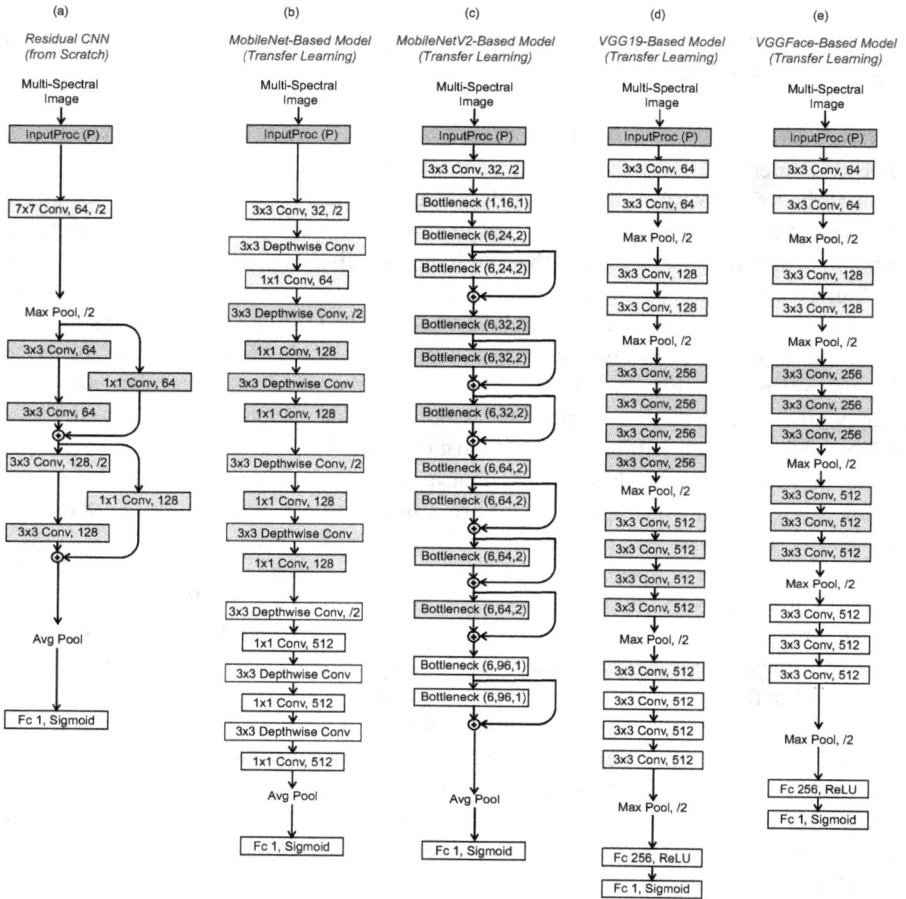

**FIGURE 5.4** CNN architectures. From left to right: (a) the residual CNN trained from scratch using only the SWIR fingerprint database (319,937 parameters); (b) the pre-trained MobileNet-based model (815,809 parameters); (c) the pre-trained MobileNetV2-based model (437,985 parameters, see Figure 5.5 for details on the bottle-necks); (d) the pre-trained VGG19-based model (20,155,969 parameters); and (e) the pre-trained VGGFace-based model (20,155,969 parameters). All pre-trained models are adapted using transfer learning techniques over the last white-background layers. Also, the first convolutional layer (purple) (i.e., "InputProc", see Figure 5.3) is trained for all networks. This figure is extracted from Gomez-Barrero and Busch (2019).

**Residual Network Trained from Scratch**: As already pointed out, the first approach is focused on training a residual CNN (He et al. 2015) from scratch. A residual connection consists of reinjecting previous representations into the downstream flow of data, by adding a past output tensor to a later output tensor. These connections help preventing information loss along the data-processing flow and allow the use of DNN architectures, decreasing their training time significantly (He et al. 2015; Szegedy, Ioffe, and Vanhoucke 2016). The five-layer ResNet utilised is depicted in Figure 5.4 (left). As it may be observed, in order to be able to train it from scratch with a small training set, it comprises only five layers. In addition, two residual connections with pointwise convolutions are added. Batch normalisation is also applied right after each convolution and before the ReLu activation in order to facilitate convergence.

**MobileNets and Transfer Learning**: The main feature of both MobileNet (Howard et al. 2017) and MobileNetV2 (Sandler et al. 2018) is the use of depthwise separable convolutions. These layers perform a spatial convolution on each channel of their input, independently, before mixing the output channels via a pointwise (i.e., $1 \times 1$) convolution. This is conceptually equivalent to separating the learning of spatial features, which will show correlations in an image, and the learning of channel-wise features, given the relative independence of each channel in an image. An additional advantage of this type of convolutions is that they require fewer parameters and computations, thereby allowing a fast training using less data. In both MobileNet networks, downsampling is directly applied by the convolutional layers that have a stride of 2 (represented by /2 in Figure 5.4), instead of adding some kind of pooling between layers.

With respect to MobileNet, the main contribution of MobileNetV2 is the use of residual connections and inverted bottlenecks (see Figures 5.4 and 5.5). These blocks model the hypothesis of the low dimensionality of the manifold of interest on which the discriminative information extracted by the internal layers of the network lies. To account for this, linear bottleneck layers are introduced in the model, and the residual connections are established between the aforementioned bottlenecks (i.e., in contrast to more common approaches where the residuals connect layers with a higher number of filters or output channels).

Finally, given the depth of both MobileNet models and the limited amount of data available, out of the 13 blocks of MobileNet, we decided to keep only eight. Similarly, out of the 16 bottlenecks of MobileNetV2, 12 are used. In addition, the last two blocks (depicted in white) are re-trained.

**VGGs and Transfer Learning**: Finally, two different VGG-based models have been studied, VGG19 (Simonyan and Zisserman 2015) and VGGFace[2] (Parkh, Vedaldi, and Zisserman 2015). These networks are older and simpler than the MobileNets; however, due to its simplicity, VGG19 is still one

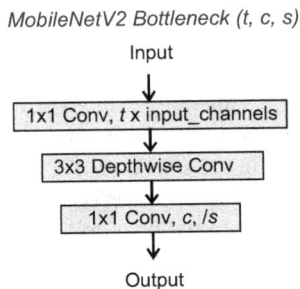

*MobileNetV2 Bottleneck (t, c, s)*

Input

↓

1x1 Conv, *t* x input_channels

↓

3x3 Depthwise Conv

↓

1x1 Conv, *c*, /*s*

↓

Output

**FIGURE 5.5**  Three-layer structure of the bottleneck residual block of MobileNetV2, where *t* denotes the expansion factor, and *c* and *s* the number of filters and stride of the last convolutional layer. This Figure is extracted from Gomez-Barrero and Busch (2019).

of the most popular network architectures, providing very good results in a wide range of competitions. In fact, VGG19 showed a superior performance with respect to MobileNet for fingerprint PAD in Tolosana et al. (2019).

Both VGG inspired models consist of blocks of two to four convolutional layers separated by max pooling layers to reduce the dimensionality of the data, and thereby facilitate convergence during the training stage. Whereas VGG19 comprises 19 different layers, VGGFace is based on the smaller VGG16 model, including 16 layers. In addition, the latter has been trained on facial databases acquired in the wild (i.e., modelling realistic scenarios in opposition to controlled environments with frontal poses and fixed illumination). Therefore, VGG19 has been pre-trained on a multi-class task (ImageNet) in contrast to the two class problem of face recognition for VGGFace. For our study, the last fully connected layers have been replaced with two fully connected layers (with a final sigmoid activation function). In addition, the last three convolutional layers, depicted in white in Figure 5.4, are re-trained in both models.

It should be finally noted that the fully connected layers trained on ImageNet (Krizhevsky, Sutskever, and Geoffrey 2012) or facial classification tasks have been removed from all MobileNet and VGG-based architectures and substituted by the corresponding fully convolutional layers with sigmoid functions for a binary classification task.

### 5.4.2.3 Score Level Fusion

As it was already observed in Tolosana et al. (2019), different CNN models are more robust to specific PAI species than others. Therefore, the fusion of the final PAD score output by several models yields a higher detection performance. In our case (see Section 5.6 for more details), we found that the optimal results are achieved fusing three different models: VGGFace, VGG19, and MobileNetV2. Therefore, we define the final PAD score as follows:

$$s = \alpha \cdot s_{\mathrm{vggF}} + \beta \cdot s_{\mathrm{vgg19}} + (1 - \alpha - \beta) \cdot s_{\mathrm{mob2}} \qquad (5.1)$$

where $\alpha + \beta \leq 1$ are the weights assigned to VGGFace and VGG19, respectively.

## 5.5  EXPERIMENTAL SETUP

### 5.5.1  DATABASE

The database used for the experimental evaluation of the proposed method was acquired in three different sessions, spanning across four months, in collaboration with the colleagues at USC within the BATL (2017) project. In total, 8,214 bona fide and 3,310 PAs, stemming from 41 different PAI species, were captured. A total of 732 different subjects participated in the data collection, from whom the ring, middle, index fingers, and thumbs of both hands were captured. This represents the largest fingerprint dataset within the SWIR spectrum in terms of both number of samples and PAI species. In addition, the proportion of PA samples has increased from 1:10 to 1:2.5, with respect to the database analysed by Tolosana et al. (2019), and the number of PA samples in the test set has been multiplied almost 10 times.

Table 5.2 presents a summary of the PAI species included in the database, classi-fied by the general type (i.e., full finger, paper print outs, and overlay) and the fabrica-tion material. With respect to the database used in Tolosana et al. (2019), a new dataset has particularly increased the number of samples captured from the most challeng-ing PAI species: overlays in general, and full fingers made of some play doh colours (the previous SWIR methods had trouble detecting some of them), and conductive materials, which pose a higher threat specially to conventional finger sensors and are therefore expected to be used by the attackers. In any case, the overall selection of PAI species follows the requirements established by the IARPA Odin program.

Table 5.3 shows the partition of the database into independent train, valida-tion, and test sets. It should be noted that some subjects participated in two dif-ferent acquisition sessions. Their samples (both the bona fide and the PA samples)

**TABLE 5.2**
**PAI Species Included in the Database and Number of Samples Considered in Our Experimental Framework**

| Type | Material | # Samples | | | |
|------|----------|-------|-------|------------|------|
| | | Total | Train | Validation | Test |
| | 3D print | 48 | 18 | 12 | 18 |
| | 3D print + silver coating | 24 | 12 | 8 | 4 |
| | Ballistic gelatine | 144 | 26 | 10 | 108 |
| | Dental material | 51 | 11 | 6 | 34 |
| | Dragonskin | 426 | 88 | 63 | 275 |
| | Dragonskin + conductive coating | 24 | 6 | 0 | 8 |
| | Dragonskin + nanotips white coating | 27 | 9 | 6 | 12 |
| | Latex + gold coating | 69 | 18 | 18 | 33 |
| | Monster latex | 78 | 28 | 11 | 39 |
| | Polydimethylsiloxane (PDMS) | 124 | 21 | 13 | 90 |
| | Playdoh black | 15 | 6 | 0 | 9 |
| Full finger | Playdoh orange | 53 | 17 | 6 | 30 |
| | Playdoh white | 24 | 6 | 6 | 12 |
| | Playdoh yellow | 24 | 3 | 9 | 12 |
| | Silicone | 147 | 47 | 38 | 62 |
| | Silicone two part | 69 | 17 | 4 | 48 |
| | Silicone + conductive coating | 18 | 8 | 4 | 6 |
| | Silicone + nanotips white coating | 54 | 12 | 13 | 29 |
| | Silicone + graphite coating | 72 | 12 | 12 | 48 |
| | Silly putty | 25 | 9 | 0 | 16 |
| | Silly putty glow in the dark | 15 | 6 | 3 | 6 |
| | Silly putty metallic | 15 | 9 | 0 | 6 |
| | Wax | 74 | 16 | 11 | 47 |
| | Finger-vein glossy paper | 37 | 18 | 4 | 5 |

(*Continued*)

**TABLE 5.2 (*Continued*)**
**PAI Species Included in the Database and Number of Samples Considered in Our Experimental Framework**

| Type | Material | # Samples | | | |
| | | Total | Train | Validation | Test |
|---|---|---|---|---|---|
| Print outs | Finger-vein matte paper | 22 | 6 | 8 | 8 |
| | Fingerprint paper | 49 | 11 | 9 | 29 |
| | Finger transparency | 64 | 16 | 8 | 41 |
| | Conductive silicone | 260 | 20 | 8 | 232 |
| | Dragonskin | 170 | 50 | 31 | 89 |
| | Dragonskin fleshtone | 10 | 4 | 2 | 4 |
| | Knox gelatine | 21 | 5 | 4 | 12 |
| | Monster latex | 34 | 15 | 8 | 11 |
| | School glue | 76 | 25 | 15 | 36 |
| Overlay | School glue white | 25 | 5 | 6 | 14 |
| | Silicone | 24 | 12 | 9 | 3 |
| | Silicone yellow | 83 | 24 | 7 | 52 |
| | Silicone fleshtone | 517 | 75 | 48 | 394 |
| | Silicone two part | 98 | 20 | 14 | 64 |
| | Urethane + Ti/Au coating | 72 | 18 | 9 | 45 |
| | Wax | 18 | 8 | 3 | 7 |
| | Wood glue | 70 | 21 | 12 | 37 |

**TABLE 5.3**
**Partition of Training, Validation and Test Datasets**

| | # Samples | # PA Samples | # BF Samples |
|---|---|---|---|
| Training set | 1,538 | 769 | 769 |
| Validation set | 940 | 470 | 470 |
| Test set | 9,046 | 2,071 | 6,975 |
| Total | 11,524 | 3,310 | 8,214 |

are consequently used only in one of the three sets. In order to achieve a balanced training and no bias towards one class, the number of bona fide and PA samples should be equal for the training and validation sets. Therefore, the limiting factor in the protocol design is the number of PA samples available: 3,310 in comparison with the 8,214 bona fide samples. We also want to maximise the number of samples in the test set to achieve more statistically relevant results. Therefore, 769 samples of each class are used for training and 470 for validation. The remaining samples (2,071 PAs and 6,975 bona fides) are used for testing purposes. The specific number of samples of each PAI species included in each set is shown in Table 5.2.

### 5.5.2 Evaluation Metrics

The performance of the PAD method is evaluated in compliance with the ISO/IEC IS 30107-3 on Biometric PAD – Part 3: Testing and Reporting (ISO/IEC JTC1 SC37 Biometrics 2017). To that end, we report Detection Error Trade-Off (DET) curves between the APCER and BPCER. In addition, the APCER at BPCER = 0.2% (denoted as $APCER_{0.2\%}$) will be also reported to evaluate systems with a high user convenience, which is the target of the Odin Program.

### 5.5.3 Experimental Protocol

As already mentioned in Section 5.4, two different deep learning approaches are considered:

- Training complete CNN models from scratch.
- Transfer learning techniques on CNN models trained on multi-class tasks (i.e., ImageNet) or on two-class problems (i.e., VGGFace).

In both cases, three different sets of experiments are carried out:

- Baseline handcrafted RGB conversion: first, a baseline detection performance is established using the handcrafted RGB conversion proposed by Tolosana et al. (2019). The results are benchmarked with Tolosana et al. (2019) and Gomez-Barrero and Busch (2019). This way, we can assess the quality of the images acquired with a new capture device and its impact on the proposed PAD method.
- Input pre-processing optimisation: then, the optimal filter size $P$ (see Section 5.4.2.1 and Figure 5.3) needs to be determined for each model, in order to obtain the best possible detection performance. This is carried out individually for each CNN model described in Section 5.4.2.2.
- Final fused system: after determining the optimal filter size and the APCEs of each CNN model, the best fusion is carried out at score level. In addition, the results are benchmarked with the state-of-the-art reported in Tolosana et al. (2019) and Gomez-Barrero and Busch (2019).

## 5.6 EXPERIMENTAL EVALUATION

### 5.6.1 Baseline: Handcrafted RGB Conversion

The DET curves for each CNN model are plotted in Figure 5.6. In all cases, the handcrafted RGB conversion is depicted in dashed dark blue lines, and different filter sizes $P$ in the range [5,50] are shown in solid thin lines. In addition, the best configuration in terms of $P$ and the corresponding $APCER_{0.2\%}$ values is highlighted with a thicker solid line.

Compared to the results presented by Tolosana et al. (2019) and Gomez-Barrero and Busch (2019), we may observe a detection performance drop for MobileNet and

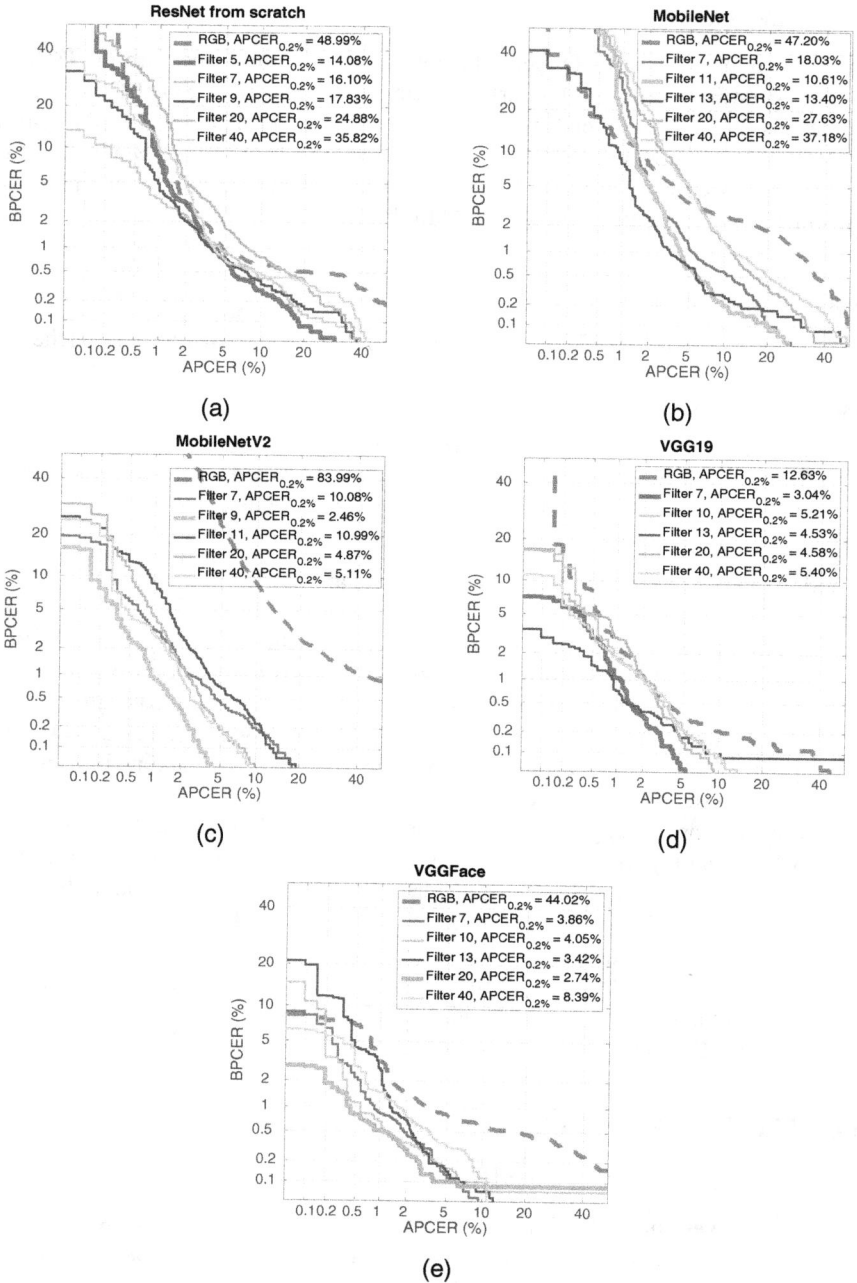

**FIGURE 5.6** DET curves for each individual CNN model approach [handcrafted (RGB) and proposed], and different filter sizes $P$. (a) ResNet from scratch; (b) MobileNet; (c) MobileNetV2; (d) VGG19; (e) VGGFace.

ResNet. This is due to increased challenge pose by the newly acquired database: the number of bona fide samples has been multiplied by two and the number of PA samples by eight. In addition, the focus on the PA fabrication has now been set on the most challenging attacks. Therefore, the $APCER_{0.2\%}$ has increased from 19.91% to 47.20% for MobileNet and from 6.79% to 48.99% for ResNet.

On the other hand, the higher resolution of the SWIR images captured with a new sensor leads to a considerable improvement for the remaining three CNN models. New data allow to use the standard image size for all CNNs: $224 \times 224$ px, instead of the reduction to $58 \times 58$ px, or $118 \times 118$ px used in the previous works. This in turn results in $APCER_{0.2\%}$ values of 12.63% for VGG19, 44.02% for VGGFace, and 83.99% for MobileNetV2, whereas a BPCER of 0.2% could not be achieved by those models in Gomez-Barrero and Busch (2019) (i.e., $APCER_{0.2\%} = \infty$ for all three models).

### 5.6.2 INPUT PRE-PROCESSING OPTIMISATION

In spite of the aforementioned enhancement, the detection rates for the handcrafted RGB conversion are far from the state-of-the-art, with the only exception of VGG19. However, this changes when the input pre-processing module described in Section 5.4.2.1 is included in the CNN models. As it may be observed in Figure 5.6, the $APCER_{0.2\%}$ are improved for all filter sizes $P$ shown, reaching values below 3% for VGGFace and MobileNetV2.

For the particular case of the ResNet trained from scratch (Figure 5.6a), the $APCER_{0.2\%}$ can be reduced to 14.08% for $P = 5$ (i.e., relative improvement of 73%). In addition, the smaller the value of $P$, the bigger the improvement. On the other hand, the best detection performance for low APCERs is reached by $P = 7$. Therefore, depending on the application scenario (i.e., convenience is preferred over security, or vice versa), different $P$ values could be selected.

Regarding MobileNet (Figure 5.6b), the $APCER_{0.2\%}$ can be further decreased to 10.61% for $P = 11$ (i.e., 78% relative improvement) at the cost of not being able to achieve APCERs under 0.5% for all $P \neq 13$. On the other hand, even if the performance of MobileNetV2 is considerable worse than that of MobileNet for the handcrafted RGB conversion, in this case, a state-of-the-art $APCER_{0.2\%}$ of 2.46% (i.e., 97% relative improvement) can be obtained for $P = 9$. In this case, the performance for low APCERs (e.g., of 0.2%) is also optimised for the same filter size.

Finally, we can see in Figure 5.6d that VGG19 achieves an $APCER_{0.2\%}$ of 3.04% for $P = 7$ (i.e., 76% relative improvement), similar to the performance reported for MobileNetV2. In addition, the BPCER remains lower for low APCERs than MobileNetV2, thereby yielding a more stable system for different operating points. This is also the case of VGGFace, which achieves an $APCER_{0.2\%}$ of 2.74% for $P = 20$ (i.e., 94% relative improvement) and an even lower BPCER around 3% for any APCER $\leq 0.2\%$. We may thus conclude that the VGG-based models achieve a higher overall performance for this particular PAD task as it was already pointed out by Tolosana et al. (2019). In addition, since VGGFace has been pre-trained on facial databases, it is able to achieve lower BPCERs than any other CNN model. It does so for a filter size $P = 20$, in comparison to the smaller filter sizes between 5 and 11 found to be optimal for the remaining models. This means that VGGFace focuses on features captured at a lower resolution and will therefore complement other models in an eventual fusion to achieve more robust results.

### 5.6.3   Final Fused System

Given the similarities between both Mobilenet models and the very superior performance of MobileNetV2, only this latter model is further considered for a score level fusion. Similarly, the ResNet trained from scratch reports, together with MobileNet, the worst results among the CNN models tested, since it has not been able to deal with larger images using only five layers. Therefore, it is also excluded from the final fused scheme.

Keeping those thoughts in mind, only MobileNetV2, VGG19, and VGGFace have been considered for the final fusion. First, the CNN models have been fused on a two by two basis, with no significant improvement of the detection accuracy. On the contrary, when the three networks are fused with $\alpha = 0.18$ and $\beta = 0.58$ (i.e., the weights are 18% for VGG19, 58% for VGGFace, and 24% for MobileNetV2), the detection performance improves, as shown in Figure 5.7a. In particular, a final $APCER_{0.2\%}$ of 1.16% can be achieved. That is, only 24 PA samples are misclassified when only two bona fide samples in 1,000 are wrongly detected as attacks. On the other hand, since for low APCERs VGGFace shows lower error rates than any of the other models, the performance in that area of the DET plot is lower for the fused scheme. As a consequence, if the deployment scenario requires very low APCERs, for instance 0.2%, a fusion of the aforementioned CNN models with different filter sizes can yield better results, as depicted in Figure 5.7b. In this case, a $BPCER_{0.2\%}$ of 1.10% is obtained – that is, only 77 bona fide and four attack presentation samples are misclassified. Therefore, depending on the application scenario, different models will be chosen and fused to optimise the performance of the system for the particular case study.

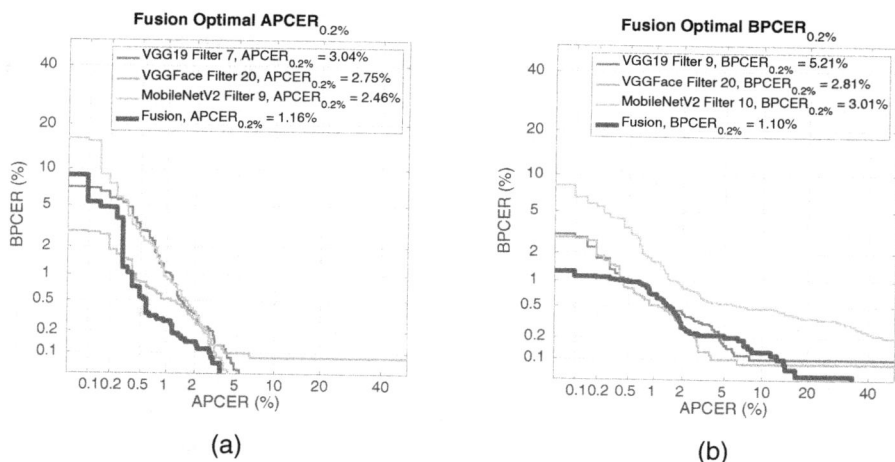

**FIGURE 5.7**   DET curves for the score level fusion of the best configurations found in Figure 5.6. (a) Fusion optimal $APCER_{0.2\%}$; (b) fusion optimal $BPCER_{0.2\%}$.

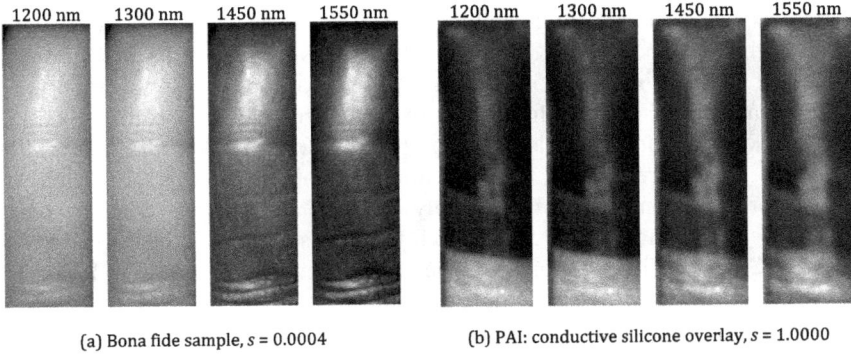

(a) Bona fide sample, $s = 0.0004$        (b) PAI: conductive silicone overlay, $s = 1.0000$

**FIGURE 5.8** (a) Bona fide sample and (b) PA sample of a conductive silicone overlay captured at all wavelengths, with the corresponding final fused PAD scores.

Since the main aim of the ODIN Program is the achievement of convenient PAD systems, we will further analyse the APCEs and BPCEs made by the optimal $APCER_{0.2\%}$ fusion. Figure 5.8 shows a bona fide and a sample of one of the most challenging PAIs for conventional fingerprint capture devices: an overlay made with conductive silicone. As it may be observed, the trend shown by the bona fide across the acquired wavelengths, with a darkening effect, is not reflected on the conductive silicone material, which thus yields the highest possible PA score: 1.

Now, in order to see to what extent the CNN models complement each other, the PAD scores of all APCEs and the lowest BPCE scores are plotted in Figure 5.9: the fused scores are included in the $x$ axis, the individual scores for each CNN model

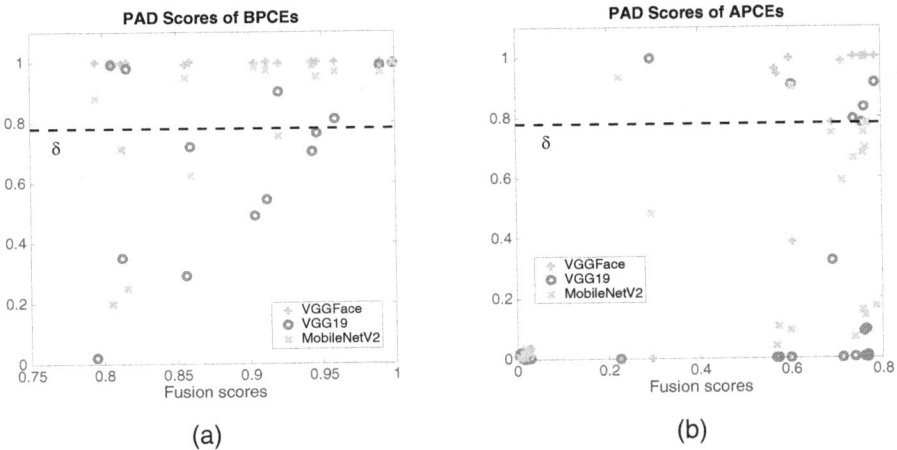

(a)        (b)

**FIGURE 5.9** Score analysis for (a) the BPCEs yielding the highest fused PAD scores, and (b) all APCEs (24). The decision threshold $\delta$ for BPCER = 0.2% is depicted with a dashed black line. The fused PAD scores are depicted on the $x$ axis, and the individual CNN scores are included in the $y$-axis.

on the $y$ axis. The decision threshold $\delta$ for a BPCER of 0.2% is depicted with a dashed horizontal line: the BPCEs show PAD scores over $\delta$ and the APCEs below $\delta$. First, we may see in Figure 5.9a for the BPCEs that the PAD scores reported by VGGFace are always higher than $\delta$ and in fact extremely close to the maximum PAD score of 1. In addition, at least one of the other CNN models also misclassifies the sample. Therefore, even if the third model is able to classify the sample as a bona fide presentation, this represents only 18%–24% of the final score. Therefore, the fused scheme is not able to correctly classify the sample. On the other hand, for the 24 APCEs (see Figure 5.9b), VGGFace reports in most cases (14) a PAD score higher than $\delta$ (i.e., correct decision). However, in almost all cases (22), MobileNetV2 outputs a PAD score below $\delta$, and in 14 cases even below 0.4. Similarly, VGG19 yields a PAD score below 0.4 for 18 of the APCE samples. Therefore, given that the threshold $\delta$ is set at 0.77 in order to achieve a low BPCER of 0.2%, those samples are not detected as attacks by the fused system. It should be also noted that for all APCEs where the fused score $s$ is lower than 0.2, all CNN models have also reported very low scores, thereby making it infeasible to detect those samples for any reasonable BPCER.

The APCEs are summarised in Table 5.4, and the corresponding samples for each PAI species are presented in Figure 5.10. A significant number of errors stems from the orange playdoh fingers: over 63% of the test samples are not detected. Furthermore, for six of them, the corresponding PAD scores remain below 0.03, and all scores $s$ below 0.2 depicted in Figure 5.9b correspond to this PAI species. In order to detect those samples, the detection threshold $\delta$ would have to be placed close to 0, thereby significantly increasing the BPCER of the system. This thus remains an open challenge for the PAD approach described in this Chapter. On the other hand, for the remaining PAI species reporting some APCEs, it is only one or two samples out of up to 275 samples included in the test set. Therefore, we may conclude that the proposed method is robust against these PAI species. Moreover, one of the main issues reported in Gomez-Barrero and Busch (2019) has now been tackled with a new capture device. In that work, out of the 222 PA samples included in the test set, three APCEs were reported for a full finger made of silicone with conductive coating and a conductive silicone overlay. Now, a total of 83 samples are included in the test set for such full fingers and 232 conductive silicone overlays. All those PA samples were correctly detected.

## TABLE 5.4
### Summary of the APCEs of the Fused Scheme Including the PAI Species

| Type | Material | # Samples | # APCEs |
|---|---|---|---|
| Full finger | Playdoh orange | 30 | 19 (63.3%) |
| | Dragonskin | 275 | 1 (0.36%) |
| Overlay | Dragonskin | 89 | 2 (2.2%) |
| | School glue white | 14 | 1 (7.1%) |
| | Silicone two part | 64 | 1 (1.6%) |

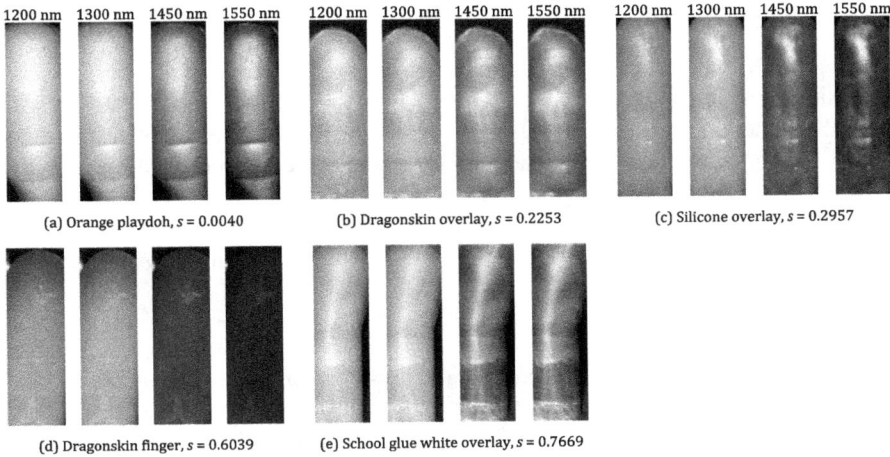

(a) Orange playdoh, $s = 0.0040$

(b) Dragonskin overlay, $s = 0.2253$

(c) Silicone overlay, $s = 0.2957$

(d) Dragonskin finger, $s = 0.6039$

(e) School glue white overlay, $s = 0.7669$

**FIGURE 5.10**   Samples acquired from all the PAI species which are partly not detected by the fused approach. (a) Orange playdoh, $s = 0.0040$; (b) Dragonskin overlay, $s = 0.2253$; (c) Silicone overlay, $s = 0.2957$; (d) Dragonskin finger, $s = 0.6039$; (e) School glue white overlay, $s = 0.7669$.

## 5.7   CONCLUSIONS AND FUTURE RESEARCH

In this chapter, we have presented a PAD method based on SWIR images and multi-spectral CNNs and evaluated its performance on fingerprint data. In particular, we have analysed both CNN models pre-trained on RGB images and training a small residual CNN from scratch. In both cases, a pre-processing module, including a convolutional layer, to transform the four acquired SWIR samples into three channel images is trained together with the rest of the model (i.e., end-to-end).

For the experimental evaluation, a database comprising 11,524 samples and 41 different PAI species has been acquired with a newly developed capture device. The higher resolution of the camera (i.e., $320 \times 256$ px in contrast to the $64 \times 64$ px of previous works) has led to a higher detection accuracy for the top performing CNN models. Specifically, an $APCER_{0.2\%}$ between 2.5% and 3% can be achieved for individual CNN models. This performance has been further enhanced by fusing three CNN models (VGG19, VGGFace, and MobileNetV2) at score level: depending on the configuration selected, an $APCER_{0.2\%} = 1.16\%$ or a $BPCER_{0.2\%} = 1.10\%$ can be attained. This yields highly secure and convenient PAD systems, tuned in for a particular scenario requiring either very high convenience (i.e., BPCER = 0.2%) or very high security (APCER = 0.2%).

In spite of the promising results, the proposed approach is still vulnerable to one of the PAI species analysed: full fingers made with orange playdoh. To tackle this issue, we will focus on our future work on the application of deep learning approaches to the newly captured LSCI and finger-vein data, which can be acquired using the same multi-modal capture device.

## ACKNOWLEDGEMENTS

This research is based upon work supported in part by the Office of the Director of National Intelligence (ODNI), Intelligence Advanced Research Projects Activity (IARPA) under contract number 2017-17020200005. The views and conclusions contained herein are those of the authors and should not be interpreted as necessarily representing the official policies, either expressed or implied, of ODNI, IARPA, or the U.S. Government. The U.S. Government is authorised to reproduce and distribute reprints for governmental purposes notwithstanding any copyright annotation therein.

## REFERENCES

BATL. 2017. "Biometric authentication with a timeless learner". www.isi.edu/projects/batl/.
    BEAT. 2012. *Biometrics Evaluation and Testing*. http://www.beat-eu.org/.
Bossen, A., R. Lehmann, and C. Meier. 2010. "Internal fingerprint identification with optical coherence tomography". *IEEE Photonics Technology Letters* 22 (7): 507–509.
Cheng, Y., and K. Larin. 2006. "Artificial fingerprint recognition by using optical coherence tomography with autocorrelation analysis". *Applied Optics* 45 (36): 9238–9245.
Cheng, Y., and K. Larin. 2007. "In vivo two-and three-dimensional imaging of artificial and real fingerprints with optical coherence tomography". *IEEE Photonics Technology Letters* 19 (20): 1634–1636.
Chug, T., and A. Jain. 2019. "OCT fingerprints: Resilience to presentation attacks". *arXiv preprint arXiv:1908.00102*.
Chugh, T., and A. Jain. 2019. "Fingerprint spoof generalization". *arXiv preprint arXiv:1912.02710*.
Fitzpatrick, T. B. 1988. "The validity and practicality of sun-reactive skin types I through VI". *Archives of Dermatology* 124 (6): 869–871.
Galbally, J., and M. Gomez-Barrero. 2017. "Presentation attack detection in iris recognition". In *Iris and Periocular Biometrics*, ed. by C. Busch and C. Rathgeb. IET, pp. 235–263.
Galbally, J., S. Marcel, and J. Fierrez. 2014. "Biometric antispoofing methods: A survey in face recognition". *IEEE Access* 2:1530–1552.
Ghiani, L., et al. 2017. "Review of the fingerprint liveness detection (LivDet) competition series: 2009 to 2015". *Image and Vision Computing* 58:110–128.
Gomez-Barrero, M., and C. Busch. 2019. "Multi-spectral convolutional neural networks for biometric presentation attack detection". In *Proc. Norwegian Conf. on Information Security (NISK)*. Submitted.
Gomez-Barrero, M., J. Kolberg, and C. Busch. 2018. "Towards fingerprint presentation attack detection based on short wave infrared imaging and spectral signatures". In *Proc. Norwegian Information Security Conf. (NISK)*.
Gomez-Barrero, M., J. Kolberg, and C. Busch. 2019. "Multi-modal fingerprint presentation attack detection: Looking at the surface and the inside". In *Proc. Int. Conf. on Biometrics (ICB)*.
Gonzalez-Soler, L. J., et al. 2019. "Fingerprint presentation attack detection based on local features encoding for unknown attacks". *arXiv preprint* https://arxiv.org/abs/1908.10163.
Goodman, J. W. 1975. "Statistical properties of laser speckle patterns". In *Laser Speckle and Related Phenomena*, ed. by J. C. Dainty, 9:9–75 Springer: Berlin, Heidelberg.
Government of India. 2012. Unique identification authority of India. https://uidai.gov.in/.
He, K., et al. 2015. "Deep residual learning for image recognition". *CoRR* abs/1512.03385.
Howard, A. G., et al. 2017. "MobileNets: Efficient convolutional neural networks for mobile vision applications". *arXiv:1704.04861*.

Huang, D., et al. 1991. "Optical coherence tomography". *Science* 254 (5035): 1178– 1181.

Hussein, M., et al. 2018. "Fingerprint presentation attack detection using a novel multi-spectral capture device and patch-based convolutional neural networks". In *Proc. Int. Wokrshop on Information Forensics and Security (WIFS)*.

ISO/IEC JTC1 SC37 Biometrics. 2016. *ISO/IEC 30107-1. Information Technology - Biometric Presentation Attack Detection*. International Organization for Standardization.

ISO/IEC JTC1 SC37 Biometrics. 2017. *ISO/IEC FIDS 30107-3. Information Technology - Biometric Presentation Attack Detection - Part 3: Testing and Reporting*. International Organization for Standardization.

Jain, A. K. 2007. "Technology: Biometric recognition". *Nature* 449: 38–49.

Kanich, O., M. Drahansky, and M. Mézl. 2018. "Use of creative materials for fingerprint spoofs". In *Proc. Int. Workshop on Biometrics and Forensics (IWBF)*.

Keilbach, P., et al. 2018. "Fingerprint presentation attack detection using laser speckle contrast imaging". In *Proc. Int. Conf. of the Biometrics Special Interest Group (BIOSIG)*.

Kolberg, J., M. Gomez-Barrero, and C. Busch. 2019. "Multi-algorithm benchmark for fingerprint presentation attack detection with laser speckle contrast imaging". In *Proc. Int. Conf. of the Biometrics Special Interest Group (BIOSIG)*, 1–10.

Kolberg, J., et al. 2019. "Presentation attack detection for finger recognition". In *Handbook of Vascular Biometrics*, ed. by S. Marcel et al. To appear, pp. 435–463.

Krizhevsky, A., I. Sutskever, and E. Geoffrey. 2012. "ImageNet classification with deep convolutional neural networks". In *Advances in Neural Information Processing Systems* 25, 1097–1105. Curran Associates, Inc.

Liu, M., and T. Buma. 2010. "Biometric mapping of fingertip eccrine glands with optical coherence tomography". *IEEE Photonics Technology Letters* 22 (22): 1677–1679.

Liu, F., G. Liu, and X. Wang. 2019. "High-accurate and robust fingerprint antispoofing system using optical coherence tomography". *Expert Systems with Applications* 130: 31–44.

Marasco, E., and A. Ross. 2015. "A survey on antispoofing schemes for fingerprint recognition systems". *ACM Computing Surveys (CSUR)* 47 (2): 28.

Marcel, S., et al., eds. 2019. *Handbook of Biometric Anti-Spoofing: Presentation Attack Detection*. Springer.

Meissner, S., R. Breithaupt, and E. Koch. 2013. "Defense of fake fingerprint attacks using a swept source laser optical coherence tomography setup". In *Proc. SPIE 8611, Frontiers in Ultrafast Optics: Biomedical, Scientific, and Industrial Applications XIII*, 8611:86110L.

Mirzaalian, H., M. Hussein, and W. Abd-Almageed. 2019. "On the effectiveness of laser speckle contrast imaging and deep neural networks for detecting known and un-known fingerprint presentation attacks". In *Proc. Int. Conf. on Biometrics (ICB)*.

Moolla, Y., et al. 2019. "Optical coherence tomography for fingerprint presentation attack detection". In *Handbook of Biometric Anti-Spoofing: Presentation Attack Detection*, ed. by S. Marcel et al., 49–70. Springer.

ODNI and IARPA. 2016. *IARPA-BAA-16-04 (Thor)*. https://www.iarpa.gov/index.php/research-programs/odin/odin-baa.

Orrú, G., et al. 2019. "LivDet in action-fingerprint liveness detection competition 2019". *arXiv preprint arXiv:1905.00639*.

Parkh, O. M., A. Vedaldi, and A. Zisserman. 2015. "Deep face recognition". In *Proc. British Machine Vision Conf. (BMVC)*.

Ratha, N., J. Connell, and R. Bolle. 2001. "Enhancing security and privacy in biometrics-based authentication systems". *IBM Systems Journal* 40: 614–634.

RESPECT. 2019. *REliable, Secure and Privacy Preserving Multi-Biometric Person Authentication*. http://www.respect-project.eu/.

Rowe, R. K., K. A. Nixon, and P. W. Butler. 2008. "Multispectral fingerprint image acquisition". In *Advances in Biometrics: Sensors, Algorithms and Systems*, ed. by N. K. Ratha and V. Govindaraju, 3–23. Springer: London.

Sandler, M., et al. 2018. "Mobilenetv2: Inverted residuals and linear bottlenecks". In *Proc. IEEE Conf. on Computer Vision and Pattern Recognition (CVPR)*, 4510–4520.

Simonyan, K., and A. Zisserman. 2015. "Very deep convolutional networks for large-scale image recognition". In *Proc. Int. Conf. on Learning Representations (ICLR)*.

Sousedik, C., and R. Breithaupt. 2017. "Full-fingerprint volumetric subsurface imaging using fourier-domain optical coherence tomography". In *Proc. Int. Workshop on Biometrics and Forensics (IWBF)*, 1–6.

Sousedik, C., R. Breithaupt, and C. Busch. 2013. "Volumetric fingerprint data analysis using optical coherence tomography". In *Proc. Int. Conf. of the BIOSIG Special Interest Group (BIOSIG)*, 1–6.

Sousedik, C., and C. Busch. 2014. "Presentation attack detection methods for fingerprint recognition systems: A survey". *IET Biometrics* 3, no. 1: 1–15. ISSN: 2047-4938.

Steiner, H., et al. 2016. "Design of an active multispectral SWIR camera system for skin detection and face verification". *Journal of Sensors*.

Szegedy, C., S. Ioffe, and V. Vanhoucke. 2016. "Inception-v4, inception-ResNet and the impact of residual connections on learning". *CoRR abs/1602.07261*.

Szegedy, C., et al. 2015. "Going deeper with convolutions". In *Proc. Conf. on Computer Vision and Pattern Recognition (CVPR)*, 1–9.

Szegedy, C., et al. 2016. "Rethinking the inception architecture for computer vision". In *Proc. Conference on Computer vision and Pattern Recognition (CVPR)*, 2818–2826.

TABULA RASA. 2010. "Trusted biometrics under spoofing attacks". http://www.tabularasa-euproject.org/.

Tolosana, R., et al. 2018. "Towards fingerprint presentation attack detection based on convolutional neural networks and short wave infrared imaging". In *Proc. Int. Conf. of the Biometrics Special Interest Group (BIOSIG)*.

Tolosana, R., et al. 2019. "Biometric presentation attack detection: Beyond the visible spectrum". *IEEE Transactions on Information Forensics and Security* 15: 1261–1275.

Vaz, P., et al. 2016. "Laser speckle imaging to monitor microvascular blood flow: A review". *IEEE Reviews in Biomedical Engineering* 9: 106–120.

# 6 AI-Based Approach for Person Identification Using ECG Biometric

*Amit Kaul*
NIT Hamirpur

*A.S. Arora*
SLIET Longowal

*Sushil Chauhan*
NIT Hamirpur

## CONTENTS

6.1   Introduction ................................................................................. 133
6.2   ECG and Related Work ................................................................ 136
    6.2.1   Advantages of ECG Biometric ..................................... 137
    6.2.2   Literature Review ............................................................ 138
6.3   Methodology Adopted ................................................................. 141
    6.3.1   Feature Extraction .......................................................... 142
6.4   Classifier ...................................................................................... 143
    6.4.1   Artificial Neural Network (ANN) ................................. 143
    6.4.2   Support Vector Machine (SVM) .................................... 144
6.5   Experiments and Results ............................................................. 145
6.6   Conclusions .................................................................................. 150
References ............................................................................................ 150

## 6.1   INTRODUCTION

The widespread computerisation has tremendously popularised the various digital payment modes, be it online transactions, e-shopping, ATMs, etc. This along with the increased security threat due to menace of terrorism has necessitated the requirement of reliable human identification systems. Biometrics, which utilise physiological (face, fingerprint, iris, etc.) or behavioural (speech, gait, keystroke, etc.) human traits, has provided the much needed solution in this regard. These systems are being extensively employed for various authentication tasks like attendance, passports, citizen registers, etc. The strength of a biometric trait is usually assessed on the basis of seven parameters, namely, uniqueness, universality, permanence, performance,

acceptability, collectability, and circumvention. However, none of the presently utilised traits possess all these qualities individually. For example, iris in certain individuals may be damaged due to some eye disease or congenital defect, thus violating
the universality trait, while speech tends to vary even in case of simple throat infection. Fingerprints also have some acceptability issues because of their linkage with
criminal identification. Moreover, as the popularity of these new age security tools
has risen, so has the tendency of the fraudsters to find means in order to deceive
or attack these systems. Attempts have been made to fool fingerprint and speech-
based biometric systems respectively by use of synthetic template of fingerprints prepared from materials such as latex gelatine etc. and pre-recorded speech utterances.
Furthermore, these type of person recognition systems need to be incorporated with
a liveness detection mechanism so as to ensure that sample has been obtained from
a living individual.

   In order to overcome the aforesaid issues, in the past decade, researchers have
explored the possibility of developing human recognition system based on bioelectric
signals like electrocardiogram (ECG), electroencephalogram (EEG), and photoplethysmogram (PPG). The bioelectric signals have been found to possess characteristics
suitable for biometric applications either in unimodal or multimodal configuration.
In this chapter, an approach related to ECG-based biometric recognition has been
described but before discussing it, an overview of biometrics and its various modes
is presented and the same is summarised in Figure 6.1.

   Design of a typical biometric system begins with an enrolment phase in which
the templates from prospective users of the technology are stored in a database.
After the enrolment phase during testing, all biometric systems can operate in two
modes – verification or identification. In the verification mode, identity claim made
by the user is either accepted or rejected, so while going through the testing phase,
the template of the individual who has claimed to be genuine user is compared with
his already stored template. Identification on the other hand is more difficult and
requires comparison with all templates stored in the database. For the closed set
case, i.e., individual to be identified has his template already stored in the database,
identification is carried out by finding the most similar template among the enrolled
subjects. Whereas for the open set case, individual to be identified may or may not be
already enrolled, so in addition to being most similar, the score between the matched
template and test sample should be below some pre-decided threshold value. Briefly
stated, in verification answer is found to the query "Am I who I claim to be", while
the question which is answered in identification is "Who I am".

   On the basis of permanence property, biometric traits can be categorised as hard
and soft. Hard biometrics are those which do not vary drastically over sufficient
duration of time and have high uniqueness, while soft biometrics are those which
provide some information about a person but lack the uniqueness and permanence
property. Hard biometrics can be further divided on the basis of type of trait used.
Physiological biometrics are those in which the sample is obtained from a physical
characteristic, e.g., fingerprint and face, and are inherent trait of a human being. On
the other hand, behavioural biometrics are the ones which are acquired by humans,
e.g., walking style and handwriting [1]. As behaviour will always influence interaction of the user with biometric sensor, so all biometric traits have some behavioural

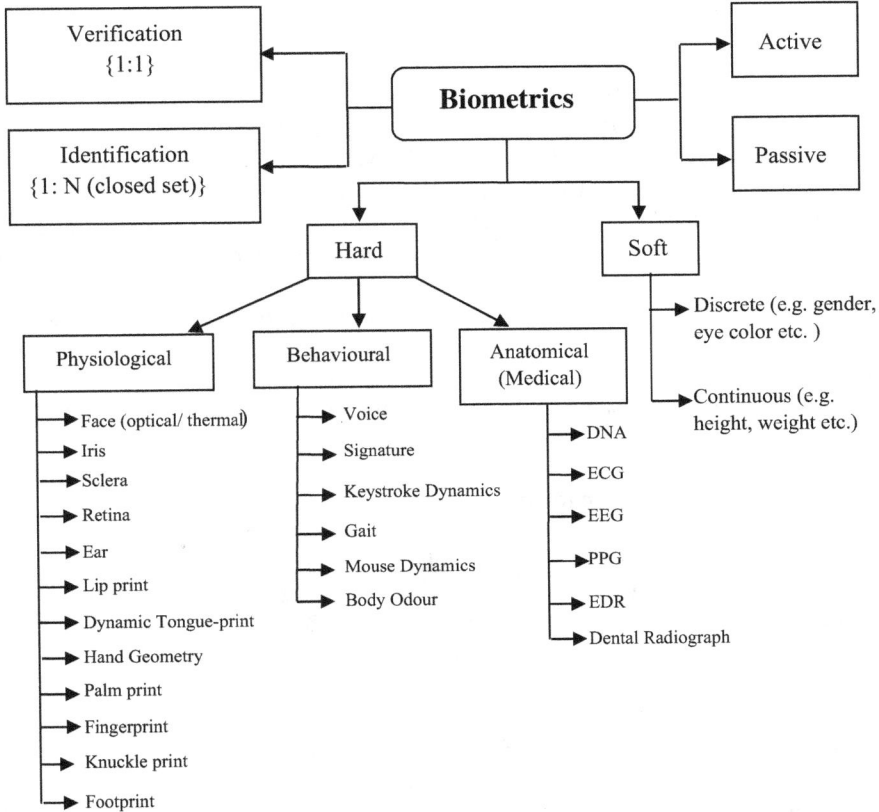

**FIGURE 6.1** An overview of biometrics.

aspect associated with them. Moreover, some researchers also consider mixed category for classifying biometric modalities like speech as it is dependent on anatomy of human vocal apparatus besides being strongly influenced by individual's behaviour. As stated earlier, physiological and behavioural traits may be absent in some humans because of a diseased state or some other reason and as such do not fulfil the universality trait. However, medical biometrics based on ECG, EEG, and PPG provide solution to these issues as they inherently possess universality trait and will ensure entire population coverage. This provides another classification for biometrics where medical biometrics along with other upcoming biometrics like odour, lip-pint, etc. are clubbed under the heading of *esoteric* biometrics, while biometrics with mature technology like fingerprint, face, etc. can be grouped under traditional biometrics.

In order to explain the approach presented here for ECG-based biometric recognition, first the description of ECG waveform, its generation, and review of the literature related to ECG biometric is given in the following section.

## 6.2 ECG AND RELATED WORK

Electrocardiogram (ECG or EKG), proposed in 1901 by William Einthoven as medical diagnostic tool, is a quasi-periodic, non-stationary signal of approximately 100 Hz, which represents the electrical activity of heart. The backbone of cardiovascular system, ensuring supply of nutrition and oxygen through blood, is human heart. It is a muscle roughly of the size of human fist comprising of four chambers, two atria for collection of blood and two ventricles for pumping out blood. From engineering point of view, it can be considered as a combination of mechanical, hydraulic, and electrical sub-systems.

It is basically a two-stage pump with its pumping action initiated by an electrical stimulus provided by the sinoatrial (SA) node (natural pacemaker of heart). The pulse produced by the SA node results in contraction of atria. The action potential generated propagates through atria, and on reaching atrioventricular (AV) node is delayed before being transmitted to the ventricles. This entire process known as cardiac cycle is composed of relaxation of ventricles for filling of blood (diastole) and contraction of ventricles for pumping the blood out of heart to pulmonary artery and aorta (systole). The contraction of so many cells at one time creates a mass electrical signal that can be detected by electrodes placed on the surface of person's chest or his extremities. ECG or EKG is a graphic recording or display of these time-varying voltages produced by heart during the cardiac cycle. Depending upon placement of electrodes, 12 lead configurations can be obtained, three limb leads (Lead I, Lead II, and Lead III), three augmented leads (avR, avL, and avF), and six chest leads ($V_1$, $V_2$, $V_3$, $V_4$, $V_5$, and $V_6$). A typical ECG waveform is depicted in Figure 6.2.

An ECG waveform consists of P-QRS-T waves; a small U wave may also be sometimes present. Each of these characteristic points in ECG is related to electrical activity in human heart during one cardiac cycle. The P-wave represents the atrial depolarisation, resulting in response to SA node triggering; PR interval indicates the AV node delay; the QRS complex characterizes the depolarisation of the ventricles, and finally the ventricular repolarisation is depicted by the T-wave.

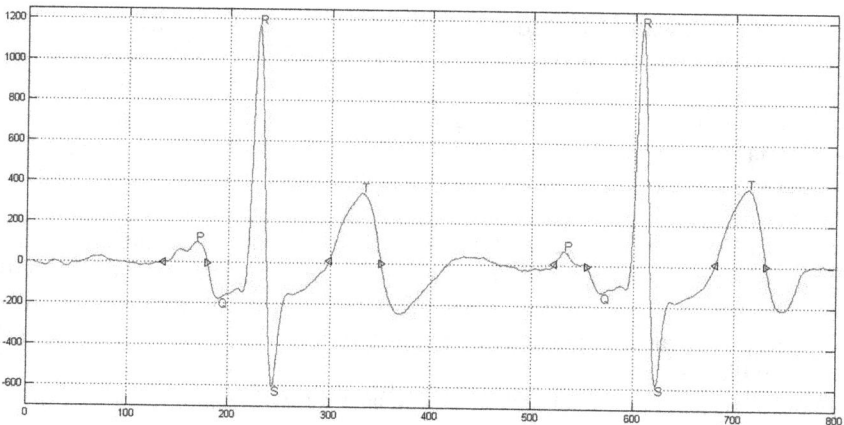

FIGURE 6.2   A typical ECG waveform showing various characteristic points.

### 6.2.1  ADVANTAGES OF ECG BIOMETRIC

ECG has been extensively used for detection of various abnormalities of the heart. In most of these diagnoses, the time duration and the amplitude of the various waves are used. As stated above, primarily ECG has been employed for detection of cardiovascular disorders but recently there has been an upsurge in activity related to its use for person recognition. The inter individual variability in ECG is attributed to the differences in position, size, and anatomy of the heart, age, sex relative body weight, chest configuration, and various other factors. Figure 6.3 shows that the ECG samples were collected from four different subjects, and the distinction in the ECG signal waveforms in these cases can be inferred easily by even visual inspection [2]. The unique and distinct advantages offered by ECG have made it a popular choice among biometric researchers. The qualities possessed by ECG include (i) inherent liveness property integrated with its universal nature, (ii) the ease of combining it with existing and established biometric modalities, (iii) difficult to copy and

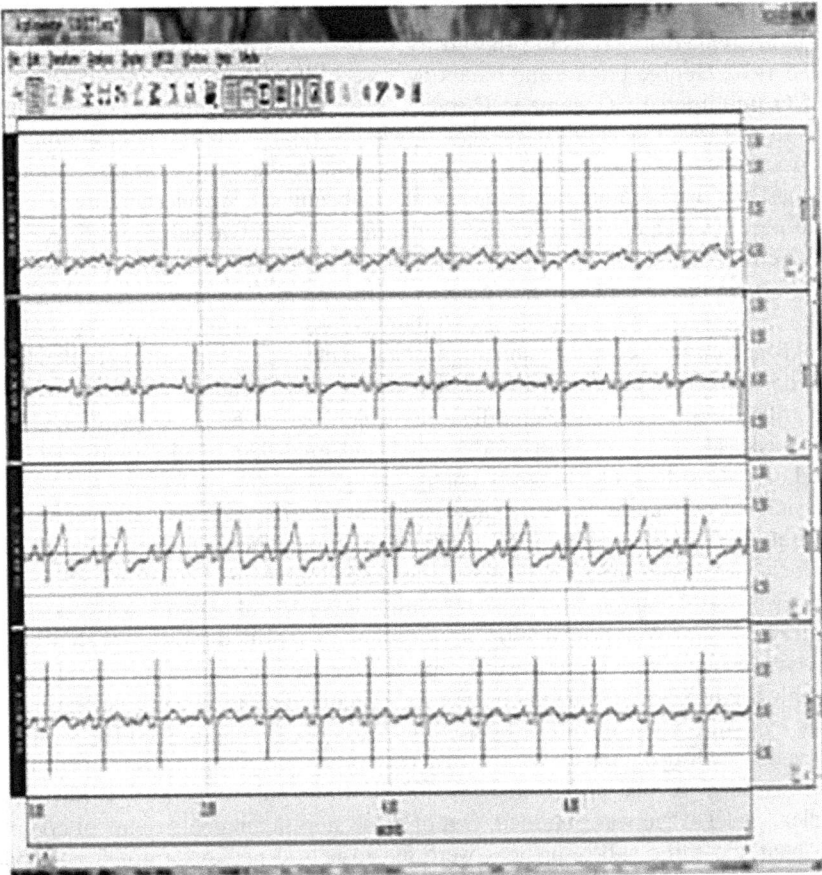

**FIGURE 6.3**   ECG waveform of four different subjects [40].

manipulate, and (iv) application of ECG for medical domain, e.g., person identification in telemedicine and patient monitoring or automatic drug delivery. Therefore, using ECG together with stable biometrics like fingerprint and iris in a multimodal framework ensures fulfilment of majority of the requirements of an ideal biometric system. The combined system due to complementary features of individual traits will be universal, stable, immune to replay attack, and liveness assured. In order to assess the ability of ECG to distinguish individual under varied conditions, a number of studies have been carried out. Some of these have been discussed in the following sub-section.

### 6.2.2 Literature Review

ECG-based biometric approaches can be divided into three main categories: (i) Time interval and amplitude-based approaches, (ii) Transform domain approaches, and (iii) model-based approaches. Some researchers consider only two main categories: fiducial-based techniques and other two grouped together under non-fiducial-based techniques.

Studies related to exploring the utility of ECG are not altogether new; in fact around 1977, George Forsen and his fellow researchers had recognised the potential of ECG (modified ECG named "C-trace") as a personal authentication modality. Then, in 1997, M. Ogawa et al. presented first practical work in this direction, when they used ECG for personal identification system in their home health monitoring system. They used features extracted with Daubechies 4 Wavelet and a three-layer neural network as a classifier for distinguishing between two subjects [3]. However, it was Lena Beil and her colleagues whose work provided the much needed boost and resulted in an upsurge in research activity in this area. They investigated the use of ECG as a biometric measure and tested their algorithm on 20 subjects. A statistical model based on time intervals and amplitudes of characteristic points (PQRST) was constructed for each person. A total of 30 features were extracted which were reduced to 12 after dropping the correlated features which were used to perform the identification tasks [4,5]. T. W. Shen used Lead I and modified Lead I (with electrodes placed on palms) for ECG-based biometric using time domain features with DBNN (Decision-Based Neural Network) for identification [6,7]. R. Palaniappan and S. M. Krishnan used three amplitude and two time interval values extracted from the ECG waveform along with the form factor of the QRS complex as the feature set for their study. Two neural network architectures – multilayer neural network with one hidden layer trained using backpropagation algorithm and simplified Fuzzy ARTMAP were used as classifiers, with better results reported for multilayer perceptron (MLP) [8]. To tackle the problem of varying heart rate, Saechia et al. used a normalisation procedure by assuming heart rate to be 80 times/minute and mapping all ECG waves to it by using scaling factor and reconstruction procedure. Fourier transform of the ECG waveform of one period and also of the three sub-sequences, P-wave, QRS complex, and T-wave was obtained. Out of these significant coefficients of complete heart beat and three sub-sequences were fed to neural network trained with backpropagation algorithms for training. The results obtained depicted that the neural network trained with three sub-sequences performed better than the network trained with Fourier coefficients from one period of ECG [9]. In order to alleviate the need of detection of characteristic points, Konstantinos and his colleagues used a non-fiducial feature extraction scheme for ECG-based authentication. In order to capture the similarity across multiple ECG cycles, the authors proposed the use of normalised autocorrelation for different lags say 'M' such that M ≪ N where "N" is the signal length chosen. The choice of the signal length is restricted by the fact that it should contain multiple cycles. Discrete Cosine Transform (DCT) coefficients of normalised autocorrelation were used as features [10–12]. In 2008, Yao and Wanput forward a wavelet-based scheme in which ECG cycles between two consecutive R peaks were selected, and wavelet features were obtained by decomposing the extracted cycles using "bior1.1" wavelet at scale 6. Principal Component Analysis (PCA) was used for classification [13]. In order to improve the classification performance, a three-layer neural network with feedforward architecture (512-64-128-1) was used. Input vector to the neural network comprised wavelet coefficient structures obtained from two different cycles either from same or different subjects. The output of network was +1 when two wavelet coefficient structures were obtained from the same subject and −1 when they were obtained from different subjects [14]. Anthony Kaveh and Wayne Chung extracted 34-dimensional feature set made up of ECG time intervals and morphological information. The features were reduced to 13 by employing PCA for dimension reduction. A multi-class support vector machine (SVM) built on "one versus all" was used for classification [15]. In 2009, S. C. Fang and H.L. Chan proposed a method of identifying individuals in phase space [16]. For this, a three-dimensional (3D) vector was built, with the averaged ECG signal as the first dimension and two other formed using a time delay $\tau$ each from the averaged signal and second component. For three leads, 3D feature vector was built from the anterior, lateral, and posterior leads after averaging each of the three leads individually as win case of a single lead. The comparison of phase portraits was carried out using spatial correlation and mutual nearest point distance. Authors also compared their work with ECG-based identification using fiducial features. Neural networks MLP, radial Basis function (RBF), and Euclidean distance were used as classifiers. The results obtained showed better performance for three leads rather than single lead ECG. Moreover, the mutual nearest point distance was better than the spatial correlation. Further results were better when only QRS complex was used rather than the complete wave. Can Ye et al. [17] employed wavelet features extracted using Daubechies wavelets clubbed with ICA (Independent Component Analysis) features extracted from two leads of ECG to form the feature vector. SVM was used for classification. Tests conducted on three public dataset showed that good identification accuracy can be achieved when information was combined using decision level fusion for the two lead ECGs. Ming Li and Shrikant Narayan came out with a technique in which two feature sets were extracted from the ECG signal. In one of them, coefficients extracted from Hermite polynomial expansion were modelled using SVM while in the second approach, cepstral features of ECG were used to model subjects using GMM (Gaussian Mixture Models) supervector to get a score. Finally, score level fusion was carried out to get the combined score. Highest accuracy was obtained for the combined scheme [18]. Extended Kalman Filter has also been applied for the ECG biometric task [19]. QRS complex is relatively stable and undergoes very little variation with a change in the heart rate. Recognising this, Tawfik and Kamal carried out normalisation of only the QT segment. DCT features were extracted from normalised QT segment and QRS complex and were fed to separate the feedforward neural network with one hidden layer for classification. Best results were obtained using features from only the QRS complex [20]. Vu Mai and his colleagues also used the QRS complex values as the feature vector and with two neural network architectures; a 2-layer MLP and RBF as classifiers [21]. Taking into note the stability of QRS complex with reference to varying heart rate, L. Hou and his fellow researchers generated spectral coefficients based on Fourier transform and DCT of the QRS complex. The dimension reduction was further carried out using PCA. A two-stage classification was carried out with feature matching as the first check and in case multiple subjects are identified, the neural network was used at the second stage [22]. Loong and his fellow researchers put forward a non-fiducial approach using 40 coefficients of linear prediction coefficients (LPC) spectrum obtained from 5 seconds ECG signal with 50% overlap and multilayer neural network with two hidden layers each having 100 nodes as the classifier. One more scheme using 50 features from wavelet packet decomposition coefficients was also investigated. Experiments conducted by them showed superior performance for the LPC method [23]. In 2010, Nahid Ghofrani and Reza Bostani studied performance of six non-conventional features for ECG biometric application, namely, autoregressive (AR) coefficients, power spectral density (PSD), Lyapunov exponent, approximation entropy, Higuchi fractal dimension and Shannon entropy. For classification, K-nearest neighbours (KNN) and two neural networks multi-layer perceptron and RBF were used. Five-dimensional feature vector formed by combining four coefficients from the fourth order AR model with mean PSD of every window, provided the best results when KNN was used as the classifier [24]. Ikenna Odinaka et al. presented an extensive study of a frequency-based method on a fairly large dataset comprising 5 minute recordings from 269 subjects with multiple session recordings. They computed spectrogram of each ECG pulse by calculating the short-time Fourier transform of the windowed sequence (Hamming window) and finding the logarithm of square of its magnitude. For classification, maximum likelihood-based model using normal distribution for feature vectors was built. Features selection was carried out based on the relative entropy approach. For verification, log likelihood ratio (LLR) was computed, and claim was accepted if its value was greater than a pre-determined threshold. Identification was based on the highest LLR value. The authors reported considerably good performance even when the suggested approach was tested on across sessions recordings [25]. In 2011, Ching-Kun Chen and his fellow workers used non-traditional six-dimensional feature vector made up of correlation dimension, four Lyapunov coefficients, and root-mean-square value and fed it to the multilayer network with two hidden layers (20 and 5 nodes). A recognition rate of 90% was achieved for database of nine subjects [26]. Andre Lourenco, Hugo Silva, Ana Fred, and their co-workers have been trying to develop a finger-based ECG biometric system. They have used fiducial features and template features with Euclidean distance, SVM, KNN, etc. as classifiers. The results reported by authors are quite encouraging [27,28]. A short-term dataset (65 subjects) and long-term dataset (63 subjects) for testing of ECG biometric methods have also been prepared by the authors [29]. Tantawi et al.

compared the performance of 36-dimensional feature vector comprising majority of fiducial features mentioned in the literature and a subset of 23features derived from the five prominent characteristic points (P, Q, R, S, T) with RBF as the classifier. It was found that the system was able to recognise the individual even with reduced features although with slightly less accuracy [30]. In 2013, Tantawi and his fellow researchers derived level six wavelet features from R-R interval of the ECG waveform by utilising bior 2.6 wavelet. The RBF neural network was employed as the classifier. Experiments were also conducted by feeding RBF with features extracted from the AC/DCT method, QRS complex, and QT interval. Testing of the proposed approach depicted high subject identification and window recognition rate. However, results were not encouraging by using features extracted from the QRS complex, QT interval, and AC/DCT method [31]. In 2014, Eduardo J. S. Luz and his colleagues have studied the viability of the biometric application of ECG signals sampled at a low frequency. For this, they used four set of features, mainly wavelet features and SVM as the classifier and found that signals obtained at low frequencies can be successfully used [32]. In the last couple of years, R. D. Labati and fellow researchers have suggested a scheme named as Deep-ECG which uses convolutional neural network (CNN) for feature extraction and perform identification by score-based matching [33]. Some authors have transformed a1-D ECG into a 2-D representation and then used CNN for subject identification [34,35]. In Ref. [36], authors have used a two-dimensional matrix created from QRS segments to obtain the model via CNN for personal identification. The experiments conducted by authors in majority of the works have been conducted on small population datasets generally less than 50. However, in Ref. [37], extensive survey of ECG-based biometric approaches has been presented, and then the comparative study of around 19 non-fiducial algorithms was carried out for authentication applications with experiments conducted on their in-house dataset of 265 individual having multiple session recordings having two weeks to six months gap between recordings. The performance of all algorithms dipped with an increase in duration between testing and training sessions, although some improvement was achieved by using multiple session recordings for training. More importantly, only three of the eighteen algorithms resulted in EER less than 10% for testing across multiple sessions. Overview of ECG biometric with challenges and opportunities related to this biometric trait has also been discussed in Ref. [38,39]. The main issues faced by ECG biometric are related to the population size of datasets used in experiments and variability of ECG with time, stress, and diseased state. Moreover, the user friendliness and convenience of the subjects is another aspect that demands the attention of the biometric researchers. For this, studies need to be carried on a datasets comprising of sufficient number of subjects and reasonable gap between recordings employed for training and testing the developed algorithms. Keeping this in mind, an approach for ECG-based human identification has been explored henceforth in this chapter.

## 6.3  METHODOLOGY ADOPTED

ECG biometrics being a pattern recognition problem comprises two blocks, a feature extractor and a classifier. The performance of the fiducial feature extraction approaches is largely dependent on the accuracy of the QRS detection algorithms as these techniques require segmentation of ECG wave and detection of various sub waves, i.e., P-wave, QRS complex, T-wave, etc. Therefore, for feature extraction, here a non-fiducial approach, as suggested by N. K. Plataniotis, has been employed. A number of classifiers have been reported in the literature. Based upon earlier studies carried by the authors [40], for the classification task, MLP, RBF, and SVM have been used. A brief review of the feature extraction technique and the classifiers is given in the following sub-sections.

### 6.3.1  FEATURE EXTRACTION

The approach suggested by N. K. Plataniotis and his colleagues is also known as the AC/DCT method as it exploits the ability of autocorrelation to extract the self-similarity in a given data sequence. The approach begins with pre-processing of the signal to remove noises in the ECG signal like the baseline wander, power line interference, electrode contact noise, etc. For this work, pre-processing has been carried out using fourth-order Butterworth band pass filter with a cutoff frequencies 1 and 40 Hz. This pre-processed ECG signal is windowed with only constraint that the windowed signal should contain at least two complete cardiac cycles; so the number of samples are accordingly chosen based on the sampling rate. This is followed by autocorrelation of the resulting sequence and its normalisation by dividing the auto-correlation with the maximum value of the autocorrelation coefficients obtained. In order to carry out the dimension reduction, utilising the energy compaction ability of DCT, DCT of the normalised autocorrelation coefficients is obtained. The main steps involved in the implementation of this technique are mentioned below:

a. Pre-process raw ECG signal to remove noise and segment it into non-overlapping windows.
b. Calculate the autocorrelation $R_{xx}(m)$ of the windowed ECG signal $x(i)$ and obtain the normalised autocorrelation coefficients by using the following expression:

$$\hat{R}_{xx}(m) = \frac{R_{xx}(m)}{R_{xx}(0)} = \frac{\displaystyle\sum_{i=0}^{N-|m|-1} x(i)x(i+m)}{R_{xx}(0)} \tag{6.1}$$

where $m$ is the time lag with values ranging from 0 to M-1 and M being very-very less in comparison to $N$, the length of the windowed signal.
c. Significant coefficients from the normalised autocorrelation coefficients are obtained by applying DCT; with many DCT coefficients having value zero or near to zero.
d. First $C$ coefficients are retained to form the feature vector of a given subject.

The three values to be chosen are, the interval of ECG signal $N$, the value of $M$ related to lag $m$, and the number of DCT coefficients to be used as the feature vector.

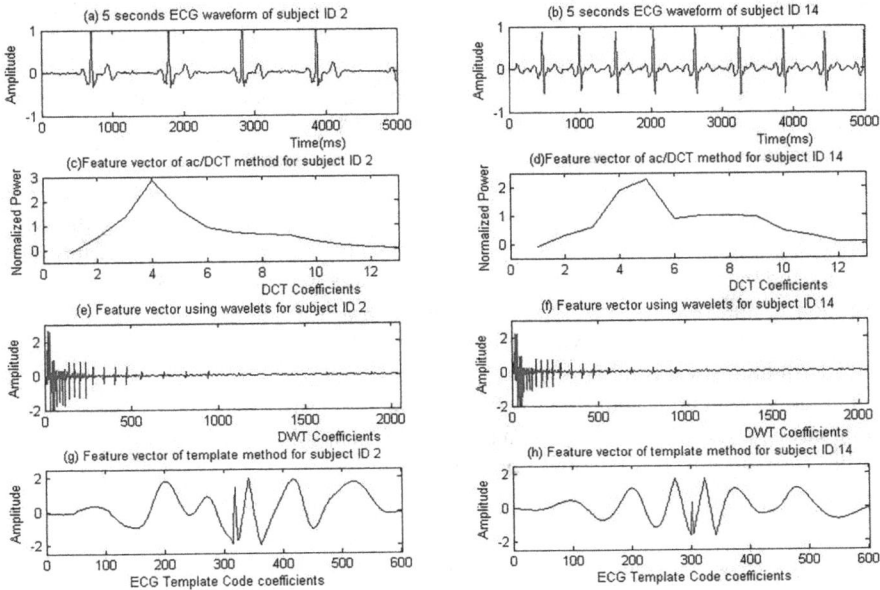

FIGURE 6.4   Comparison of AC/DCT features of two subjects.

Based on the results reported in the literature and further experimental evaluation, the values of these parameters were chosen as $N = 10{,}000$, corresponding to 10 seconds of signal, $M = 180$ and $C = 13$. The outputs of the various steps of the AC/DCT method listed above are shown in Figure 6.4 for ECG signals of two subjects. The difference in feature vectors extracted from ECG signals of two subjects is evident from the inspection of Figure 6.4g and h.

## 6.4   CLASSIFIER

Artificial Neural Networks and Support Vector Machines have been used for biometric recognition scheme explained in this chapter. An overview of these two classifiers is given below.

### 6.4.1   ARTIFICIAL NEURAL NETWORK (ANN)

Usually the requirement is that the system has the ability to learn from known samples of pattern and then adapt itself to take decision for unseen patterns; somewhat similar to what humans go on to do. In order to replicate the human learning capability many techniques have been put forward, out of these ANNs tries to computationally model the fundamental building block of the brain, i.e., a neuron.

ANN is an interconnection of neurons in layered manner with the output of a neuron dependent on the input weighted by connecting weights and non-linearly mapped by transfer function or activation function. Neuron layers between input and output layers are known as hidden layers. The manner in which the various neurons are arranged is known as architecture of the neural network. The architectures can be broadly classified as single layer feedforward networks, multilayer feedforward networks, and recurrent networks. Some of the popular networks reported in the literature are MLP, RBF, learning vector quantization (LVQ), self-organising map (SOM), and Hopfield neural network. On the other hand, the algorithm used to compute the weights and other parameters is known as the learning rule which can be supervised or unsupervised. The training of the MLP network is carried out by using the backpropagation algorithm which is a supervised learning scheme. In this algorithm, the weights and biases of the different layers are updated by propagating in the backward direction of the sensitivities obtained after feeding the feature vector at the input of the network. Among the various variations of backpropagation, the Levenberg Marquardt algorithm is most popular [41].

The development of neural networks followed a heuristic path, with theory developing later on. On the other hand, an approach which is based on the sound theoretical background and has become popular among pattern recognition community is SVM.

### 6.4.2 Support Vector Machine (SVM)

SVM proposed by Vladimir Naumovich Vapnik and Alexey Yakovlevich Chervonenkis has emerged as a powerful tool for binary classification. It separates the two classes by constructing an optimal separating hyperplane and ensures that the margin between two classes is maximised. "Support Vectors" are the bounds between datasets and the optimal separating hyperplane. The objective of support vector is to maximise the distance or the margin between the support vectors. In fields such as handwritten digit recognition, text categorisation, and information retrieval, SVMs hold records in performance benchmarks [42].

For any $N$-dimensional feature, vector $f_i$ can be considered as a point in the $N$-dimensional plane belonging to a class $c_i \in \{-1, 1\}$. The optimal separating hyperplane, in the case of linear classification is obtained for the two classes in the following manner:

$$w \cdot f_i + b \geq 1, \quad c_i = 1 \tag{6.2}$$

$$w \cdot f_i + b \leq -1, \quad c_i = -1 \tag{6.3}$$

The objective here is that $d = \dfrac{2}{\|w\|}$, i.e., the distance between the support vectors is maximised. For this, a Lagrange function is formulated and solved for minimisation of $w$ and $b$. Usually linear classification is not always attained. Therefore, the input vector is mapped to higher dimension using a kernel function. A wide variety of kernel functions have been proposed by the researchers like polynomial, RBF kernel. In the experiments discussed in next section, some of these kernels have been used for the classification task.

## 6.5 EXPERIMENTS AND RESULTS

As mentioned earlier in majority of the studies conducted for studying utility of ECG as a biometric measure, experiments have been conducted on public databases (MIT–BIH–Arrhythmia database and PTB Database) which were collected with focus on heart ailments. Normally single session records are available which are partitioned into two sets, out of which one is used for training or template creation and other for testing. This does not reflect the true performance of the methods used for a biometric like ECG as its permanence property is not tested through such experimentation. To investigate long duration stability of ECG for biometric applications, sufficient gap should exist between samples used for training and testing.

Keeping this in mind, two ECG datasets were collected as a part of multimodal biometric database, and experiments have been conducted on these datasets [43]. For the session I, the Lead II recordings of duration 5 minutes for 229 subjects were collected. The remaining three session records were collected after a gap of at least two to three months between each subsequent session. However, the population size in the later sessions decreased due to non-availability of the same subjects. The second database comprises two sessions of Lead I recordings of 3 minutes duration for 110 subjects. A summary of the two datasets is given in Table 6.1.

All ECG records were collected using a physiological data acquisition system (Biopac, MP-150) at a sampling rate of 1,000 Hz. Majority of the subjects belonged to the age group of 17–30 years, with none of them reporting any cardiovascular abnormality.

In order to study the efficacy of the artificial intelligence tools in solving pattern recognition problems like the one discussed in this chapter, identification task was performed by computing distance between template and feature vectors from different sessions using different distance measures. The best results obtained for the two datasets are presented in Table 6.2.

The results in the above table depict that the identification rate is reasonably low for all distance measures. However, for authentication tasks, the distance measures have good accuracy, as reported in Ref. [40].

**TABLE 6.1**
**Summary of Two in-House ECG Datasets**

| | Lead Configuration | | | | | |
|---|---|---|---|---|---|---|
| | Lead II | | | Lead I | | |
| Session | No. of Subjects | Gap b/w Sessions | Recording Duration (Minutes) | No. of Subjects | Gap b/w Sessions | Recording Duration (Minutes) |
| I | 229 | - | 5 | 110 | - | 3 |
| II | 200 | 2–4 months | 5 | 110 | 1–20 weeks | 3 |
| III | 155 | 6 months | 5 | - | - | - |
| IV | 82 | 1½–2 years | 5 | - | - | - |

**TABLE 6.2**

**Across-Session Identification Rate (%) for ECG Datasets Using Distance Measures**

| Dataset | Euclidean | Manhattan | Canberra | Square Chord | Square Chi-Squared | Bray Curtis |
|---|---|---|---|---|---|---|
| | | | Distance Measure | | | |
| Lead II | 41 | 42.5 | 8.5 | 28 | 28.5 | 42.5 |
| Lead I | 52.73 | 50.91 | 29.09 | 50 | 50 | 50.91 |

For the AI techniques studied in this work, features extracted from first 2-minute record have been used for training of the neural networks and SVM. For each subject, a separate MLP network, as shown in Figure 6.5, was built with one hidden layer having 26 neurons and one neuron in the output layer. The output of the network was set to be equal to 1 for feature vectors from the same subject and 0 for feature vectors of other subjects.

For all three classifiers, two training schemes were adopted. In the first training scheme, feature vectors extracted from first 3 minutes of session 1 of Lead II datasets were used for training. This comprised 18 feature vectors per subject. Whereas for Lead I ECG dataset, feature vectors from first 2 minutes of recording were used for training. In order to embed information from multiple sessions in the classifier, second training format was adopted. For this, scheme training vectors were the features extracted from first 3 minutes of ECG data for both session 1 and session 2 records. As only two records are available for Lead I datasets, so multiple session training was not performed for this dataset. In addition to this, SVM performance was evaluated using three combinations of kernel function and learning method, namely, linear kernel with sequential minimal optimisation (SMO), quadratic with least square (LS) and linear with LS. The performance evaluation was carried out within session analysis followed by across-session testing.

In the tables listed henceforth, Lead A – X, Y denotes that experiments were conducted on Lead A dataset by training the classifiers using feature vectors from X session, and testing was carried out with feature vectors extracted from session Y. Table 6.3 provides the classification accuracy in terms of true match rate (TMR) and false match rate (FMR) for within session analysis. MLP and RBF neural networks and SVMs were trained with the first methodology, i.e., feature vectors from single

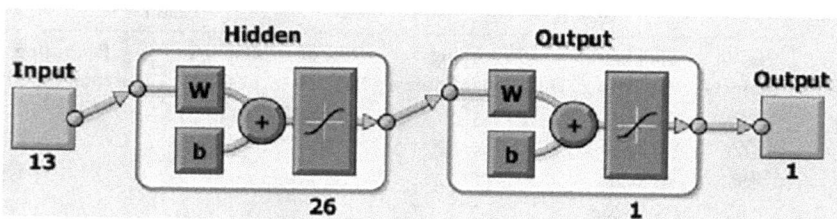

**FIGURE 6.5** Architecture of the feedforward network.

**TABLE 6.3**

**Within Session Classification Performance for ANNs and SVM**

| Session | | MLP | RBF | Classifier SVM Linear SMO | Quadratic LS | Linear LS |
|---|---|---|---|---|---|---|
| Lead II-S1,S1 | TMR | 99 | 99.5 | 99 | 99.5 | 100 |
| | FMR | 3.47 | 0.86 | 3.4 | 2.11 | 8.95 |
| Lead I-S1,S1 | TMR | 99.09 | 99.09 | 98.18 | 100 | 100 |
| | FMR | 3.71 | 0.8 | 3.14 | 2.73 | 8.92 |

session records were used for training. An average identification rate of more than 99% was achieved with all three classifiers.

The lowest and highest FMR was respectively obtained with the RBF network and with SVM having linear kernel and the LS learning method. The results achieved with these classifiers were significantly better than those attained with AC/DCT features and simple distance measures as classifiers.

Extending the investigations further for across-session analysis, the networks and SVMs trained earlier and employed for within session analysis were fed with feature vectors extracted from ECG records of other sessions. Table 6.4 provides the summary of results for across-session testing of both Lead II and Lead I dataset for various sessions with MLP and RBF networks. For all three sessions of Lead II dataset, identification accuracy of around 55%–60% was achieved with the MLP network, while for RBF, network accuracy in the 60%–70% range was achieved. Although the accuracy for the both the classifiers drastically decreased with reference to within session analysis, it is about 20% higher than the accuracy achieved earlier when across-session analysis was carried for the AC/DCT method using simple distance measures, as depicted in Table 6.2. Even for Lead I dataset, the identification accuracy of nearly 70% achieved with both MLP and RBF networks is better than the performance achieved with various distance measures.

In addition to this when feature vectors for two datasets were tested using SVM as a classifier, the identification accuracy was found to be better than MLP for different combinations of kernel functions and learning methods used in this work. On the other hand, in comparison to RBF networks, a higher recognition rate was achieved with SVM only when the LS learning method was utilised for quadratic and linear kernels. However, the misclassification rate, i.e., FMR was little higher for both these cases. For Lead I dataset, also mean identification rate of more than 85% was obtained with one of the combinations of the kernel function and learning method. In addition to this, accuracy as high as 90% was achieved for Lead I dataset. A summary of the across-session results of the various tests conducted on the two in-house datasets is presented in Table 6.5 and clearly indicates the superiority of SVM as a classifier.

**TABLE 6.4**

**Across Session Classification Performance for MLP & RBF (Single Session Training)**

| Session | | Classifier | | | |
|---|---|---|---|---|---|
| | | MLP | | RBF | |
| | | TMR | FMR | TMR | FMR |
| Lead II-S1,S2 | Mean | 61.97 | 2.39 | 71.42 | 1.93 |
| | Min | 56 | 2.32 | 68 | 1.8 |
| | Max | 65 | 2.53 | 74 | 2.02 |
| | SD | 2.15 | 0.06 | 1.66 | 0.05 |
| Lead II-S1,S3 | Mean | 56.47 | 2.41 | 61.83 | 1.99 |
| | Min | 52.9 | 2.25 | 58.71 | 1.89 |
| | Max | 59.35 | 2.64 | 65.16 | 2.2 |
| | SD | 1.56 | 0.1 | 1.62 | 0.06 |
| Lead II-S1,S4 | Mean | 57.24 | 2.86 | 68.37 | 2.08 |
| | Min | 52.44 | 2.66 | 63.41 | 1.85 |
| | Max | 63.41 | 3.06 | 73.17 | 2.32 |
| | SD | 2.74 | 0.1 | 2.42 | 0.12 |
| Lead I-S1,S2 | Mean | 68.64 | 4.29 | 70 | 2.35 |
| | Min | 64.55 | 4.12 | 66.36 | 2.16 |
| | Max | 72.73 | 4.5 | 76.36 | 2.54 |
| | SD | 1.88 | 0.1 | 2.35 | 0.1 |

Besides this, another set of testing was carried out by using ANNs and SVMs trained with feature vectors extracted from first two sessions of Lead II dataset, leading to a multiple session training. This was carried out to enable the classifiers to handle variations occurring in feature vectors due to time variability in a more efficient way. Feature vectors extracted from session 3 and 4 were used for testing. All three classifiers showed considerable improvement with reference to the results obtained with classifiers trained by feature vectors derived from recordings of single session. With MLP, an average identification rate of around and above 75% was achieved, while with RBF networks, the recognition rate was nearly 80% and that too at relatively lower FMR. An overview of the test results obtained with multiple session trained MLP and RBF networks is presented in Table 6.6.

Moreover, as depicted in Table 6.7, performance showed considerable improvement for SVMs trained with training vectors from multiple sessions. For both sessions 3 and 4 of Lead II ECG dataset, mean identification accuracy of about 93% was achieved when SVMs with linear kernel and LS learning methods were used as classifiers. However, FMR in this case was slightly high and was around 10%.

**TABLE 6.5**

**Across Session Classification Performance for SVM (Single Session Training)**

| | | Classifier | | | | | |
|---|---|---|---|---|---|---|---|
| | | SVM | | | | | |
| | | Linear SMO | | Quadratic LS | | Linear LS | |
| Session | | TMR | FMR | TMR | FMR | TMR | FMR |
| Lead II-S1,S2 | Mean | 66.08 | 3.85 | 73.52 | 3.54 | 87.13 | 9.06 |
| | Min | 60 | 3.61 | 70 | 3.35 | 81.5 | 8.71 |
| | Max | 69.5 | 4.1 | 76.5 | 3.81 | 90 | 9.36 |
| | SD | 2.43 | 0.1 | 1.94 | 0.13 | 1.85 | 0.19 |
| Lead II-S1,S3 | Mean | 57.38 | 3.92 | 64.77 | 3.56 | 83.68 | 9.14 |
| | Min | 52.26 | 3.61 | 60.65 | 3.28 | 80 | 8.89 |
| | Max | 60 | 4.23 | 69.03 | 3.84 | 86.45 | 9.56 |
| | SD | 2.03 | 0.13 | 2.12 | 0.14 | 1.74 | 0.16 |
| Lead II-S1,S4 | Mean | 65.57 | 4.25 | 76.06 | 4.1 | 87.32 | 9.72 |
| | Min | 59.76 | 3.93 | 70.73 | 3.75 | 81.71 | 9.29 |
| | Max | 71.95 | 4.5 | 81.71 | 4.62 | 91.46 | 10.27 |
| | SD | 3.07 | 0.16 | 2.23 | 0.21 | 1.88 | 0.23 |
| Lead I-S1,S2 | Mean | 63.69 | 4.03 | 70.15 | 5.74 | 85.51 | 9.81 |
| | Min | 59.09 | 3.82 | 65.45 | 5.31 | 82.73 | 9.64 |
| | Max | 67.27 | 4.45 | 75.45 | 6.26 | 90 | 10.22 |
| | SD | 1.96 | 0.16 | 2.42 | 0.27 | 2.17 | 0.16 |

**TABLE 6.6**

**Across-Session Classification Performance for MLP & RBF (Multiple Session Training)**

| | | Classifier | | | |
|---|---|---|---|---|---|
| | | MLP | | RBF | |
| Session | | TMR | FMR | TMR | FMR |
| Lead II-S1S2,S3 | Mean | 76.19 | 3.98 | 78.06 | 3.01 |
| | Min | 72.26 | 3.75 | 74.19 | 2.77 |
| | Max | 80 | 4.1 | 83.23 | 3.23 |
| | SD | 1.9 | 0.09 | 2.01 | 0.1 |
| Lead II-S1S2,S4 | Mean | 74.19 | 4.48 | 77.28 | 3.6 |
| | Min | 69.51 | 4.13 | 70.73 | 3.34 |
| | Max | 80.49 | 4.88 | 81.71 | 3.85 |
| | SD | 2.58 | 0.16 | 2.63 | 0.13 |

**TABLE 6.7**

**Across Session Classification Performance for SVM (Multiple Session Training)**

| | | Classifier | | | | | |
| | | SVM | | | | | |
| | | Linear SMO | | Quadratic LS | | Linear LS | |
| Session | | TMR | FMR | TMR | FMR | TMR | FMR |
|---|---|---|---|---|---|---|---|
| Lead II-S1S2,S3 | Mean | 79.55 | 5.32 | 84.37 | 4.31 | 93.48 | 10.33 |
| | Min | 74.19 | 5.03 | 81.29 | 4.06 | 91.61 | 10.03 |
| | Max | 83.23 | 5.63 | 87.74 | 4.52 | 96.13 | 10.78 |
| | SD | 1.74 | 0.12 | 1.75 | 0.12 | 1.14 | 0.17 |
| Lead II-S1S2,S4 | Mean | 80.81 | 5.47 | 85.16 | 4.7 | 92.89 | 10.63 |
| | Min | 75.61 | 4.83 | 80.49 | 4.35 | 89.02 | 10.07 |
| | Max | 86.59 | 5.84 | 90.24 | 5.3 | 96.34 | 11.19 |
| | SD | 2.66 | 0.19 | 1.81 | 0.19 | 1.76 | 0.28 |

## 6.6   CONCLUSIONS

The utility of ECG as a biometric trait for the identification task was explored by conducting experiments on large population datasets. The results obtained clearly depict the effectiveness of AI tools over simple distance measures used in this study. In addition to this, the results also clearly demonstrate that the performance of ECG-based biometrics can further be improved by using training data from multiple sessions and classifiers like SVM.

However, relatively high FMR can be handled by utilising ECG-based biometric systems in conjunction with some other stable biometric modality. These results show that training the classifiers with feature vectors from recordings of multiple sessions aid in enhancing the performance. This study provides results for sufficiently large population size ECG datasets with a significant gap between the training and the test session recordings. Future work in this area can be carried out by testing the suggested approach and other algorithms on multiple session recordings.

## REFERENCES

1. Jain, A. K., Ross, A., and Prabhakar, S. (2004). An introduction to biometric recognition, *IEEE Transactions on Circuits and Systems for Video Technology*, Vol. 14, No. 1, pp. 4–20.
2. Chauhan, S., Arora, A.S., and Kaul, A. (2010) A survey of emerging biometric modalities. In *Proceedings of the International Conference and Exhibition on Biometrics Technology*.
3. Ogawa, M., Tamura, T., Yoda, M., and Togawa, T. (1997). Fully automated biosignal acquisition system for home health monitoring. In *Engineering in Medicine and Biology Society, 1997. Proceedings of the 19th Annual International Conference of the IEEE*, Vol. 6, pp. 2403–2405.

4. Biel, L., Pettersson, O., Philipson, L., and Wide, P. (1999). ECG analysis: A new approach in human identification in IMTC-99. *Proceedings of the 16th IEEE Instrumentation and Measurement Technology Conference*, Vol. 1, pp. 557–561.
5. Biel, L., Pettersson, O., Philipson, L., and Wide, P. (2001). ECG analysis: A new approach in human identification. *IEEE Transactions on Instrumentation Measurement*, Vol. 50, pp. 808–812.
6. Shen, T. W., Tompkins, W. J., and Hu, Y. H. (2002). 'One-lead ECG for identity verification' in EMBS/BMES '02. *Proceedings of the 2nd Joint Engineering in Medicine and Biology, 24th Annual Conference and the Annual Fall Meeting of the IEEE Biomedical Engineering Society*, Houston, TX, USA Vol. 1, pp. 62–63.
7. Shen, T.W. (2005). Biometric identity verification based on electrocardiogram (ECG), PhD Dissertation, University of Wisconsin, Madison, WI.
8. Palaniappan, R., and Krishnan, S. M. (2004). Identifying individuals using ECG beats. *Proceedings of International Conference on Signal Processing and Communications*, pp. 569–572.
9. Saechia, S., Koseeyaporn, J., and Wardkein, P. (2005). Human identification system based ECG signal. in *IEEE Region 10 TENCON* 2005, IEEE, pp. 1–4.
10. Plataniotis, K. N., Hatzinakos, D., and Lee, J. K. (2006). ECG biometric recognition without fiducial detection. in *Special Session on Research at IEEE Biometric Consortium Conference*, pp. 1–6.
11. Wang, Y., Plataniotis, K. N., and Hatzinakos, D. (2006).Integrating analytic and appearance attributes for human identification from ECG signals. in *Proceedings of Special Session on Research at Biometric Consortium Conference*, 2006 Biometrics Symposium, pp. 1–6.
12. Wang, Y., Agrafioti, F., Hatzinakos, D., and Plataniotis, K. N. (2008, January). Analysis of human electrocardiogram for biometric recognition. *EURASIP Journal on Advances in Signal Processing* Vol. 2008, 148658.
13. Yao, J., and Wan, Y. (2008). A wavelet method for biometric identification using wearable ECG sensors. *5th International Summer School and Symposium on Medical Devices and Biosensors, 2008*. ISSS-MDBS 2008, IEEE, pp. 297–300.
14. Wan, Y., and Yao, J. (2008). A neural network to identify human subjects with electrocardiogram signals. *Proceedings of the World Congress on Engineering and Computer Science, WCECS 2008*.
15. Kaveh, A., and Chung, W. (2013). Temporal and spectral features of single lead ECG for human identification. *IEEE Workshop on Biometric Measurements and Systems for Security and Medical Applications (BIOMS)*, 2013, IEEE, pp. 17–21.
16. Fang, S.-C., and Chan, H.-L. (2009). Human identification by quantifying similarity and dissimilarity in electrocardiogram phase space. *Pattern Recognition*, Vol. 42, no. 9, pp. 1824–1831.
17. Ye, C., Miguel Tavares, C., and Vijaya Kumar, BVK (2010). Investigation of human identification using two-lead electrocardiogram (ECG) signals. *4th IEEE International Conference on Biometrics: Theory Applications and Systems (BTAS)*, 2010, IEEE, pp. 1–8.
18. Li, M., and Narayanan, S. (2010). Robust ECG biometrics by fusing temporal and cepstral information. *20thInternational Conference on Pattern Recognition (ICPR)*, 2010, IEEE, pp. 1326–1329.
19. Ting, C.-M., and Salleh, S.-H. (2010). ECG based personal identification using extended Kalman filter. *10th International Conference on Information Sciences Signal Processing and their Applications (ISSPA)*, 2010, IEEE, pp. 774–777.
20. Tawfik, M. M., Selim, H., and Kamal, T. (2010). Human identification using time normalized QT signal and the QRS complex of the ECG. *7th International Symposium on Communication Systems Networks and Digital Signal Processing (CSNDSP)*, 2010, IEEE, pp. 755–759.

21. Mai, V., Khalil, I., and Meli, C. (2011). ECG biometric using multilayer perceptron and radial basis function neural networks. *2011 Annual International Conference of the IEEE Engineering in Medicine and Biology Society*, EMBC, IEEE, pp. 2745–2748.
22. Hou, L. S., Subari, K. S., and Syahril, S. (2011). QRS-complex of ECG-based biometrics in a two-level classifier. *TENCON 2011-IEEE Region 10 Conference*, IEEE, pp. 1159–1163.
23. Loong, J. L. C., Subari, K. S., Besar, R., and Abdullah, M. K. (2010). A new approach to ECG biometric systems: A comparative study between LPC and WPD systems. *World Academy of Science, Engineering and Technology*, Vol. 68, pp. 759–764.
24. Ghofrani, N., and Bostani, R. (2010). Reliable features for an ECG-based biometric system. *17th Iranian Conference of Biomedical Engineering (ICBME)*, 2010, IEEE, pp. 1–5.
25. Odinaka, I., Lai, P.-H., Kaplan, A. D., O'Sullivan, J. A., Sirevaag, E. J., Kristjansson, S. D., Sheffield, A. K., and Rohrbaugh, J. W. (2010). ECG biometrics: A robust short-time frequency analysis. *IEEE International Workshop on Information Forensics and Security (WIFS)*, 2010, IEEE, pp. 1–6.
26. Chen, C.-K., Lin, C.-L., and Chiu, Y.-M. (2011). Individual identification based on chaotic electrocardiogram signals. *6th IEEE Conference on Industrial Electronics and Applications (ICIEA)*, 2011, IEEE, pp. 1771–1776.
27. Lourenço, A., Silva, H., and Fred, A. (2011). Unveiling the biometric potential of finger-based ECG signals. *Computational Intelligence and Neuroscience*, Vol. 5, pp. 1–8.
28. Da Silva, H. P., Fred, A., Lourenco, A., and Jain, A. K. (2013).Finger ECG signal for user authentication: Usability and performance. *Sixth IEEE International Conference on Biometrics: Theory, Applications and Systems (BTAS)*, 2013, IEEE, pp. 1–8.
29. Da Silva, H. P, Lourenco, A., Fred, A., Raposo, N., and Aires-de-Sousa, M. (2013). Check your biosignals here: A new dataset for off-the-person ECG biometrics. *Computer Methods and Programs in Biomedicine*, Vol. 113, pp. 503–514.
30. Tantawi, M., Revett, K., Tolba, M. F., and Salem, A. (2012). A novel feature set for deployment in ECG based biometrics. *Seventh International Conference on Computer Engineering & Systems (ICCES)*, 2012, IEEE, pp. 186–191.
31. Tantawi, M. M., Revett, K., Salem, A.-B., and Tolba, M. F. (2013). A wavelet feature extraction method for electrocardiogram (ECG)-based biometric recognition. *Signal, Image and Video Processing*, Vol. 9, pp. 1–10.
32. Luz, E. J. d. S., Menotti, D., and Schwartz, W. R. (2014). Evaluating the use of ECG signal in low frequencies as a biometry. *Expert Systems with Applications*, Vol. 41, no. 5, pp. 2309–2315.
33. Labati, R.D., Muñoz, E., Piuri, V., Sassi, R., and Scotti, F., (2019). Deep-ECG: Convolutional neural networks for ECG biometric recognition. *Pattern Recognition Letters*, Vol. 126, pp.78–85.
34. Kim, M.G., Ko, H., and Pan, S.B., (2019). A study on user recognition using 2D ECG based on ensemble of deep convolutional neural networks. *Journal of Ambient Intelligence and Humanized Computing*, Vol. 11, no. 5, pp. 1859–1867.
35. Zhang, Q., Zhou, D., and Zeng, X., (2017). PulsePrint: Single-arm-ECG biometric human identification using deep learning. *8th IEEE Annual Ubiquitous Computing, Electronics and Mobile Communication Conference (UEMCON)* in October 2017, pp. 452–456.
36. Xu, J., Li, T., Chen, Y., and Chen, W. (2018).Personal identification by convolutional neural network with ECG signal. *IEEE International Conference on Information and Communication Technology Convergence (ICTC)* in October, 2018, pp. 559–563.
37. Odinaka, I., Lai, P.-H., Kaplan, A. D., O'Sullivan, J. A., Sirevaag, E. J., and Rohrbaugh, J. W. (2012).ECG biometric recognition: A comparative analysis. *IEEE Transactions on Information Forensics and Security*, Vol.7, no. 6, pp. 1812–1824.

38. Fratini, A., Sansone, M., Bifulco, P., and Cesarelli, M., (2015). Individual identification via electrocardiogram analysis. *Biomedical Engineering Online*, Vol. 14, no. 1, p.78.

39. Pinto, J.R., Cardoso, J.S., and Lourenço, A., (2018). Evolution, current challenges, and future possibilities in ECG biometrics.*IEEE Access*, Vol.6, pp.34746–34776.

40. Gautam, N., Kaul, A., Nath, R., Arora, A.S., and Chauhan, S., (2012). Multi-algorithmic approach for ECG based human recognition. *Journal of Applied Security Research*, Vol.7 no. 4, pp.399–408.

41. Hagan, M. T., Demuth, H. B., and Beale, M. H. (1996). Neural network design. *Cengage Learning*, 2nd edition.

42. Burges, C. J. (1998). A tutorial on support vector machines for pattern recognition. *Data Mining and Knowledge Discovery*, Vol. 2, pp. 121–167.

43. Kaul, A., Gautam, N., Jain, R., Choudhary, D., Nath, R., Arora, A.S., and Chauhan, S. (2012). NITH_MBD-multimodal biometric database. In *2nd Inter-National Conference on Biomedical Engineering and Assistive Technologies (BEATs–2012)*.

# 7 Cancelable Biometric Systems from Research to Reality

## The Road Less Travelled

*Harkeerat Kaur*
IIT Jammu

*Pritee Khanna*
PDPM IIITDM Jabalpur

## CONTENTS

7.1 Introduction ........................................................................................ 155
7.2 Cancelable Biometric Systems: Introduction and Review .......................... 157
    7.2.1 Conventional Template Transformation Techniques ...................... 158
    7.2.2 Role of Deep Learning in Biometrics and Need for Privacy ............ 160
    7.2.3 Neutral Network-Based Template Transformation Techniques ........ 162
7.3 Experimental Reporting ......................................................................... 164
7.4 Real-Life Challenges for Applications of Cancelable Biometric Systems ... 167
7.5 Conclusions and Foresights .................................................................... 169
References ...................................................................................................... 170

## 7.1 INTRODUCTION

Biometric systems have become the most convenient and useful means of providing secure access and authentication. The evolution in the artificial intelligence and Internet of Things (IoT) has inflicted the biometric authentication technology to become an integral part of common mans' life where he is able to imply controlled access over anything and everything ranging from mobile phone to bank locker all with the help of single touch. As this technology is soon gaining the status of 'implied security' with various commercial products and solutions entering the market, biometrics are soon becoming a kind of definitive identity credentials. It therefore becomes imperative to analyse the security loopholes and privacy concerns that come along with its widespread usage. Like any other system, biometric systems are also prone to attacks at various levels where an intruder is able to gain illegitimate access [1]. The implications arising from the loss of a biometric identity are more serious and far fetching as compared to the loss of a PIN/password [2]. Biometric identity once comprised cannot be used to safeguard another application. Storage of biometric information over centralised platform (server/cloud) makes them most vulnerable to hacking and other malicious activities. Covert tracing and tracking of individuals by cross-matching their biometric database is another privacy invading issue. The effect of biometric loss at some common and less secure application may affect its usability at some other security critical application.

Biometric template protection suggests use of some auxiliary/helper data to transform the reference biometric into a new format to curb unintended use of biometric templates. At the same time, these transformed templates must not compromise the ability to identify/verify individuals, maintain discriminability as well as inter-user variability, and address various attack scenarios. Among biometric template protection techniques proposed in this regard, the most popular ones being, Biometric Cryptosystems (1996–1998) [3,4], Fuzzy Commitment and Fuzzy Vaults (1998–2004) [5,6], and Cancelable Biometrics (2001) [2]. While all the other techniques successfully imparted template protection, cancelable biometrics also imparted biometric templates with *revocability*, i.e., the ability to be cancelled and revoked like passwords. Apart from template protection, the concept of cancelable biometrics provides a useful mechanism of enhancing *biometric data privacy*. In biometrics, privacy refers to an individual's personal control over the collection, use, and disclosure of recorded information about them, as well as an organisation's responsibility for data protection and safeguarding of personally identifiable information in its custody or control. By enforcing use of only pseudo-biometric identity (PI) during authentication, cancelable biometrics prevents any unintended use, cross-matching, or learning any important personal information linked with a biometric template of a user such as gender, ethnicity, race, or medical information. Moreover, it links the template generation process with a user-specific token which adds as an extra security factor and provides more user control over the collection and use of his personal information. In spite of these very inspiring features which allow one to conveniently regenerate a new biometric template and enhance privacy and security, the technology has still not come into potential usage among masses. There has been tremendous research in this regard ever since 2001 to shape the design paradigms and address template protection requirements of the cancelable biometric system, yet its public interaction is still awaited.

This work aims to provide a situation awareness and preparedness for biometrics and deep learning application which are gaining significant public outreach in almost all applications requiring authentication [7]. This extension to smart technologies and applications expects to impact numerous other applications in near future. Section 7.2 presents an overview of the concept of biometric privacy offered by the cancelable biometric system. Various schemes proposed in the cancelable biometric domain are mentioned here. Recent advances of deep neural networks (DNN) in biometrics and biometric template protection are discussed in Section 7.3 with research alignments of DNN and cancelable biometrics followed by reporting of experimental outcomes of the discussed techniques in Section 7.4. Section 7.5 systematically outlines the implementation challenges for cancelable systems that prevent its practical usage in real life. Some design issues can be addressed if these two technologies can be merged for a greater experience, as concluded in Section 7.6.

## 7.2  CANCELABLE BIOMETRIC SYSTEMS: INTRODUCTION AND REVIEW

The concept of cancelable biometric proposes that a biometric template/feature should never be used in its raw format for storing and matching purposes. Unlike other schemes which transform biometric using encryption and invertible vaults, cancelable biometric follows the one-way, non-invertible approach to map an 'original biometric' identity into a 'pseudo-biometric' identity (PI) with the help of some auxiliary data (AD). Figure 7.1 shows this distortion affect where an original template 'M' is transformed to a pseudo-biometric identity 'PI' using a transformation function which takes user-specific AD or key as its input arguments. An essential property of this transform is that it must be non-invertible and must preserve the discriminability of the original features after distortion. It implies that after distortion, the biometric features belonging to the same user must have a similar distribution and those belonging to the different users must have distinct distributions, indicating that inter-user and intra-user variations must be maintained in the transformed domain.

The basic cancelable biometric setup is shown in Figure 7.2. The transformation function is incorporated as an intermediate step in conventional biometric authentication systems, where only 'pseudo-biometric' identity is generated at enrollment or authentication, while the AD is provided to user in a tokenised manner (e.g., smart card). At enrollment, the original biometric identity 'B' of a user is transformed with the help of some secret key/AD to generate a transformed feature/PI which is stored as a reference template. At authentication, the probe biometric (B') of the same user is transformed in a similar way to generate transformed query template (PI'). Transformed reference and query templates are matched to determine access.

Cancelable biometric systems are characterised by their ability to provide four important template protection requirements specified as *discriminability, revocability, diversity,* and *non-invertibility*. These characteristic can be followed from Figure 7.1 as (a) the transformed identity PI must preserve the discriminating characteristics of original biometric template M (*discriminability*); (b) if a PI is compromised, it can be regenerated from the same template M by changing the transformation function or AD (*revocability*). Also, the same template can be mapped as different PIs for the diverse usage of biometric over different applications (*diversity*); and (c) in the case of compromise, the original template is not revealed due to non-invertible nature of transform (*non-invertibility*).

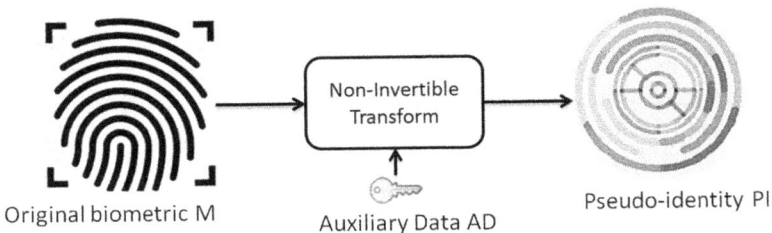

Original biometric M          Auxiliary Data AD          Pseudo-identity PI

Non-Invertible Transform

**FIGURE 7.1**   Cancelable biometric transformation process.

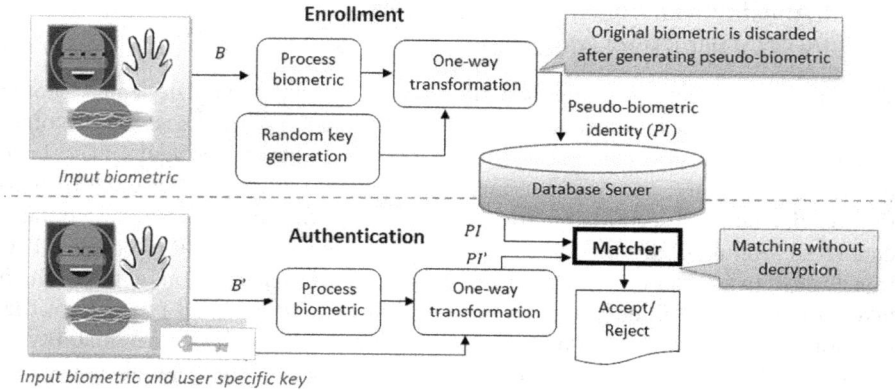

**FIGURE 7.2** Enrolment and authentication processes with cancelable biometrics.

While it is utmost important that any template protection scheme must deliver these important requirements, the challenge is to design a transformation paradigm which distorts the biometric features and at the same time not to the extent in which the discriminability is compromised. The balance between discriminability and non-invertibility is important to claim the security of the system. The next section studies the conventional template transformation schemes followed by the effect of technology shift on these transformations due the use of neural networks.

## 7.2.1 CONVENTIONAL TEMPLATE TRANSFORMATION TECHNIQUES

The template transformation paradigms are broadly classified as *biometric salting* and *non-invertible transforms*. Biometric salting techniques distort the data by mixing it with some random noises followed by some many to one mapping. AD are obtained externally and it interacted directly with the biometric to increase the entropy of the template, which makes it difficult for an adversary to make a guess. The salting operation is generally followed by some many-to-one mapping in order to impart non-invertibility. The techniques under this category can be further classified as *Random Projection, Random Convolution*, and *Random Noise*, and *Random Mapping*-based transforms. The techniques under these categories are summarised in Figure 7.3 and are discussed below.

**Random Projection** (RP)-based transformations are most widely used biometric salting techniques. RP transforms biometric data by projecting it over a random subspace defined by a user-specific key. Teoh et al. (2004) proposed the most popular biometric salting technique known as BioHashing [8]. Here, the biometric features are salted by projecting those on a random subspace defined by orthonormal random matrices. It is later quantised into binary codes via thresholding operations to achieve many-to-one mapping and non-invertibility. Although the approach is well known to preserve discriminability, it is also susceptible to inverse operations if the transformed biometric and projection matrix are leaked [9,10]. Various techniques, such as Random Multi-space Quantisation (RMQ) in BioHash [11], Multispace Random

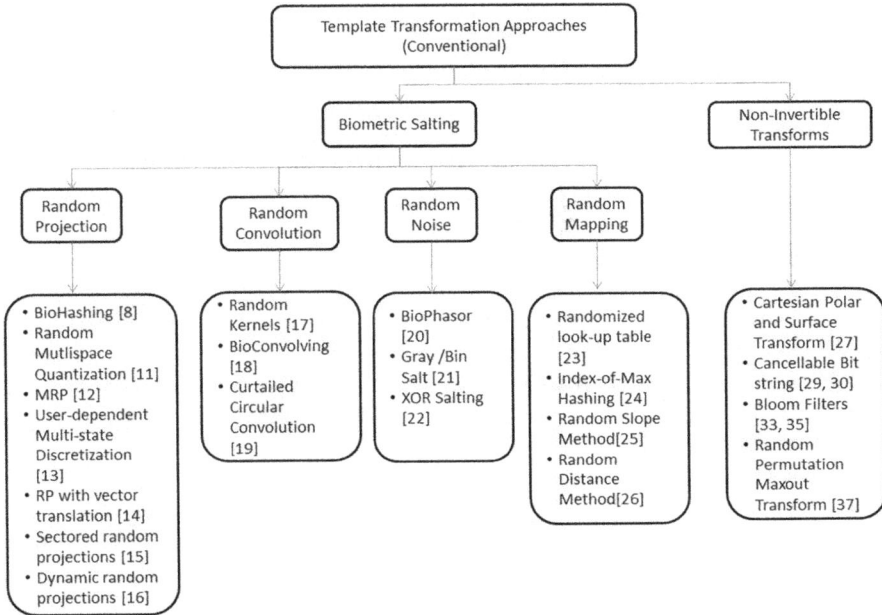

**FIGURE 7.3**   Categorywise depiction of conventional template transformation techniques.

Projections (MRP) [12], User-dependent Multi-state Discretisation (Ud-MsD) BioHash [13], RP with vector translation [14], Sectored Random Projections [15], and Dynamic Random Projections [16] are proposed to improve upon the drawbacks.

**Random Convolution**-based transformations convolve biometric signal with some random kernel to generate transformed templates. Savvides et al. (2004) transformed face images by convolving those with random kernels [17]. However, deconvolution can be attempted to recover features if random kernel is known. Maiorana et al. (2010) proposed BioConvolving, which uses random user-specific key to divide the original feature into fixed sized segments that are later convolved to generate transformed templates [18]. However, discriminability and non-invertibility properties are not justified in stolen token scenario. Wang et al. (2014) used curtailed circular convolution in which binary fingerprints features are convolved with random binary strings in circular manner to impart non-invertibility [19].

**Random Noise**-based transformations distort biometric templates by adding random noise patterns. Teoh et al. (2006) proposed BioPhasoring to generate a set of complex vectors where the original features form real part and the user-specific random vectors form imaginary part [20]. The phase/arctangent of the complex vector is used as non-invertible transformed template. Leng et al. (2011, 2013) improvised BioHashing and BioPhasoring techniques for palmprint modality. The transformation algorithm is extended to 2D for both the techniques to generate templates with reduced computational complexity and storage cost. Zuo et al. (2008) proposed GRAY salting (template-based salting) and BIN salting (code-based salting) for generating cancelable iris templates [21]. These techniques add unique random noise or synthetic textures to underlying Gabor features. Kaur and Khanna (2017) XORed original features with random patterns that is followed by median filtering to ensure non-invertibility [22].

**Random Mapping Transform** initially maps biometric features to other values in transform domain like decimal values, indices, distance, or slope. Dwivedi et al. (2016) proposed randomised look-up table mapping to generate cancelable iris templates [23]. Consistent bits are extracted from features to generate randomly mapped decimal value. But the mapping can be inverted, if look-up table and transformation parameters are known. Another scheme proposed by Jin et al. (2018) maps real-valued iris features into discrete index (max ranked) hashed codes. It is based on locality sensitive hashing (LSH) also known as 'Index-of-Max (IoM)' hashing [24]. Kaur and Khanna (2018) proposed a method that maps biometric features and some random user-specific data as points on the Cartesian space. The slopes and intercepts of the lines passing through these features and random points are calculated to generate transformed features [25]. In another work, instead of computing slopes the distances between the feature points and random points are used for the same [26].

**Non-invertible Transforms** map biometric features to a new random subspace such that the inverse mapping is not possible. Ratha et al. (2007) proposed three concrete functions that randomly map fingerprint minutiae points to a new subspace using Cartesian, polar, and surface folding transforms [27]. In spite of many-to-one mappings used by these transform, Quan et al. (2008) proved that the transforms are invertible when transformed templates and parameters are simultaneously known [28]. Similarly, Farooq et al. (2007) and Lee and Kim (2010) proposed a many-to-one mapping of minutiae features onto a predefined 3D array based on some user-specific key and reference minutia's position and orientation [29,30]. However, the mapping used here tends to compromise discriminability. Also, inverse attacks are possible if user-specific key are revealed. Recently, Alam et al. (2018) have proposed improvisations to preserve the discriminability and non-invertibility using minutiae-based bit strings methods [31]. Yang et al. (2013) extracted local structures of minutiae features using Delaunay triangulation which were subjected to non-invertible polar transformation [32]. Rathgeb et al. (2013) proposed a template protection approach which mapped input binary iris features to hashed vectors consisting of only zeros and ones using the concept of bloom-filters [33]. However, the irreversibility of bloom-filters was identified shortly by Herman et al. (2004), and it was also observed that the technique is vulnerable to cross-matching attacks [34]. Barrero et al. (2016) discussed an improvement which was built upon the original concept of bloom-filter-based template protection followed by an additional feature rearrangement technique to provide unlinkability and irreversibility [35]. Wang et al. (2017) used partial discrete Fourier transform to get good performance as the local structures of minutiae points preserve discriminability after non-invertible distortions [36]. Teoh and Wang (2018) proposed random permutation maxout transform, which maps a real-valued face feature vector into a discrete index code used as transformed template [37].

## 7.2.2 Role of Deep Learning in Biometrics and Need for Privacy

Deep learning techniques have almost replaced the conventional feature extraction techniques to become the future of the state-of-the art techniques where the upcoming systems will be based on DNN models only. Ongoing works in DNN are contributing significantly towards improving the performance and exhibiting superior recognition accuracy of biometric systems. Several convolutional neural networks (CNN)-based models have been developed for face, speech, finger-vein, and other characteristics which allow feature extraction to be completely automated. DNN have also been in trend for multi-instance and multi-biometric fusion [38]. However, developing the training model for CNN requires learning from plenty of labeled datasets and millions of network train parameters. Li et al. (2018) proposed a technique for robust face recognition which combines normalised CNN with probabilistic max-pooling so that the feature information can be preserved while maintaining feature invariance, thus eliminating conventional principal component analysis (PCA) and linear discriminant analysis (LDA) techniques [7].

Designing the layers and activation functions for deep learning models is important for achieving good accuracy. DNN require significant amount of training, which imparts them with ability to learn complex relations and subtle modalities like palm-vein and finger-vein. Apart from authentication, DNN and biometrics are coming together for variety of purposes. There are advances in cognitive system design with biometric technology decision-making on the underlying DNN technology. Not only this, the face features and posture have also been useful towards emotion and anxiety detection, mood, and satisfaction. Some biometrics like heart rate can also tell about medical fitness of the person. Postures biometrics can be used by physiotherapists to understand if a person is in pain. We even have social biometrics nowadays which may identify a person's age and gender on the basis of digital activities. While the complete automation offered by DNN models gives greater convenience, it abstracts the details of inner workings. The ignorance about the details of the inner systems may put the system to a great risk. In case of multi-biometric instance, it may even become difficult to predict which component of the traits will be used for recognition purposes. While the DNN-based components are increasingly becoming significant part of our laptops, smartphones, smart TVs, banking systems, it introduced a new challenge to the development of suitable biometric template protection techniques to preserve the privacy of our biometric data [38,39].

It is important to understand privacy enhancement and template protection in the context of DNN-based authentication architecture which is establishing itself as an important component of next-generation devices. The evolution of DNN has made it possible to identify and verify a user by using the traits like electro-cardiogram (ECG) and electroencephalogram (EEG) signals. Significant research studies has been reported in the last few years exploring various DNN models like Boltzmann machines, Deep Belief Networks (DBNs), Stacked Auto Encoders, CNN, Generative Adversarial Network (GAN), Recurrent Models, etc. for various biometric traits like face, palmprint, finger-vein, gait, voice, and key stroke, etc. [40]. Amongst these, CNN has been one of the most successful models for face and speaker verification. However, apart from learning features only for face recognition, the models can also be designed for extracting some more personal information about the person, such as age, gender, and ethnicity. Similar things are predictable with voice samples for speaker verification. Much of the recorded ECG and EEG signals can reveal information about medical health of a person. With DNN proliferating into IoT devices, there is imposed risk of these applications invading users' privacy [41]. As discussed above, amongst the various template protection schemes, cancelable biometrics are the most promising and privacy enhancing solutions by mapping biometric template into non-invertible domain. Thereafter, entire storing and matching are performed in the transformed domain. A number of approaches have been proposed in the recent years (2017–2019) which combines DNN techniques with cancelable transforms to meet the real-world requirement of high performance and privacy preservation.

### 7.2.3 NEUTRAL NETWORK-BASED TEMPLATE TRANSFORMATION TECHNIQUES

Similar to the conventional techniques discussed above, category wise depiction of neural network-based template transformation techniques is given in Figure 7.4 and discussed below.

**Random Projection-Based DNN Techniques**: Several recent approaches proposed in the literature used cancelable transformations like RP to first map the image into a random subspace and then learning features on transformed domain using CNN networks or vice versa. Liu et al. (2018) proposed a method named as FVR-DLRP for secure authentication of finger-vein templates using deep learning and RP methods [42]. They first extracted finger-vein features and projected those on orthonormal random matrices for template transformation and dimensionality reduction. The transformed finger-vein features were then trained over DBN consisting of three layers of the restricted Boltzmann machine for authentication purposes. The DBN is a multilayer network structure, which can learn the complex mapping relationship between input and output.

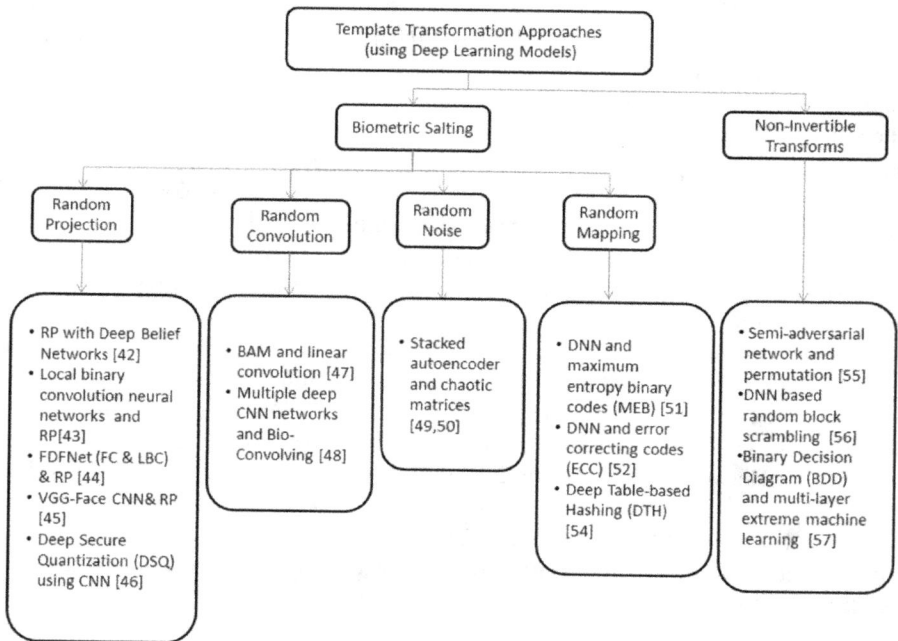

**FIGURE 7.4** Categorywise depiction of neural network-based template transformation techniques.

Singh et al. (2018) developed a cancelable knuckleprint recognition system. The authors used local binary convolution (LBC) neural networks for feature extraction which reduces large number of learnable parameters in comparison to the standard convolution layer. The LBP-CNN features are transformed using BioHashing to impart cancelability to the deep knuckle features [43]. Another work proposes cancelable Finger Dorsal Feature Extraction Net (FDFNet) for extracting discriminative features from major and minor finger knuckle biometric templates paired up with BioHashing for cancelability [44]. The underlying network consists of five layers of which first three are fully connected layers using bubble ordinal pattern filters, difference filters, and last two are LBC layer comprises of a set of fixed sparse predefined and non-trainable binary convolutional filters.

Jindal et al. (2019) proposed CNN-based face recognition with privacy enhancement using RP techniques. They used a pre-trained VGG-Face CNN (trained over 2.6 M images for 2.6 K people) for extraction of face features. The extracted features are subjected to RP to provide revocability, diversity, and eliminate redundancy [45]. The project feature is again subjected to a set of fully connected CNN layers in order to learn a robust mapping and minimise the intra-user variations.

It has been observed from the above techniques that integration of RP improved the performance of the above techniques. Apart from enhancing matching performance, recent works also explored the application of DNN to improve the security of RP-based techniques. Chen et al. (2019) developed a technique called Deep Secure Quantisation (DSQ) to protect random project-based hashing against similarity-based attacks (SA) [46].

**Random Convolution-Based DNN Techniques**: Tarek et al. (2016) generated cancelable iris features using bi-directional associative memory (BAM) neural network and linear convolution to prevent correlation attacks [47]. They utilised BAN to bind original iris templates to random bit strings, which are further subject to convolution-based cancelable transform. The scheme also enhances the security of transformation key and biometric template by hiding then in an encoded form using BAM. The process consist of two stages where in the first stage user-specific transformation key parameters and its association with biometric traits are memorised using BAM models' weight. In the second stage, they both are binded using convolution and subjected to binarisation using thresholding operation to impart non-invertibility. They also present an analysis against correlation attack for the proposed scheme.

Abdellatef et al. (2019) proposed an instance for securing face templates where they extracted features from different regions of the face separately using multiple deep CNN networks, which are later fused using a fusion network. After fusion the final descriptors are subjected to BioConvolving-based cancelable transform [48].

**Random Noise-Based DNN Techniques**: No specific works have been introduced until now which can be added to this category. However there exists some suggestive work that can be used on the same line to provide cancelability to biometric templates. Fei et al. (2017) suggested stacked autoencoders for generating chaotic matrices similar to logistic maps [49]. The chaotic behaviour of logistic map can be exerted useful means of generating random matrices and transformation schemes for revocable templates [50].

**Random Mapping-Based Techniques**: Pandey et al. (2016) used deep CNNs to learn mappings which transform face image templates to maximum entropy binary (MEB) codes [51]. These codes are then hashed using any hash function that follows the random oracle model (like SHA-512) to generate protected face templates. Apart from improving matching performance, their approach offers enhanced privacy and cancelability to the templates. Talreja et al. (2017) proposed a multi-biometric system for face and iris modalities using DNN and error correcting codes [52]. They used dedicated CNNs to first map templates of both the modalities into a common feature space by extracting domain-specific features which are then fused with the help of a fully connected or bilinear joint representation layer. It is followed by a feature selection process to reduce template dimension. Finally, a cancelable binary vector is generated that is within a certain distance from a codeword of an error-correcting code. Nasir et al. (2018) proposed some recent image protection techniques where pre-trained CNN is used to extract features which are then transformed to compact binary codes using a deep autoencoder [53]. Jang and Cho (2019) proposed a cancelable authentication system using CNN-based face image retrieval system [54]. The authors develop a novel Deep Table-based Hashing (DTH) framework which encodes CNN-based features into a binary code by utilising the index of the hashing table. Their distortion process included noise embedding and intra-normalisation to fulfil the essential requirement of non-invertibility.

**Non-invertible and Other Transforms**: Apart from conventional template transformation schemes techniques, various techniques have been developed using DNN in its own way for imparting biometric privacy and cancelability. Vahid et al. (2018) proposed privacy enhancing transforms based on convolutional autoencoders which perturbs an input face image such that the transformed image can be successfully used for face recognition but not for gender classification but can be sued by matcher for classification purposes [55]. Shen et al. (2018) proposed a deep CNN-based random block scrambling method to impart privacy to face templates [56]. The face images and key parts are subject to Arnold random scrambling which is fed into CNN models for training and verification. Yang et al. (2019) developed template protection algorithm for deep learning-based finger-vein biometric system using the binary decision diagram (BDD) [57]. The approach first transforms the templates using BDD in a non-invertible manner. Later it is fed into multi-layer extreme machine learning model for further processing.

## 7.3  EXPERIMENTAL REPORTING

Important experimental observations made for the various template transformation approaches under conventional and deep learning-based categories are summarised in Tables 7.1 and 7.2. The databases and modalities under evaluation are also mentioned. For many approaches multiple modalities and databases are used for experimentation. Also various parameter selections are defined. Due to brevity of space, results are reported for the best parameters defined in the manuscript and for selected modalities. In the case of DNN-based transformation scheme, comparison of original templates are performed using many DNN architectures, and performance is reported in the transformed domain.

## TABLE 7.1
## Experimental Observations for Conventional Template Transformation Approaches

| Category | Technique | Modality | Database | Performance in Transformed Domain | Base Line Performance |
|---|---|---|---|---|---|
| Random Projection | BioHashing [8] | Fingerprint | FVC2002 DB1 | 0% EER | - |
| | | Face | ORL | 0% EER | - |
| | | Palmprint | Private | 0% EER | EER |
| | | Knuckleprint | PolyU FKP | 25.9% EER | 30.5% EER |
| | Multispace Random Quantisation [11] | Face | FERET | 7.09% EER | 4.52% EER |
| | Multispace Random Projection [12] | Face | ORL | 25.77% EER | 25.11% EER |
| | User-dependent Multi-state Discretisation [13] | Fingerprint | FVC2002 DB1 | 3.42% EER | 14.84% EER |
| | RP with vector translation [14] | Face | FERET+AR+ Aging+PIE | 18.68% EER | 17.54% EER |
| Random Convolution | Random Kernel [17] | Face | CMU-PIE | 100% RI | 100% RI |
| | BioConvolving [18] | Signature | MYCT | 7.95% EER | 6.33% EER |
| | Curtailed circular convolution [19] | Fingerprint | FVC2002 DB1, DB2, DB3 | 2%, 2.3%, 6.12% EER respectively | - |
| Random Noise | BioPhasor [20] | Palmprint | PolyU | 0.13% EER | - |
| | Gray Salting [21] | Iris | MMU 1 | 95.6% GAR | 98% GAR |
| | XOR-based salting [22] | Palmprint | CASIA | 0.55% EER | 0.50% EER |
| Random Mapping | Table indices [23] | Iris | CASIA-V1 | 0.37% EER | 0.28% EER |
| | Index-of-Max [24] | Iris | CASIA-v3 | 0.54% EER | 0.38% EER |
| | Random Slope-V1 [25] | Palmprint | PolyU | 0.48% EER | 0.42% EER |
| | Random Distance [26] | Palmprint | CASIA | 0.53% EER | 0.50% |

*(Continued)*

**TABLE 7.1 (*Continued*)**
**Experimental Observations for Conventional Template Transformation Approaches**

| Category | Technique | Modality | Database | Performance in Transformed Domain | Base Line Performance |
|---|---|---|---|---|---|
| Non-invertible | Polar transforms [32] | Fingerprint | FVC2002 DB1,DB2 | 5.93%, 4.0% EER | 5.41%, 2.82% EER |
| | Bloom filters [33,35] | Iris | CASIA-v3 | 1.49% EER | - |
| | Random Permutation Maxout Transform [37] | Face | AR, FERET | 6.95%, 3.65% EER | 8.53%, 4.52% EER |

**TABLE 7.2**
**Experimental Observations for Neural Networks Based Template Transformation Approaches**

| Category | Technique | Modality | Database | Performance in Transformed Domain | Base Line Performance |
|---|---|---|---|---|---|
| Random Projection | Deep Belief Network [42] | Finger-vein | FV_NET64 | 91.2% GAR | 91.8% GAR |
| | Local binary convolution neural networks [43] | Finger Knuckle | PolyU FKP | 0.125% FAR | - |
| | FDFNet (FC & LBC) & RP [44] | Finger Knuckle | PolyU FKI | 0.002% EER (minor knuckle) | - |
| | VGG-Face CNN& RP [45] | Face | CMU-PIE | 99.95% GAR | - |
| | Deep Secure Quantisation (DSQ) [46] | IRIS | CASIA-v4 | EER ≤ 1% | - |
| Random Convolution | BAM and linear convolution [47] | IRIS | CASIA-IrisV3-Interval | 3.56% EER | 1.78% EER |
| | Multiple deep CNN networks and Bio-Convolving [48] | Face | FERET, LFW, PaSc | 97.14%, 98.93%, 97.38% accuracy | - |
| Random Noise | Stacked autoencoder and chaotic matrices [49,50] | General | - | - | - |

(*Continued*)

**TABLE 7.2** (*Continued*)
**Experimental Observations for Neural Networks Based Template Transformation Approaches**

| Category | Technique | Modality | Database | Performance in Transformed Domain | Base Line Performance |
|---|---|---|---|---|---|
| Random Mapping | DNN and maximum entropy binary codes (MEB) [51] | Face | PIE, YALE, Multi-PIE | 1.14%, 0.71%, 0.90% EER, | - |
| | DNN and error correcting codes (ECC) [52] | Face & Iris | Casia-Webface & CASIA-Iris-Thousand1 | 99.99% GAR | - |
| | Deep Table-based Hashing (DTH) [54] | Face | YouTube Faces +Face Scrub | 0.0048% | - |
| Non-Invertible and other Transforms | SAN and permutation [55] | Face | Celeb A, MCT, LFW | 39.3%, 39.2%, 72.5%, EER | 19.7%, 8.0%, 33.4%, 16.9% EER |
| | DNN-based random block scrambling [56] | Face | - | 96.72% RI | |
| | Binary Decision Diagram (BDD) and multi-layer extreme machine learning [57] | Finger-vein | SDUMLA, MMCBNU_6000, UTFVP | 93.09%, 98.70%, 98.61% CIR | |

## 7.4 REAL-LIFE CHALLENGES FOR APPLICATIONS OF CANCELABLE BIOMETRIC SYSTEMS

In spite of sufficient proof-of-concept, the present situation lags a necessary proof-of-work concept. The above section provides an exhaustive set of template transformation approaches and their results on matching performance. The contribution of DNN is towards improved matching performance. However, only evaluation on matching performance is not sufficient for designing a system that meets the practical implementation scenario. In order to meet the gap between research and reality certain implementation challenges must be addressed. This section highlights the drawback of implementing cancelable system which limits its important applications:

  a. **Multiple Identity Registrations Scenario:** The cancelable system enrols a user '$i$' only on the basis of its transformed $PI_i$ generated using a user-specific key $K_i$. There may be cases when a same person may enrol again with different key $K_j$. In that case the system outs a new pseudo-identity $PI_j$. It becomes imperative to identify the person re-enrolling with the same biometric but different key. This is challenging as the system enrols diverse templates of the same person as two entirely different pseudo-identities. An attacker may fool the system this way to mask him and have multiple access accounts or even obtain multiple keys. A possible solution to this may be multi-level non-invertible transforms, where all users are transformed using the same key at the first level and then transformed using different user-specific keys at the second level. While it becomes difficult to differentiate between templates at the first level conventionally, there is tremendous scope for neural networks for learning to improve this learning process.

b. **Generating Random Keys and Their Mixing:** Another important aspect is related with generating random keys (AD) and fixing the range of values to which they belong. For most salting techniques, determining the range and distribution of random keys is important. If the random keys range is pre-dominating, it will have more effect while generating the transformed template. The range and distribution of the values generated for random key must be in accordance with the extracted feature vector. The specified range for a proposed work must not be used directly as it entirely depends on the type of feature extraction technique and the biometric trait to which it belongs. Some recent works also suggest the use of one biometric to generate random inputs to be used for feature distortion, like brain signals, or voice [58]. Feng et al. (2018) have generated techniques for generating and revoking brain passwords for head gear devices [59]. Again DNN forms the backbone of defining such extraction and their mixing here.

c. **Insider Attack Scenario:** The cancelable systems enhance user privacy by storing only transformed identities that are revocable and do not reveal any information about the original template due to non-invertible nature of the transform. However, the system still remains susceptible to insider attacks, where a malicious insider may uplift the transformed template from database to intercept the system. This can be prevented by using secret sharing techniques like [60,61]. The transformed pseudo-identity may be divided into two or more shares, which are distributed over multiple database servers and user-token. In that case, a malicious insider will only have a share which does not reveal the actual referenced transformed pseudo-identity until the remaining shares are available.

d. **Designing the Storage over User-Specific Token:** As a deviation to the name, the user-specific token must not store the entire key, but may be a hash or index to it. If the entire key a stored on the user-token it becomes an easy target for the attackers to read the information. Also a multi-secret sharing scheme like [61,62] may be useful here which shall input both transformed template and key to output distributed shares.

e. **Re-enrollment Scenario in Case of Token Compromise:** The cancelable systems allow user to re-enroll if the template is compromised. However, if the token is compromised then generation of new key is easy. User may re-enroll to cast the effect of the changed key. This is an important design issue and need clever tricks that allow only issuance of new token without inputting the biometric again. Some works like [62] address this issue by again applying secret sharing and designing a separate enrolment, authentication, and revocable modules. Still it remains an open issue and design challenge to be addressed.

f. **Entire Database Compromise Scenario:** If the entire database is compromised, then it will require re-enrolment of all users, which is the main limiting factor to practical implementation of the system. Again solutions which transcends over multi-level transformation or cascading non-invertible and invertible mappings in presence of user-specific key might be useful; but multi-level transformation may have reduction in performance. Again a secret sharing-based approach might be helpful in this case. The entire database is transcended into three or more shares stored over distributed database servers and user token. If one of the database server is compromised it shall not reveal any original information and can be traced back to generate new shares without user re-enrolment [62]. This designing also needs more improvisations in future.

g. **Exhaustive Evaluation for Performance:** Most of the performance evaluations are either performed in the worst- and best-case scenarios. In the worst-case scenario, the discriminability of the transformation is analysed by assigning the same transformation key to each user. The then generated transformed templates are matched against each other to measure the performance in the transformed domain. It is expected to be comparable to the original, yet a little degradation is observed. In the best-case scenario, the templates are transformed by assigning different user-specific key to each user, which significantly increases the inter-user variations to give almost 0% *false accept rate*. The practical implementations need more detailed analyses on matching scores to set the system threshold for a match or non-match. The actual testing to qualify is the combination of both worst and best case testing. The combination can be defined as, initially generate a set of reference transformed templates by assigning each user a different user-specific keys say, $K_1...K_n$ (best case). Then generate probe templates by first transforming all users $u_1,...u_n$ using key $K_1$, to match against reference template of user $u_1$. This outputs a set of genuine and impostor matching scores for all users against user $u_1$. By repeating this process for all users, one may be able to map the overall scores obtained for genuine and impostor analysis. Segregation can be easily defined over the overall set of genuine and impostor mappings to set the system threshold.

## 7.5   CONCLUSIONS AND FORESIGHTS

The current scenario with respect to prevalence of DNN in biometric authentication technology is discussed in this chapter. The importance and need of biometric security and privacy is highlighted. It is suggested that the upcoming frameworks must include means to address privacy by design and not as an aftermath. The models must be backed with privacy enhancing concepts to look for public acceptance in order to protect their privacy rights. The cancelable biometric-based template protection scheme appears to be a useful means to be incorporated in the authentication framework. It provides privacy enhancing abilities of revocability, diversity, and non-invertibiliy; which only operates over PIs and not the original ones. Until now only few concrete works have been proposed as the alignment of these two technologies.

Some important research studies based on CNN, autoencoders, and adversarial learning are provided here for combination of these technologies in future. The following advantages of integrating cancelable-based template protection solution to DNN networks are highlighted:

1. **Privacy Enhancing Solutions**: Learning on the transformed space, i.e., 'pseudo-biometric identity' instead of 'original biometric identity' prevents unwanted learning, for example, age and gender in the case of face and voice biometrics.
2. **Increase in Performance**: The use of DNN may increase the matching performance of the cancelable systems for the existing transformation approaches, thereby reducing the performance gap between the original and transformed domain.
3. **Same Biometric Regeneration for Various Applications**: It provides viable solution for remote authentication and multi-server access, where the same biometric sample can be safely transformed into diverse transformed templates for usage of different applications or multi-server applications, such that these applications are unable to share the data amongst themselves for tracking, linking, cross-matching, and other personal data mining attacks.
4. **Increase in User-Acceptance**: Improved user control over the use and disclosure of biometric information increases public acceptance and removes the fear of covert surveillance, thereby enhancing their right to privacy.

Also, the major design problems while implementing a practical cancelable system are presented here. This research leads can be followed in order to have a highly robust and privacy preserving framework for the next-generation biometric authentication framework.

## REFERENCES

1. Roberts, Chris. "Biometric attack vectors and defences." *Computers & Security* 26, no. 1 (2007): 14–25.
2. Ratha, Nalini K., Jonathan H. Connell, and Ruud M. Bolle. "Enhancing security and privacy in biometrics-based authentication systems." *IBM Systems Journal* 40, no. 3 (2001): 614–634.
3. Soutar, Colin, Danny Roberge, Alex Stoianov, Rene Gilroy, and B.V.K. Vijaya Kumar. "Biometric encryption." In *ICSA guide to Cryptography*, vol. 22. McGraw-Hill, 1999.
4. Soutar, Colin, Danny Roberge, Alex Stoianov, Rene Gilroy, and Bhagavatula Vijaya Kumar. "Biometric encryption using image processing." In *Optical Security and Counterfeit Deterrence Techniques II*, vol. 3314, pp. 178–188. US: International Society for Optics and Photonics, 1998.
5. Juels, Ari, and Martin Wattenberg. "A fuzzy commitment scheme." In *Proceedings of the 6th ACM Conference on Computer and Communications Security*, pp. 28–36, 1999.
6. Uludag, Umut, and Anil K. Jain. "Fuzzy fingerprint vault." In *Proceedings of the Workshop: Biometrics: Challenges Arising from Theory to Practice*, pp. 13–16, 2004.
7. Li, Jing, Tao Qiu, Chang Wen, Kai Xie, and Fang-Qing Wen. "Robust face recognition using the deep C2D-CNN model based on decision-level fusion." *Sensors* 18, no. 7 (2018): 2080.

8. Jin, Andrew Teoh Beng, David Ngo Chek Ling, and Alwyn Goh. "Biohashing: Two factor authentication featuring fingerprint data and tokenised random number." *Pattern Recognition* 37, no. 11 (2004): 2245–2255.

9. Cheung, King Hong, Adams Wai-Kin Kong, Jane You, and David Zhang. "An analysis on invertibility of cancelable biometrics based on biohashing." *CISST* 2005: 40–45, 2005.

10. Lacharme, Patrick, Estelle Cherrier, and Christophe Rosenberger. "Preimage attack on biohashing." In *International Conference on Security and Cryptography (SECRYPT)*, pp. 1–8. IEEE, 2013.

11. Teoh, Andrew BJ, Alwyn Goh, and David C.L. Ngo. "Random multispace quantization as an analytic mechanism for biohashing of biometric and random identity inputs." *IEEE Transactions on Pattern Analysis and Machine Intelligence* 28, no. 12 (2006): 1892–1901.

12. Teoh, Andrew Beng Jin, and Chong Tze Yuang. "Cancelable biometrics realization with multispace random projections." *IEEE Transactions on Systems, Man, and Cybernetics, Part B (Cybernetics)* 37, no. 5 (2007): 1096–1106.

13. Teoh, Andrew Beng Jin, Wai Kuan Yip, and Kar-Ann Toh. "Cancellable biometrics and user-dependent multi-state discretization in BioHash." *Pattern Analysis and Applications* 13, no. 3 (2010): 301–307.

14. Wang, Yongjin, and Konstantinos N. Plataniotis. "An analysis of random projection for changeable and privacy-preserving biometric verification." *IEEE Transactions on Systems, Man, and Cybernetics, Part B (Cybernetics)* 40, no. 5 (2010): 1280–1293.

15. Pillai, Jaishanker K., Vishal M. Patel, Rama Chellappa, and Nalini K. Ratha. "Sectored random projections for cancelable iris biometrics." In *2010 IEEE International Conference on Acoustics, Speech and Signal Processing*, pp. 1838–1841. IEEE, 2010.

16. Yang, Bian, Daniel Hartung, Koen Simoens, and Christoph Busch. "Dynamic random projection for biometric template protection." In *2010 Fourth IEEE International Conference on Biometrics: Theory, Applications and Systems (BTAS)*, pp. 1–7. IEEE, 2010.

17. Savvides, Marios, B.V.K. Vijaya Kumar, and Pradeep K. Khosla. "Cancelable biometric filters for face recognition." In *Proceedings of the 17th International Conference on Pattern Recognition, 2004. ICPR 2004.*, vol. 3, pp. 922–925. IEEE, 2004.

18. Maiorana, Emanuele, Patrizio Campisi, and Alessandro Neri. "Bioconvolving: Cancelable templates for a multi-biometrics signature recognition system." In *2011 IEEE International Systems Conference*, pp. 495–500. IEEE, 2011.

19. Wang, Song, and Jiankun Hu. "Design of alignment-free cancelable fingerprint templates via curtailed circular convolution." *Pattern Recognition* 47, no. 3 (2014): 1321–1329.

20. Teoh, Andrew B.J., and David C.L. Ngo. "Biophasor: Token supplemented cancellable biometrics." In *2006 9th International Conference on Control, Automation, Robotics and Vision*, pp. 1–5. IEEE, 2006.

21. Zuo, Jinyu, Nalini K. Ratha, and Jonathan H. Connell. "Cancelable iris biometric." In *2008 19th International Conference on Pattern Recognition*, pp. 1–4. IEEE, 2008.

22. Kaur, Harkeerat, and Pritee Khanna. "Cancelable features using log-Gabor filters for biometric authentication." *Multimedia Tools and Applications* 76, no. 4 (2017): 4673–4694.

23. Dwivedi, Rudresh, Somnath Dey, Ramveer Singh, and Aditya Prasad. "A privacy-preserving cancelable iris template generation scheme using decimal encoding and look-up table mapping." *Computers & Security* 65 (2017): 373–386.

24. Lai, Yen-Lung, Zhe Jin, Andrew Beng Jin Teoh, Bok-Min Goi, Wun-She Yap, Tong-Yuen Chai, and Christian Rathgeb. "Cancellable iris template generation based on Indexing-First-One hashing." *Pattern Recognition* 64(2017): 105–117.

25. Kaur, Harkeerat, and Pritee Khanna. "Random slope method for generation of cancelable biometric features." *Pattern Recognition Letters* 126(2019): 31–40.

26. Kaur, Harkeerat, and Pritee Khanna. "Random distance method for generating unimodal and multimodal cancelable biometric features." *IEEE Transactions on Information Forensics and Security* 14, no. 3 (2018): 709–719.

27. Ratha, Nalini, Jonathan Connell, Ruud M. Bolle, and Sharat Chikkerur. "Cancelable biometrics: A case study in fingerprints." In *18th International Conference on Pattern Recognition (ICPR'06)*, vol. 4, pp. 370–373. IEEE, 2006.

28. Quan, Feng, Su Fei, Cai Anni, and Zhao Feifei. "Cracking cancelable fingerprint template of Ratha." In 2008 *International Symposium on Computer Science and Computational Technology*, vol. 2, pp. 572–575. IEEE, 2008.

29. Farooq, Faisal, Ruud M. Bolle, Tsai-Yang Jea, and Nalini Ratha. "Anonymous and revocable fingerprint recognition." In 2007 *IEEE Conference on Computer Vision and Pattern Recognition*, pp. 1–7. IEEE, 2007.

30. Lee, Chulhan, and Jaihie Kim. "Cancelable fingerprint templates using minutiae-based bit-strings." *Journal of Network and Computer Applications* 33, no. 3 (2010): 236–246.

31. Alam, Badiul, Zhe Jin, Wun-She Yap, and Bok-Min Goi. "An alignment-free cancelable fingerprint template for bio-cryptosystems." *Journal of Network and Computer Applications* 115 (2018): 20–32.

32. Yang, Wencheng, Jiankun Hu, Song Wang, and Jucheng Yang. "Cancelable fingerprint templates with delaunay triangle-based local structures." In *Cyberspace Safety and Security*, pp. 81–91, Springer, Cham, 2013.

33. Rathgeb, Christian, Frank Breitinger, and Christoph Busch. "Alignment-free cancelable iris biometric templates based on adaptive bloom filters." In 2013 *International Conference on Biometrics (ICB)*, pp. 1–8. IEEE, 2013.

34. Hermans, Jens, Bart Mennink, and Roel Peeters. "When a bloom filter is a doom filter: Security assessment of a novel iris biometric template protection system." In 2014 *International Conference of the Biometrics Special Interest Group (BIOSIG)*, pp. 1–6. IEEE, 2014.

35. Gomez-Barrero, Marta, Christian Rathgeb, Javier Galbally, Christoph Busch, and Julian Fierrez. "Unlinkable and irreversible biometric template protection based on bloom filters." *Information Sciences* 370 (2016): 18–32.

36. Wang, Song, Wencheng Yang, and Jiankun Hu. "Design of alignment-free cancelable fingerprint templates with zoned minutia pairs." *Pattern Recognition* 66 (2017): 295–301.

37. Teoh, Andrew Beng Jin, Sejung Cho, and Jihyeon Kim. "Random permutation Maxout transform for cancellable facial template protection." *Multimedia Tools and Applications* 77, no. 21 (2018): 27733–27759.

38. Alpar, Orcan. "Intelligent biometric pattern password authentication systems for touch-screens." *Expert Systems with Applications* 42, no. 17–18 (2015): 6286–6294.

39. Brostoff, George. "How AI and biometrics are driving next-generation authentication." *Biometric Technology Today* 2019, no. 6 (2019): 7–9.

40. Sundararajan, Kalaivani, and Damon L. Woodard. "Deep learning for biometrics: A survey." *ACM Computing Surveys (CSUR)* 51, no. 3 (2018): 1–34.

41. Gautam, Kalpna, Vikram Puri, Jolanda G. Tromp, Nhu Gia Nguyen, and Chung Van Le. "Internet of Things (IoT) and deep neural network-based intelligent and conceptual model for smart city." In *Frontiers in Intelligent Computing: Theory and Applications*, pp. 287–300. Springer, Singapore, 2020.

42. Liu, Yi, Jie Ling, Zhusong Liu, Jian Shen, and Chongzhi Gao. "Finger vein secure biometric template generation based on deep learning." *Soft Computing* 22, no. 7 (2018): 2257–2265.

43. Singh, Avantika, Shreya Hasmukh Patel, and Aditya Nigam. "Cancelable knuckle template generation based on LBP-CNN." In *Proceedings of the European Conference on Computer Vision (ECCV)*, 2018.

44. Singh, Avantika, Ashish Arora, Shreya Hasmukh Patel, Gaurav Jaswal, and Aditya Nigam. "FDFNet: A secure cancelable deep finger dorsal template generation network secured via. bio-hashing." In *2019 IEEE 5th International Conference on Identity, Security, and Behavior Analysis (ISBA)*, pp. 1–9. IEEE, 2019.

45. Jindal, Arun Kumar, Srinivasa Rao Chalamala, and Santosh Kumar Jami. "Securing face templates using deep convolutional neural network and random projection." In *2019 IEEE International Conference on Consumer Electronics (ICCE)*, pp. 1–6. IEEE, 2019.

46. Chen, Yanzhi, Yan Wo, Renjie Xie, Chudan Wu, and Guoqiang Han. "Deep Secure Quantization: On secure biometric hashing against similarity-based attacks." *Signal Processing* 154 (2019): 314–323.

47. Tarek, Mayada, Osama Ouda, and Taher Hamza. "Robust cancellable biometrics scheme based on neural networks." *IET Biometrics* 5, no. 3 (2016): 220–228.

48. Abdellatef, Essam, Nabil A. Ismail, Salah Eldin S.E. Abd Elrahman, Khalid N. Ismail, Mohamed Rihan, and Fathi E. Abd El-Samie. "Cancelable multi-biometric recognition system based on deep learning." *The Visual Computer* (2019): 1–13.

49. Hu, Fei, Jingyuan Wang, Xiaofei Xu, Changjiu Pu, and Tao Peng. "Batch image encryption using generated deep features based on stacked autoencoder network." *Mathematical Problems in Engineering* 2017 (2017): 1–12.

50. Soliman, Randa F., Noha Ramadan, Mohamed Amin, Hossam Eldin H. Ahmed, Said El-Khamy, and Fathi E. Abd El-Samie. "Efficient cancelable Iris recognition scheme based on modified logistic map." *Proceedings of the National Academy of Sciences, India Section A: Physical Sciences*, pp. 1–7, 2018.

51. Pandey, Rohit Kumar, Yingbo Zhou, Bhargava Urala Kota, and Venu Govindaraju. "Deep secure encoding for face template protection." In *2016 IEEE Conference on Computer Vision and Pattern Recognition Workshops (CVPRW)*, pp. 77–83. IEEE, 2016.

52. Talreja, Veeru, Matthew C. Valenti, and Nasser M. Nasrabadi. "Multibiometric secure system based on deep learning." In 2017 *IEEE Global conference on signal and information processing (globalSIP)*, pp. 298–302. IEEE, 2017.

53. Rahim, Nasir, Jamil Ahmad, Khan Muhammad, Arun Kumar Sangaiah, and Sung Wook Baik. "Privacy-preserving image retrieval for mobile devices with deep features on the cloud." *Computer Communications* 127 (2018): 75–85.

54. Jang, Young Kyun, and Nam Ik Cho. "Deep face image retrieval for cancelable biometric authentication." In *2019 16th IEEE International Conference on Advanced Video and Signal Based Surveillance (AVSS)*, pp. 1–8. IEEE, 2019.

55. Mirjalili, Vahid, Sebastian Raschka, Anoop Namboodiri, and Arun Ross. "Semi-adversarial networks: Convolutional autoencoders for imparting privacy to face images." In *2018 International Conference on Biometrics (ICB)*, pp. 82–89. IEEE, 2018.

56. Shen, Wei, Zhendong Wu, and Jianwu Zhang. "A face privacy protection algorithm based on block scrambling and deep learning." In *International Conference on Cloud Computing and Security*, pp. 359–369. Springer, Cham, 2018.

57. Yang, Wencheng, Song Wang, Jiankun Hu, Guanglou Zheng, Jucheng Yang, and Craig Valli. "Securing deep learning based edge finger vein biometrics with binary decision diagram." *IEEE Transactions on Industrial Informatics* 15, no. 7 (2019): 4244–4253.

58. Bajwa, Garima, and Ram Dantu. "Neurokey: Towards a new paradigm of cancelable biometrics-based key generation using electroencephalograms." *Computers & Security* 62 (2016): 95–113.

59. Lin, Feng, Kun Woo Cho, Chen Song, Wenyao Xu, and Zhanpeng Jin. "Brain password: A secure and truly cancelable brain biometrics for smart headwear." In *Proceedings of the 16th Annual International Conference on Mobile Systems, Applications, and Services*, pp. 296–309. 2018.

60. Ogiela, Lidia, Makoto Takizawa, and Urszula Ogiela. "Impact of sharing algorithms for cloud services management." In *International Conference on Broadband and Wireless Computing, Communication and Applications*, pp. 423–427. Springer, Cham, 2019.
61. Yan, Bin, Yong Xiang, and Guang Hua. "Basic visual cryptography algorithms." In *Improving Image Quality in Visual Cryptography*, pp. 15–33. Springer, Singapore, 2020.
62. Kaur, Harkeerat, and Pritee Khanna. "Privacy preserving remote multi-server biometric authentication using cancelable biometrics and secret sharing." *Future Generation Computer Systems* 102 (2020): 30–41.

# 8 Gender Classification under Eyeglass Occluded Ocular Region

## An Extensive Study Using Multi-spectral Imaging

*Narayan Vetrekar*
Goa University

*Raghavendra Ramachandra and Kiran Raja*
Norwegian University of Science and Technology (NTNU)

*R. S. Gad*
Goa University

## CONTENTS

| | | |
|---|---|---|
| 8.1 | Introduction | 176 |
| | 8.1.1 Our Contributions | 177 |
| 8.2 | Related Works | 178 |
| | 8.2.1 Visible Spectrum | 180 |
| | 8.2.2 Near-Infra-Red Spectrum | 182 |
| | 8.2.3 Visible and Near-Infra-Red Spectrum | 183 |
| | 8.2.4 Multi-Spectral Imaging | 183 |
| 8.3 | Database | 184 |
| | 8.3.1 Data Preprocessing | 186 |
| 8.4 | Proposed Method | 186 |
| | 8.4.1 Spectral Bands Selection | 186 |
| | 8.4.2 Feature Extraction | 189 |
| | 8.4.3 Classification | 190 |
| 8.5 | Experiments and Results | 190 |
| | 8.5.1 Experimental Evaluation Protocol | 191 |
| | 8.5.2 Evaluation 1: Without-Glass v/s Without-Glass | 192 |
| | 8.5.2.1 Individual Band Comparison | 192 |
| | 8.5.2.2 Fused Band Comparison | 194 |
| | 8.5.3 Evaluation 2: Without-Glass v/s With-Glass | 196 |

        8.5.3.1   Individual Band Comparison............................................ 197
        8.5.3.2   Fused Band Comparison.................................................. 197
8.6   Conclusions................................................................................................. 199
Acknowledgement ................................................................................................200
References..............................................................................................................200

## 8.1   INTRODUCTION

Soft biometric traits such as age, gender, ethnicity, weight, height, and colour of skin have been considered useful in biometric applications and forensics. Among all soft biometric traits, classifying gender has been widely studied for various applications such as identity verification, surveillance, retrieval system, and human computer interaction (Bekios-Calfa et al., 2011; Vetrekar et al., 2017a, 2017b; Raghavendra et al., 2018). Due to its stability and permanence in the features compared to other soft biometric traits, gender is predominantly used as stable auxiliary information for biometric identification and verification system (Lyle et al., 2010). In another application domain, gender information has also been used in categorising the larger set of biometric data in two sub-bins for biometric database management (Jain, 2004). Gender information not only reduces the time required to search the legitimate user from an enrollment dataset (or template dataset) but also improves the overall accuracy of the biometric system (Moeini & Mozaffari, 2017).

Although facial features have shown great potential in predicting gender, it can be noted that face information may not be fully available due to clothing preferences where face is covered by masks. Despite the clothing preferences, especially in semicooperative biometric data capture, ocular information can be easily obtained. An ocular region consists of a small region that surrounds the eye having essential information such as textural and geometric details compared to other facial parts such as nose, forehead, chin, and cheeks (Burge & Bowyer, 2013). The use of ocular information has been well demonstrated in many biometric applications in classical setting to recent smartphone biometrics (Park et al., 2009; Raja et al., 2015). Motivated by earlier works on ocular information in biometrics, one can also deduce the potential of the ocular region for classifying the gender (Vetrekar et al., 2017a, 2017b; Raghavendra et al., 2018).

While we note the applicability of the ocular region for classifying the gender, we also acknowledge a number of challenges in this direction. Similar to other biometric characteristics, the ocular region can also suffer from few challenges due to capture conditions in unconstrained and unsupervised environmental factors (Proença & Alexandre, 2005). Similarly, ocular information cannot be fully available when the subject is wearing the eyeglasses leading to decreased biometric performance (Drozdowski et al., 2018; Lee et al., 2001). Not only does the eyeglasses occlude the information but also present specular and ambient reflections that further degrade the performance of the biometric system (Drozdowski et al., 2018; Lee et al., 2001; Vetrekar et al., 2018). It is also recommended as per the biometric standards (ISO/IEC JTC1 SC37 Biometrics, 2015) to remove the eyeglasses while data acquisition. The recent survey also concluded that the more than 50% of the world population wear eyeglasses ("Data on optometry and optics in Europe", 2017; "Vision-watch-Council", 2016), especially the rise of shortsightedness in east Asia and in general around the world. A similar impact on classifying the gender using the ocular region can be hypothesised under non-ideal data and further with the presence of eyeglasses (Bowyer et al., 2008; Bharadwaj et al., 2010; Proença & Alexandre, 2010).

### 8.1.1 OUR CONTRIBUTIONS

Considering the wide population using eyeglasses across the world-wide population ("Data on optometry and optics in Europe", 2017; "Vision-watch-Council", 2016) and the adverse effect on the performance of the biometric system, it can be noted that classification of gender is not well addressed in the earlier works. The previous studies are limited to analysis of ocular recognition and gender classification in data without the presence of the glasses. In this work, we address the problem of gender classification from the ocular region under the presence of eyeglasses. We present a systematic analysis to establish the effect of eyeglasses covering the ocular region for gender classification. The idea of employing multi-spectral imaging based on facial features have been very well addressed in the recent works (Vetrekar et al., 2017a, 2017b; Raghavendra et al., 2018), thereby extracting spatio-spectral details across the electromagnetic spectrum. In principle, multi-spectral imaging exploits the complementary image information in the form of reflectance and/or emittance to extract discriminative features for better performance accuracy. Motivated by such works, we explore multi-spectral imaging for gender classification using ocular data captured with multi-spectral sensors unlike the works focusing only on visible (VIS) and near-infra-red (NIR) spectrum.

We assert the presence of discriminative spectral band information due to the inherent properties of multi-spectral imaging across male and female class, which can help in classifying the gender in a robust manner despite the presence of glasses. Based on our earlier works on multi-spectral imaging for biometrics (Vetrekar et al., 2017a, 2017b; Raghavendra et al., 2018), we present in this work ocular gender classification using multi-spectral images collected in eight different narrow spectrum bands such as $530\,nm$, $590\,nm$, $650\,nm$, $710\,nm$, $770\,nm$, $890\,nm$, $950\,nm$, and $1000\,nm$ spanning from $530\,nm$ to $1000\,nm$ wavelength range. Further, to explore the inherent characteristics of multi-spectral imaging, we propose an approach that selects four discriminative ocular band images based on the highest entropy value. The selected images are further processed independently for feature extraction using banks of Gabor filters, and the features are used to learn a classifier model using Probabilistic Collaborative Representation Classifier (ProCRC) for predicting the gender. To validate the proposed approach, we present two sets of experimental evaluations based on two protocols on 104 ocular instances corresponding to a total of 16640 sample images for our gender classification study. In the first protocol, we evaluate the classification accuracy when training and testing correspond to the same category – "Without-Glass" and in the second protocol, we evaluate the classification accuracy when training and testing set correspond to "Without-Glass" and "With-Glass," respectively. Both protocols are designed to demonstrate the effect of wearing eyeglasses on the performance accuracy of gender classification.

We further present a fair comparison against the multiple approaches used in gender classification across individual spectral bands and fusion of bands with five different state-of-the-art methods employing Local Binary Pattern (LBP), Local Phase Quantization

(LPQ), Histogram of Oriented Gradients (HOG), GIST, and Binarized Statistical Image Feature (BSIF) independently with Support Vector Machine (SVM) Classifier. However, in the case of fusion, we have employed three different fusion methods such as Image Matting Fusion (IMF) (Li, Kang, Hu, & Yang, 2013), Guided Filtering-Based Fusion (GFF) (Li, Kang, & Hu, 2013), and 2-Discrete Wavelet Fusion (2-DFT) (Amolins et al., 2007) to demonstrate the applicability of our proposed band selection approach. All the evaluation results carried out in this work are presented in the form of average classification accuracy obtained over 10-fold cross-validation to select training and testing samples in random manner such that both the sets belonging to training and testing set are separated disjointly for the analysis. In the due course of this work, we present a number of contributions in this chapter which can be summarised as follows:

- Presents an analysis of gender classification using 104 unique ocular images captured using multi-spectral imaging sensor with eight narrow spectrum bands such as $530\,nm$, $590\,nm$, $650\,nm$, $710\,nm$, $770\,nm$, $890\,nm$, $950\,nm$, and $1000\,nm$ spanning from $530\,nm$ to $1000\,nm$ wavelength range.
- Proposes a new approach of selecting four most discriminative spectral band images based on the highest entropy value, followed by feature extraction using banks of Gabor filters and ProCRC classification for gender classification.
- The approach is further evaluated for gender classification even under the occlusion of eyeglasses. The approach is analysed on ocular data captured with "Without-Glass" and "With-Glass" to establish the robustness of our proposed approach.
- Further, to present a fair comparison, the performance of our proposed method is compared against the five different state-of-the-art feature extraction methods such as LBP, LPQ, HOG, GIST, and BSIF, independently along with SVM classifier, performed on individual spectrum band and fusion of bands.

In the remainder of this chapter, Section 8.2 introduces the literature review on gender classification based on the ocular region and the related work discussed in this section is divided into VIS, NIR, VIS and NIR, and Multi-spectral imaging categories. Along with the detail literature survey, this section also presents the abstract literature in the tabulated form (Table 8.1) for better comparison of previous works. Section 8.3 presents the detailed description of multi-spectral imaging database collected with eight bands across VIS and NIR spectrum. Section 8.4 provides the detailed description of our proposed method for gender classification, and Section 8.5 presents the experimental evaluation along with the protocols. With a set of analysis on evaluation results, Section 8.6 presents the conclusive remarks and lists the potential future works.

## 8.2  RELATED WORKS

The presence of eyeglasses is considered as one of the major noise factors for degraded biometric performance, as shown by various studies (Bowyer et al., 2008; Bharadwaj et al., 2010; Proença & Alexandre, 2010). Another set of works have specifically

## TABLE 8.1
## Summary of Most Relevant Gender Classification Research from Ocular Images

| Authors | Database | Features | Classification | Accuracy |
|---|---|---|---|---|
| | | **Visible Spectrum** | | |
| Merkow et al. (2010) | Proprietary | LBP | LDA-NN PCA-NN SVM | 85.00% |
| Lyle et al. (2010) | FRGC | LBP | SVM | 93.00% |
| Kumari et al. (2012) | FERET | ICA | BPNN RBFNN PNN | 90.00% |
| Castrillón-Santana et al. (2016) | GROUPS | LBP HOG LTP WLD LOSTB | SVM | 92.46% |
| Rattani et al. (2017) | VISOB | LBP HOG LTP LPQ BSIF | SVM MLP | 92.00% |
| Rattani et al. (2018) | VISOB | RPI | VGG ResNet | 90.00% |
| Tapia et al. (2019a) | CSIP MICHE MODBIO INACAP | SRCNNs | RF | 90.00% |
| | | **Near-Infra-Red Spectrum** | | |
| Bobeldyk and Ross (2016) | BioCOP | BSIF | SVM | 85.70% |
| Kuehlkamp et al. (2017) | GFI | LBP GF RPI | MLP CNN | 66.00% |
| Tapia & Aravena (2018) | ND-GFI | RPI | CNN | 87.26% |
| Viedma et al. (2019) | 5 Public DBs | ULBP HOG RPI | SVM NECA | 89.22% |
| | | **Visible and Near-Infra-Red Spectrum** | | |
| Dong & Woodard (2011) | FRGC MBGC | GSF LAF CPF | MD LDA SVM | FRGC:97.00% MBGC:96.00% |
| Lyle et al. (2012) | FRGC MBGC | LBP HOG DCT LCH | ANN SVM | FRGC:97.30% MBGC:90.00% |

*(Continued)*

**TABLE 8.1 (*Continued*)**
**Summary of Most Relevant Gender Classification Research from Ocular Images**

| Authors | Database | Features | Classification | Accuracy |
| --- | --- | --- | --- | --- |
| Tapia et al. (2019b) | 10 Public DBs | RPI | CNN | 86.60% |
| | **Multi-spectral Imaging** | | | |
| Raja et al. (2020) | Proprietary | GIST | CRC | 81.00% |

FRGC, Face Recognition Grand Challenge; FERET, Face Recognition Technology; GROUPS, The images of Groups; VISOB, Visible Light Mobile Ocular Biometric; CSIP, Cross-sensor Iris and Periocular; MICHE, Mobile Iris Challenge Evaluation; MODBIO, A multimodal database captured with a portable handheld device; INACAP, Hand-made Periocular Iris Image Database Captured from Cellphones; BioCOP, FBI Biometric Collection of People; GFI, Gender from Iris; ND-GFI, University of Notre Dame Iris Image Dataset; MBGC, Multiple Biometric Grand Challenge; GSF, Global Shape Features; LAF, Local Area Features; CPF, Critical Point Feature; GF, Gabor Filter; RPI, Raw Pixel Intensity; RF, Random Forest; NECA, Nine Ensemble Classifier Algorithm.

studied the impact of wearing eyeglasses to establish the performance degradation in ocular biometrics (Lee et al., 2001; Vetrekar et al., 2018). The works further conclude that the presence of eyeglasses on the ocular region seriously deteriorates the overall accuracy. Despite the serious problem posed by influence of eyeglasses on the performance of ocular biometrics, the set of related works available is very limited. We therefore present a set of related works in the subsequent section listing out the approaches and open challenges, specifically for gender classification based on ocular information.

Earlier works on ocular region-based gender classification have focused on using VIS spectrum (Merkow et al., 2010; Lyle et al., 2010; Kumari et al., 2012; Castrillón-Santana et al., 2016; Rattani et al., 2017, 2018; Tapia et al., 2019a), NIR spectrum (Bobeldyk & Ross, 2016; Kuehlkamp et al., 2017; Tapia & Aravena, 2018; Tapia et al., 2019a; Viedma et al., 2019), VIS and NIR spectrum (Dong & Woodard, 2011; Lyle et al., 2012; Tapia et al., 2019b) and more recently the multi-spectral imaging operated in nine narrow spectrum bands using the VIS and NIR wavelength range (Raja et al., 2020). A list of all related works in this direction is listed out in Table 8.1.

## 8.2.1 VISIBLE SPECTRUM

One of the early works by Merkow et al. (2010) studied the gender classification using the ocular region cropped from the face images to analyse the reliability of the ocular region for gender classification. The facial database employed in this work is of low-resolution Joint Photographic Experts Group (JPEG) face images acquired using web crawler from Flikr [Inc., 2010 (available online)]. In their work, ocular gender classification was performed using three different classification methods such as Linear Discriminant Analysis along with 1 Nearest Neighbor (LDA-1NN) classifier, Principal Component Analysis along with 1 Nearest Neighbor (PCA-1NN) classifier, and SVM Classifier. Each of these classification methods was used in conjunction with LBP feature descriptor. Authors reported an accuracy of 85.00% for classifying gender, while considering the images with frontal face pose with minimal occlusion and pitch.

In another work by Lyle et al. (2010), the ocular features was employed to classify soft biometric information such as gender and ethnicity. With the use of LBP texture descriptor on the grey-scale images, the work illustrated the effectiveness in predicting soft biometric traits. The gender classification accuracy of 93.00% with SVM was reported using Face Recognition Grand Challenge (FRGC) database (Phillips et al., 2005) while demonstrating the improvement in performance accuracy of the existing ocular-based authentication system when combined with soft biometric informations.

Kumari et al. (2012) attempted to classify gender from the poor quality grayscale ocular region using FERET face database (Phillips et al., 2000). The authors used Independent Component Analysis (ICA) on the high dimensional data and classified the gender using Convolutional Neural Network (CNN) methods such as Back Propagation Neural Network (BPNN), Radial Basis Function Neural Network (RBFNN), and Probabilistic Neural Network (PNN). Although the evaluation was performed on low-quality images, the reported ocular gender classification of 90.00% demonstrated satisfactory applicability.

Castrillón-Santanae et al. (2016) studied exhaustively the problem of gender classification based on ocular information on most challenging dataset in wild. The purpose of this work was to demonstrate the validity of using ocular region in a large population and the use of complementary information of different feature descriptors to improve the overall accuracy. Features including LBP, HOG, Local Ternary Patterns (LTP), Weber Local Descriptor (WLD), and Local Oriented Statistics Information Booster (LOSIB) were employed in their system along with a SVM (Radial Basis Function (RBF) Kernel) classifier for classifying the gender. The classification results computed on the GROUPS database (Gallagher & Chen, 2009) showed 92.46% gender classification accuracy.

Further, Rattani et al. (2017) explored the problem of gender estimation using the ocular images acquired using three different smartphones such as iPhone 5s, Samsung Note 4, and Oppo N1. The texture descriptor methods such LBP, LTP, HOG, LPQ, and BSIF were used in conjunction with SVM and Multi-layer Perceptron (MLP) classifier on publicly available VISOB database of periocular images (Rattani et al., 2016). A maximum of 80.00% classification accuracy was obtained with SVM classifier on LPQ descriptor, while 91.60% accuracy was obtained with MLP classifier on HOG descriptor on the ocular data captured from smartphone. Latter, in an extended work by Rattani et al. (2018), the authors performed extensive evaluation based on deep learning on the ocular image collected using smartphone. The pre-trained CNNs such as very deep convolutional network for large image recognition (VGG) and residual network (ResNet) were employed for gender classification with an accuracy of 90.00%.

On similar lines, selfie images collected using smartphone also used for ocular gender classification, especially by cropping the ocular region from selfie images of individual faces which was demonstrated by Tapia et al. (2019a). The authors have employed Super Resolution Convolutional Neural Networks (SRCNNs) to improve the overall quality of the ocular region cropped from selfie images to 2X and 3X in their work prior to gender classification. The results were obtained on three existing databases: CSIP (Santos et al., 2015), MICHE (Marsico et al., 2015), MOBBIO (Sequeira et al., 2014), and their in-house INACAP database. The study demonstrated a gender classification accuracy of 90.15% using Random Forest by employing SRCNNs on ocular images.

Overall, the works on ocular gender classification using VIS spectrum are significantly dependent on the use of prominent features such as eyebrow (structural information) (Dong & Woodard, 2011). While we also note that the ocular regions cropped from the holistic facial region collected specifically for facial database are available in the public domain for academic research (Tapia & Aravena, 2018; Bobeldyk & Ross, 2016; Lyle et al., 2010), dedicated datasets with equal gender balance are not available. Finally, it was noted that the performance accuracy decreased when eyebrows were not considered in analysis (Dong & Woodard, 2011).

## 8.2.2 Near-Infra-Red Spectrum

Bobeldyk and Ross (2016) investigated the gender prediction based on ocular images collected in NIR spectrum. The purpose of this work was to explore Iris or ocular region for gender classification. The work focused on classifying gender using four different regions: only iris region, normalised iris-only region, ocular region, and iris occluded ocular region. In the context of this work, a statistical feature extraction method (BSIF) along with SVM classifier was employed to predict gender using BioCOP database (BioCOP, Database Available Online) collected using NIR sensor. The experimental evaluation results have demonstrated the better classification accuracy using the ocular region compared to the iris region for gender prediction.

The effect of cosmetics on eyelashes was examined for gender classification by Kuehlkamp et al. (2017). The authors used mascara on the subjects such that the eyelashes appeared more thicker and darker, to increase the artefacts to make the extraction of the texture details challenging. The works have explored the use of hand-crafted features such as LBP and Gabor filter and data-driven features (raw pixel intensity), while the MLPs and CNNs as the classification approach used in their work. The result of this work based on Gender from Iris (GFI) dataset (Tapia et al., 2016) indicated 66.00% ocular gender classification accuracy.

Tapia et al. (2018) have also considered the usefulness of CNN in providing the competitive gender prediction results for the ocular region, rather than relying on textural information. The authors demonstrated the results by training the CNN model on left and right eyes and merging the models of left and right eye to explore the benefits of merging two models. The experimental classification accuracy of 87.26% was obtained in their work using publicly available ND-GFI database (Dame, n.d.).

Also, in the recent study by Viedma et al. (2019), it was indicated that the ocular region contributes significant information than iris for gender classification. In general, authors have analysed and demonstrated the location of relevant features in the ocular region for gender classification. The features such as raw pixel intensity, texture [Uniform Local Binary Patterns (ULBP)], and shape (HOG) were used for gender classification using ocular information. However, to estimate the relevance of each feature, the Gini Index with the XgBoost algorithm was used, while the classification accuracy was obtained with SVM and nine ensemble classifiers.

The experimental result obtained with five publicly available database suggested the highest classification accuracy of 89.22%.

We summarise that the ocular gender classification based on NIR spectrum focus heavily on iris pattern (Tapia & Aravena, 2018; Bobeldyk & Ross, 2016) for robust performance, while we note that the recent study also suggests the contribution of the ocular region than iris alone for gender classification (Viedma et al., 2019). Further, texture-based methods such as LBP, HOG, BSIF along with strong SVM and CNN classifier have been used independently in various studies (Tapia et al., 2019a). A noted limitation in the NIR spectrum is that it requires dedicated sensor for iris image capture and high degree of subject cooperation during data collection (Tapia et al., 2019b).

### 8.2.3 Visible and Near-Infra-Red Spectrum

Recent studies have also evaluated both VIS and NIR spectrum images for ocular-based gender classification. We briefly present some of these works in this section for the brevity of reader.

Dong et al. (2011) investigated the use of eyebrow shape features from the ocular region for gender classification. Global shape features (GSF), local area features (LAF), and critical point features, which mainly represent the eyebrow shape features, have been extracted as major features for gender classification from the ocular region. Further, classification results were obtained using three different classifier such as Minimum Distance (MD), Linear Discriminant Analysis (LDA), and SVM classifier. The best gender prediction accuracy obtained on MBGC database and FRGC database was 96.00% and 97.00% respectively, suggesting the applicability of eyebrow shapes for efficiently classifying soft biometrics trait of gender.

Lyle et al. (2012) also demonstrated the effectiveness of various feature descriptors such as LBP, HOG, DCT, and LCHE for appearance-based ocular region and gender classification using Artificial Neural Network (ANN) and SVM. The paper reported 90.00% and 97.30% ocular gender prediction accuracy using MBGC and FRGC database.

Further, Tapia et al. (2019a) have explored the generalisability of the deep neural network-based CNN algorithm for ocular gender classification. Specifically, the authors in their work have obtained the competitive classification accuracy for various scenarios such as cross-sensor, cross-spectral, and multi-spectral data. The investigation of this work on multiple publicly available database using VIS and NIR spectrum indicated the 86.60% gender classification accuracy.

### 8.2.4 Multi-Spectral Imaging

With multi-spectral imaging gaining more attention in the recent times in biometrics, a recent work by Raja et al. (2020) also investigated the ocular gender classification using multi-spectral images collected in eight bands across VIS and NIR spectrum. The relevance of this work relies on fusing the spectral band feature using GIST features in kernalised space to fully leverage the individual band features. Further classification preformed using Collaborative Representative Classifier (CRC) have demonstrated 81.00% average ocular gender classification, signifying the potential of multi-spectral imaging for gender prediction.

## 8.3 DATABASE

We first summarise the multi-spectral ocular database employed for gender classification based on the ocular region in this section (Figure 8.1). The database for this experimental evaluation is collected using our custom-built multi-spectral imaging sensor (Vetrekar et al., 2016) to acquire the images in eight narrow spectrum bands corresponding to 530nm, 590nm, 650nm, 710nm, 770nm, 890nm, 950nm and 1000nm spanning from VIS to NIR wavelength range. The acquired ocular database for gender classification consists of 104 unique ocular images from 52 subjects with a distribution of 32 male and 20 female participants.

As the goal of this study is to perform the gender classification when the ocular region is occluded with eyeglasses, we have collected multi-spectral images for the ocular region with and without the presence of glasses over two sessions. The first set of data collection comprises multi-spectral ocular instances when subjects are not wearing eyeglasses and in a simplified manner, we call this set of images by an acronym "Without-Glass" data. On similar lines, the second set of data collection comprises multi-spectral ocular instances when subjects are asked to wear eyeglasses and in a simplified manner, we call this set of images by an acronym "With-Glass" data. Table 8.2 presents the description of acronyms used for two different categories of data collection.

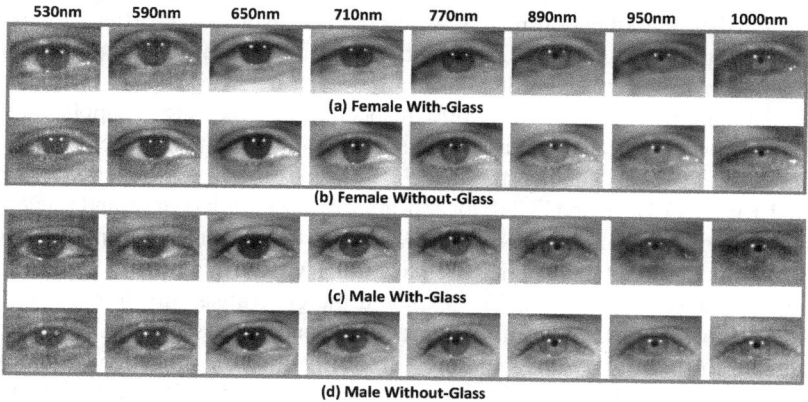

FIGURE 8.1 Multi-spectral ocular instances collected using *eight* narrow bands across visible and near-infra-red spectrum. Figure illustrates the sample ocular images for male and female class for two classes of "Without-Glass" and "With-Glass".

## TABLE 8.2
## Description of Acronyms Used for Two Different Categories of Multi-Spectral Images Collected for Ocular Region

| Acronym | Summary |
|---|---|
| Without-Glass | Multi-spectral images acquired when subject are not wearing eyeglasses |
| With-Glass | Multi-spectral images acquired when subject are wearing eyeglasses |

Further, each category of ocular image is captured in five instances in two different sessions, and the time difference between each of these sessions was 3–4 weeks. The sample ocular images were collected at a stand-off distance of one meter between the multi-spectral imaging sensor and subject in a controlled indoor illumination conditions. Proper care has been exercised while selecting the eyeglasses for this experimental data collection, in order to avoid any kind of reflections due to the illuminations from the indoor lighting. Under each category of ocular instances for male and female, a total of 8320 ocular instances [corresponding to (64 Male Ocular Instances × 2 Sessions × 5 Samples × 8 Bands) + (40 Female Ocular Instances × 2 Sessions × 5 Samples × 8 Bands) = 8320 total ocular instances] are obtained. The detailed summary representing the total number of ocular instances corresponding to two different categories across male and female class is illustrated in Table 8.3.

Further, the database employed in this work has following unique points compared to existing database employed in the ocular biometric for gender classification:

- The database available in the public domain are mainly in VIS spectrum or NIR spectrum, while these databases, especially VIS spectrum, do not contain the distinct information due to the integration process that exist in the formation of image over broad spectrum [except our recent work (Raja et al., 2020)]. However, the database presented in this work addresses these limitations by capturing the ocular images in narrow spectrum bands in discrete and disjoint manner to obtain intrinsic features.
- Further, the use of narrow spectrum band in discrete and disjoint manner allows to extend the inherent characteristic details in the spatial spectral domain without any redundancy in the data.
- Majority of previous works have used the facial images to crop the ocular region for gender classification, while we have build our database specially for ocular biometric only.
- Also most of the work uses the eyebrow information in their database, however this work eliminates the effect of eyebrow in the database to avoid any kind of biasing in the proposed work

**TABLE 8.3**

**Summary Illustrating the Total Number of Ocular Instances Collected for Male and Female Using Multi-Spectral Imaging**

| | Categories of Data Collection | | | | | | | |
|---|---|---|---|---|---|---|---|---|
| | Without-Glass | | | | With-Glass | | | |
| Multi-spectral Ocular Database | Ocular Instances | Sessions | Samples | Bands | Ocular Instances | Sessions | Samples | Bands |
| Male | 64 | 2 | 5 | 8 | 64 | 2 | 5 | 8 |
| Female | 40 | 2 | 5 | 8 | 40 | 2 | 5 | 8 |
| Total ocular instances | | 8,320 | | | | 8,320 | | |

### 8.3.1 Data Preprocessing

The sample ocular instances collected for "With-Glass" category of data collection consists of eyeglass. To avoid any bias in just classifying the frames alone, we process the images to remove frames from the ocular images by cropping the ocular region, as shown in Figure 8.2. As the original multi-spectral images collected for the ocular region is of spatial resolution of $1280 \times 1024$ pixels, the images result in $270 \times 160$ pixels after cropping. On similar lines, the ocular sample images corresponding to "Without-Glass" category of database is also preprocessed for the analysis. Figure 8.2 illustrates the cropping of male and female ocular regions corresponding to the left and right eye region. Further, these sample multi-spectral images collected in *eight* different wavelengths, hence to maintain the similar illumination across all the *eight* spectrum bands, we have employed histogram enhancement technique to enhance the contrast uniformly across individual spectral bands.

## 8.4 PROPOSED METHOD

This section explains in detail the proposed approach employed in this chapter for ocular gender classification using the multi-spectral images collected in eight narrow spectrum bands across VIS and NIR spectrum. The ocular instances collected across individual bands consist of discriminative information (refer Section 8.3), and hence to improve the robustness in the classification, we present our proposed scheme to efficiently utilise the spectral band information for gender classification in the robust manner. The proposed approach therefore first selects the four most discriminative spectral band images based on the highest entropy value. The selected spectral band images are then processed using the bank of Gabor filters (Haghighat et al., 2015) independently to extract the local and global features. Further, the histogram features obtained for selected spectral band images using Gabor filters are concatenated to learn the classifier model using an efficient ProCRC for gender classification. The schematic representation of our proposed approach for gender classification using multi-spectral images collected for the ocular region occluded with eyeglasses is illustrated in Figure 8.3. However, to present our approach in more detail, we divide this section of our chapter in three subsections: (i) Spectral Band Selection, (ii) Feature Extraction, and (iii) Classification. Details related to each of these sections are discussed in the following section.

### 8.4.1 Spectral Bands Selection

Let $Q_\lambda \in \mathfrak{R}^{m \times n}$ represent the set of preprocessed ocular spectral band images corresponding to eight different spectral bands and can be expressed using (Equation 8.1) as follows:

$$Q_\lambda = \{Q_1, Q_2, Q_3, ..., Q_8\} \in \mathfrak{R}^{m \times n} \tag{8.1}$$

where the individual eight spectral bands comprising $530\,nm$, $590\,nm$, $650\,nm$, $710\,nm$, $770\,nm$, $890\,nm$, $950\,nm$, and $1000\,nm$ are represented by using the notation $\lambda = \{1, 2, 3,.., 8\}$, and $m \times n$ represents the spatial dimension of each ocular instances. To perform the gender classification based on ocular images is always a challenging task in situations when the details such as eyebrows are not present (important geometric feature useful for gender classification). The result of which, it is even difficult to discriminate between male and female class with the naked eyes, as can be seen from the sample images of database shown in Figure 8.1. Further, the inclusion of eyeglasses makes it even more challenging task when the classifier is trained with ocular instances without wearing eyeglasses and tested with ocular instances with eyeglasses. Now, the individual spectral band leverage the discriminative band information which varies among themselves. As a result of different photometric reflectance and transmittance properties of individuals, some bands may not lead to discriminative information. Hence, to extract the dominant features from the bands, we have selected four characteristic ocular spectral band images based on the highest entropy value to enhance the gender classification accuracy. Further, the idea here is to select the complementary ocular band images, at the same time to reduce the computational expenses involved in the processing.

For the given set of ocular spectral band images $Q_\lambda \in \Re^{m \times n}$ belongs to eight bands, the entropy value for each individual bands can be computed using (Equation 8.2) as follows:

$$E_\lambda = \sum_k P_k \log_2 P_k \qquad (8.2)$$

where $P_k$ represents the $k^{th}$ probability of difference between the two adjacent pixels, and $E_\lambda$ represents the entropy value corresponding to $Q_\lambda$ individual spectral bands.

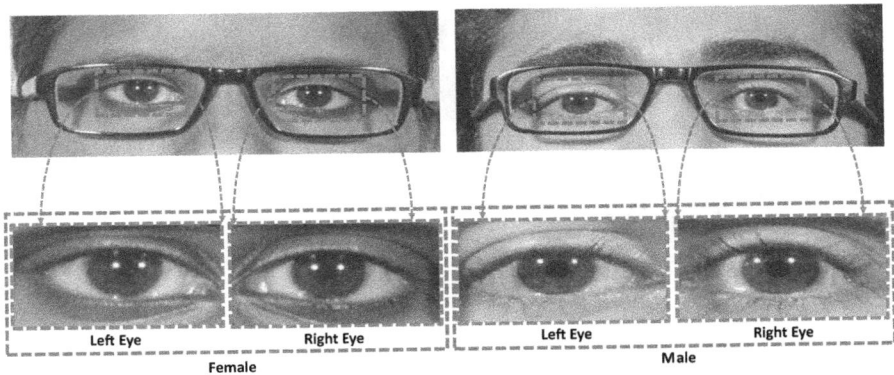

**FIGURE 8.2** Cropping of ocular instances from left eye and right eye for male and female class, performed during the preprocessing of data.

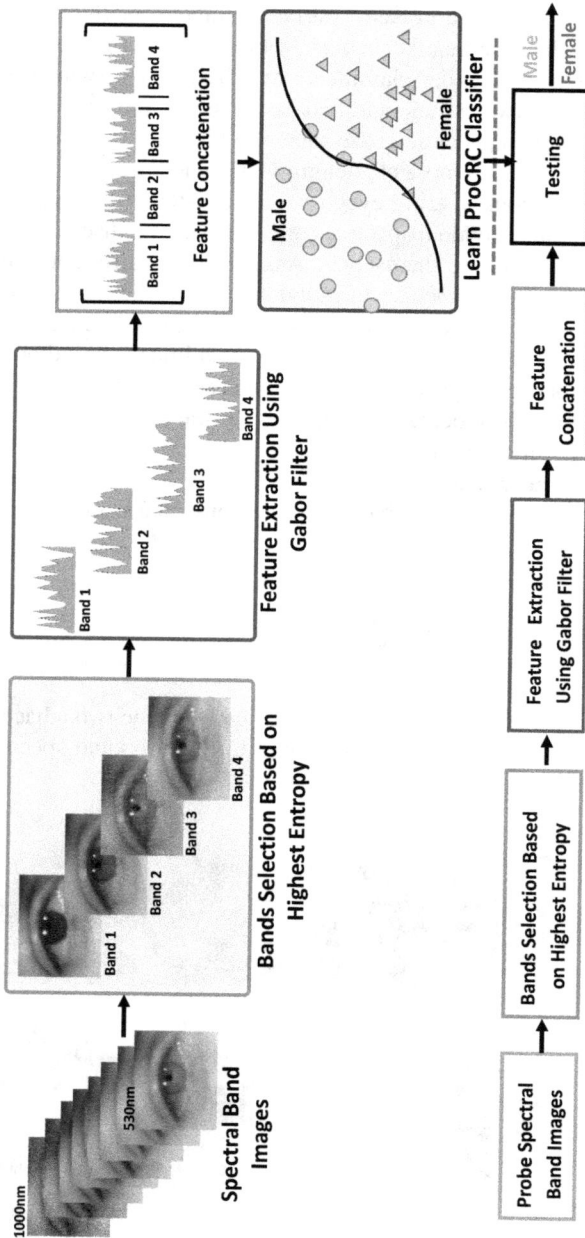

**FIGURE 8.3** Proposed approach for gender classification using ocular images collected using multi-spectral images collected across eight spectrum bands spanned from 530*nm* to 1000*nm* spectrum. The approach learns concatenated Gabor features corresponding to four different bands selected based on highest entropy value for the classification.

The entropy value computed for the individual eight spectral bands can be expressed using (Equation 8.3)

$$E_\lambda = \{E_1, E_2, E_3, \ldots, E_8\} \tag{8.3}$$

As described earlier in this section, for efficient processing, we have employed the band selection approach in this work for gender classification. The four selected ocular spectral bands $\{\varrho_1, \varrho_2, \varrho_3, \varrho_4\}$, corresponding to the highest entropy value, are represented in Equation (8.4):

$$\{\varrho_1, \varrho_2, \varrho_3, \varrho_4\} = \{Q_{\max 1}, Q_{\max 2}, Q_{\max 3}, Q_{\max 4}\} \tag{8.4}$$

where {max1, max2, max3, max4} represents the selected spectral band images corresponding to the maximum entropy, i.e., max $\{E_1, E_2, E_3,.., E_8\}$. The selected spectral bands $\varrho_1, \varrho_2, \varrho_3,$ and $\varrho_4$ are processed independently for feature extraction in the following section.

### 8.4.2 Feature Extraction

The dominant features are obtained in the form of local and global features using the bank of Gabor filters (Haghighat et al., 2015) separately on the selected spectral bands ocular instances (selection process of spectral band is explained in Section 8.4.1) before performing classification. The strength of Gabor filters has been widely utilised in biometrics due to its high performance. The significance of Gabor feature descriptor is that it employs the bank of Gabor filters in different orientation and scale to obtain the characteristic feature information in the highest frequency region for a given image. Hence, in this work, we obtain Gabor features of individual selected spectral bands to extract discriminative band information. The transfer function for 2-D Gabor function defined in the space domain can be expressed in Equation (8.5):

$$g(u, v) = g(u, v; f_c, \theta) = \frac{f_c^2}{\pi \sigma \psi} \exp^{-\frac{f_c^2 u_p^2}{\sigma^2} + \frac{f_c^2 v_q^2}{\psi^2}} \exp^{2\pi i f_c u_p} \tag{8.5}$$

where $(u, v)$ represents the spatial coordinates; $u_p = u \cos \theta + v \sin \theta$, $v_p = -u \sin \theta + v \cos \theta$ and $\theta$ represents the rotation angle; $f_c$ represents the central frequency; $\sigma$ and $\psi$ control the bandwidth of Gabor filters across $u$ and $v$ axis, respectively. The performance of Gabor filters is based on the selection of $\theta$ (orientation) and $f_c$ (scale) which is set to orientation 4 and scale 4 empirically.

Using Gabor function (Equation 8.5), the Gabor feature vector corresponding to four selected spectral bands $\{\varrho_1, \varrho_2, \varrho_3, S \varrho_4\}$ (Equation 8.4) can be expressed as $\{g_1, g_2, g_3, g_4\}$, respectively. Further, we combine these Gabor feature vectors corresponding to selected spectral bands to obtain the final histogram feature vector $h$ to process in a classifier using Equation (8.6):

$$h = \{g_1(u, v) \| g_2(u, v) \| g_3(u, v) \| g_4(u, v)\} \tag{8.6}$$

### 8.4.3 CLASSIFICATION

In order to efficiently classify the histogram features corresponding to male and female class, we employ ProCRC (Cai et al., 2016) to improve the classification accuracy of predicting gender. The idea of ProCRC is to jointly maximise the likelihood ratio of test samples belonging to each of the classes, and the final classification is performed by computing the maximum likelihood ratio of test sample with each of the classes.

In this work, the set of histogram features $h$ obtained in the above section corresponding to each class (male or female) forms the training set. The histogram features corresponding to two different classes are further processed to learn probabilistic collaborative representation subspace $\xi$. Now to compute the likelihood ratio of the test sample, a regularised least square regression is employed on the learnt histogram feature belonging to training set and probe histogram feature vector belonging to each of the classes, as can be expressed in Equation (8.7):

$$\partial = \underset{\hat{\alpha}}{\arg\min}\left\{\| \omega + \xi\alpha \|_2^2 + \beta \| \alpha \|_2^2 + \frac{\mu}{M}\sum_{m=1}^{M} \| \xi\alpha - \alpha_k\xi_k \|_2^2\right\} \qquad (8.7)$$

where the first two terms $\| \omega + \xi\alpha \|_2^2 + \beta \| \alpha \|_2^2$ of (Equation 8.7) form the collaborative representation framework and last term $\sum_{m=1}^{M} \| \xi\alpha - \alpha_k\xi_k \|_2^2$ attempts to find a common point inside each subspace of $k$ class (in this work, $k = 2$, i.e., two class: male and female). Further, balancing role of the three term is carried out by $\alpha$, $\beta$, and $\mu$, regularisation parameters. The obtained comparison scores $\partial$ is then used as performance analysis parameters for gender classification.

## 8.5 EXPERIMENTS AND RESULTS

This section presents in detail the evaluation protocol employed for this study and related results on gender classification based on using ocular instances. The classification accuracy is computed on the multi-spectral ocular images collected in eight narrow spectrum bands across the VIS and NIR wavelength range. The goal of this work is to examine the influence of wearing eyeglasses on the performance accuracy of ocular gender classification. To present the robustness in the classification accuracy, we present the results based on our proposed multi-spectral ocular gender classification approach. Specifically, we select four discriminative ocular band images corresponding to the highest entropy value and process independently using the bank of Gabor filters to extract local and global features. Finally, the Gabor feature vectors corresponding to selected spectral band images are concatenated to learn the classifier model using ProCRC for gender classification problem. We present the extensive set of classification results in the form of average classification accuracy on the larger dataset of 16640 ocular images collected using multi-spectral imaging sensor. To present the average classification accuracy, we have performed 10-fold cross-validation experiment to randomly select the ocular sample images for training and testing set in a disjoint manner without overlap. The results are presented using our proposed method, and comparison is provided against five different state-of-the-art methods to present its significance.

## 8.5.1 EXPERIMENTAL EVALUATION PROTOCOL

In this section of the chapter, we present the experimental evaluation protocol employed in this work for ocular gender classification. To present the gender classification using multi-spectral images collected for ocular instances, we present our experimental evaluation protocol where images are partitioned into training and testing set. The training set comprises an equal number of 20 male and 20 female ocular instances including their samples from "Without-Glass" data (Table 8.4). The total number of sample images in the training set consists of 3200 ocular images [corresponding to (20 Male ocular instances × 2 Sessions × 5 Samples × 8 Bands) + (20 Female ocular instances × 2 Sessions × 5 Samples × 8 Bands) = 3200 images]. The testing set comprises 44 male and 20 female ocular instances including their samples from "With-Glass" data. The total number of sample images in the testing set consists of 5,120 ocular images [corresponding to (44 Male ocular instances 2 Sessions 5 Samples 8 Bands) + (20 Female ocular instances × 2 Sessions × 5 Samples × 8 Bands) = 5120 images].

To present the significance of this work, we have employed two sets of experimental evaluation. *Evaluation 1* corresponds to Without-Glass *v/s* Without-Glass, and *Evaluation 2* corresponds to Without-Glass *v/s* With-Glass gender classification. The details of experimental results related to each of these evaluations are provided in the next sections. Further to present the fair comparison with our proposed approach, we have compared the classification accuracy with the performance of each individual spectral bands and across three different fusion methods. In the case of fusion, we have employed three different fusion methods corresponding to IMF (Li, Kang, Hu, & Yang, 2013), GFF (Li, Kang, & Hu, 2013), and 2-Discrete Wavelet Transform (DCT) (Amolins et al., 2007) to present our results. The gender classification results obtained across individual spectral bands, and the fusion of bands are performed independently using five different feature extraction algorithms such as LBP (Ojala et al., 2002), LPQ (Ojansivu & Heikkil¨a, 2008), HOG (Dalal & Triggs, 2005), GIST (Oliva & Torralba, 2001), and BSIF (Kannala & Rahtu, 2012). The performance evaluation results are obtained by processing independently these

---

## TABLE 8.4
## Experimental Evaluation Protocol Summarising the Total Number of Multi-spectral Ocular Instances Partitioned Under Training and Testing Set

| | Training Set | | | | | | | |
|---|---|---|---|---|---|---|---|---|
| **Database** | **Male** | | | | **Female** | | | |
| | **Ocular Instances** | **Sessions** | **Samples** | **Band** | **Ocular Instances** | **Sessions** | **Samples** | **Band** |
| *Without-Glass* | 20 | 2 | 5 | 8 | 20 | 2 | 5 | 8 |
| Total sample images | | 1,600 | | | | 1,600 | | |
| *With-Glass* | 44 | 2 | 5 | 8 | 18 | 2 | 5 | 8 |
| Total sample images | | 3,520 | | | | 1,600 | | |

five different feature descriptor methods along with SVM classifier (Raghavendra et al., 2018; Vetrekar et al., 2017a, 2017b). Use of these feature extraction methods along with SVM classifier has recently been used in gender classification studies conducted on multi-spectral imaging data.

## 8.5.2 Evaluation 1: Without-Glass v/s Without-Glass

In this section of the chapter, we present in detail the experimental evaluation results, when the multi-spectral ocular instances corresponding to "Without-Glass" category of data are employed to learn the two class model during training, and comparison is performed against the ocular multi-spectral images corresponding to the same category of data, i.e., "Without-Glass" during testing, based on our proposed approach discussed in this chapter (refer Section 8.4) for gender classification. The purpose of this set of evaluation is to present the benchmark results when the same categories of data are learned and tested. The classification accuracy results obtained based on the proposed method is compared across the performance accuracy of individual bands and the fusion of bands. The performance of individual bands and fusion of bands is carried out for gender classification using five state-of-the-art methods such as LBP-SVM, LPQ-SVM, HOG-SVM, GIST-SVM, and BSIF-SVM. Based on the experimental evaluation protocol, we first present the results based on individual spectral bands (Section 8.5.2.1) and fusion of bands (Section 8.5.2.2) in the following sections.

### 8.5.2.1 Individual Band Comparison

We present in this section the performance analysis of individual spectral bands for ocular gender classification. Table 8.5 tabulates the average gender classification accuracy after 10-fold cross-validation, and Figure 8.4 illustrates the mean and variance plot describing the classification accuracy of individual spectral bands. The overall results have shown the reasonable average classification accuracy across the

**TABLE 8.5**

**Average Gender Classification Accuracy (in %) across Individual Bands and Proposed Method, When Training Ocular Sample Images Belongs to *Without-Glass* and Testing Ocular Sample Images Belongs to *Without-Glass* Category of Data**

| Algorithm | Spectral Bands | | | | | | | |
|---|---|---|---|---|---|---|---|---|
| | 530 nm | 590 nm | 650 nm | 710 nm | 770 nm | 890 nm | 950 nm | 1,000 nm |
| LBP-SVM | 52.65 | 56.74 | 58.40 | 62.13 | 61.19 | 63.45 | 58.37 | 60.64 |
| •LPQ-SVM | 58.82 | 61.03 | 63.37 | 68.43 | 70.40 | 68.68 | 62.41 | 65.82 |
| •HOG-SVM | 53.27 | 54.80 | 70.57 | 69.07 | 66.53 | 67.26 | 60.23 | 62.21 |
| •GIST-SVM | 68.97 | 69.58 | 74.66 | 72.00 | 73.66 | 74.39 | 71.37 | 67.27 |
| •BSIF-SVM | 61.97 | 64.74 | 71.22 | 72.80 | 74.79 | 73.81 | 71.78 | 68.05 |
| •Proposed approach | | | | 75.72 | | | | |

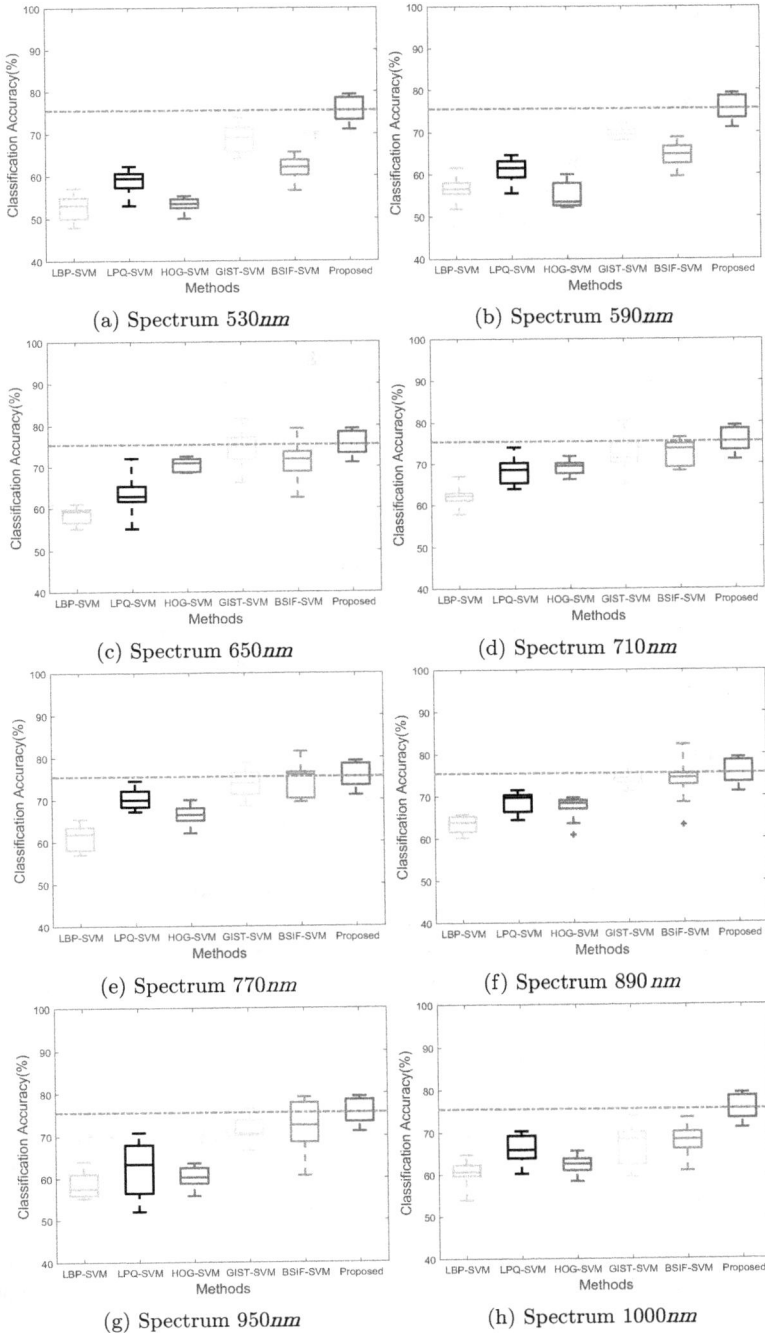

**FIGURE 8.4** Average classification accuracy (%) illustrated in terms of mean and variance plot for Without-Glass *v/s* Without-Glass evaluation for gender prediction.

individual spectral bands for gender prediction. Further, based on the evaluation results obtained for this set, we summarise our major observations as below:

- The highest average gender classification accuracy obtained across the individual band is 74.79% for 710 nm spectrum band using BSIF along with SVM classifier (BSIF-SVM). On the other hand, the lowest average gender classification accuracy obtained across the individual band is 52.65% for 530 nm spectrum band using LBP along with SVM classifier (LBP-SVM).
- Of the eight spectral bands employed in this work, bands such as 650 nm, 710 nm, 770 nm, 890 nm, and 950 nm demonstrated consistently an higher classification accuracy for most of the algorithms used, while the bands such as 530 nm, 590 nm, and 1000 nm have indicated lower performance using state-of-the-art approaches. The better performance across 650 nm, 710 nm, 770 nm, 890 nm, and 950 nm could be attributed to better Signal-To-Noise Ratio, as compared to the other spectral bands. However, their performance can be improved by using robust algorithms such as BSIF-SVM and GIST-SVM, as seen from the Table 8.5.
- Among the five feature descriptor algorithms used in this chapter, BSIF and GIST have indicated the highest performance accuracy compared to other algorithms such as LBP, LPQ, and HOG used in this work. The same can be very well observed from the mean and variance plot illustrated in Figure 8.4.

### 8.5.2.2   Fused Band Comparison

To provide the significance of individual band for classification accuracy, we also present the evaluation across three different fusion methods such as IMF, GFF, and 2-DWT for gender prediction. All the three methods were then analysed for the robustness by independently using five different feature extraction methods described above, followed by SVM classifier. Table 8.6 presents the average gender classification accuracy after 10-fold cross-validation. Figure 8.5 illustrates the mean and variance plot describing the classification accuracy of three different fusion methods using five different feature descriptor methods. A similar observation was

**TABLE 8.6**

**Average Gender Classification Accuracy (in %) across Fusion of Bands and Proposed Method, When Training Ocular Samples Belongs to *Without-Glass* and Testing Samples Belongs to *Without-Glass* Category**

| Fusion Method | Algorithm | | | | | |
|---|---|---|---|---|---|---|
| | LBP-SVM | LPQ-SVM | HOG-SVM | GIST-SVM | BSIF-SVM | Proposed Approach |
| IMF | 57.97 | 60.21 | 58.81 | 69.57 | 64.39 | |
| GFF | 58 | 58.38 | 59.65 | 71.-53 | 64.96 | 75.72 |
| 2-DWT | 59.78 | 68.36 | 73.57 | 72.65 | 72.88 | |

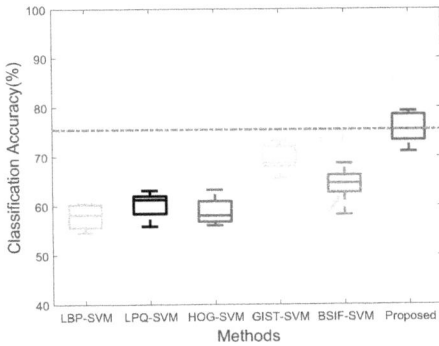

(a) IMF

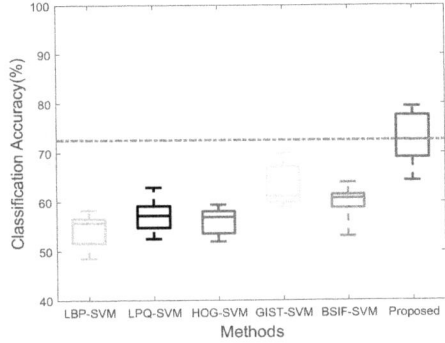

(b) IMF

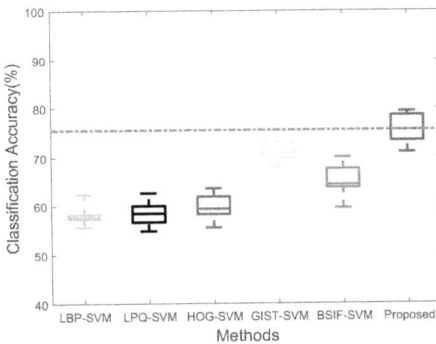

(c) GFF

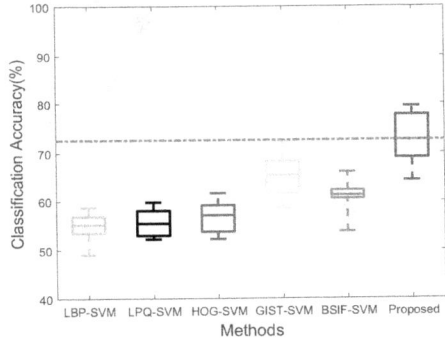

(d) GFF

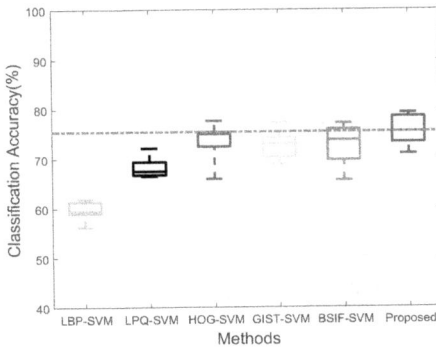

(e) 2-DWT

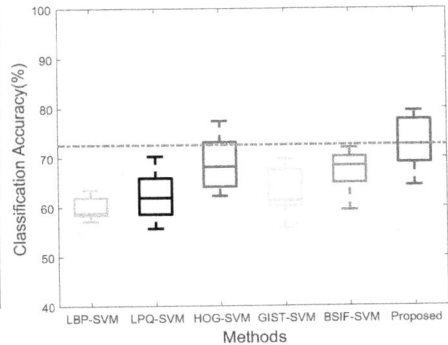

(f) 2-DWT

**FIGURE 8.5**   Average classification accuracy (%) illustrated in terms of mean and variance plot for three different fusion methods such as IMF, GFF, and 2-DWT. From the figure, the results corresponding to (a), (c), (e), represents the classification accuracy related to Without-Glass *v/s* Without-Glass evaluation and the results corresponding to (b), (d), and (f), represents the classification accuracy related to Without-Glass *v/s* With Glass evaluation.

made in terms of average classification across the fusion methods in comparison with the individual band performance, as illustrated in Table 8.5. Further, based on the results obtained, we summarise our specific observations as below:

- The highest average classification accuracy of 73.57% is obtained with the 2-DFT fusion method using the HOG-SVM algorithm, while the lowest average classification accuracy of 57.97% is obtained for the IMF fusion method using the LBP-SVM algorithm.
- Out of three different fusion approaches employed in this work, spectral band fusion based on 2-DFT demonstrates the better performance accuracy across all the five different state-of-the-art feature extraction methods. On the other hand, the fusion methods such as GFF, IMF have shown slightly poor classification accuracy compared to 2-DFT, but with the help of robust algorithms such as GIST, BSIF, their results are also comparable with 2-DFT.

Based on the benchmark results obtained in the above subsections for individual spectral bands and fusion of eight spectral bands, we can present the classification accuracy results based on our proposed approach. Tables 8.5 and 8.6 illustrate the average ocular gender classification accuracy, and Figures 8.4 and 8.5 illustrate the mean-variance plot describing the performance analysis of our proposed approach in comparison with individual spectral band and fusion of spectral bands performed using state-of-the-art gender classification techniques. The proposed approach has outperformed the state-of-the-art feature descriptor methods employed in this work for gender classification. Specifically, the new approach used for gender classification has obtained a maximum of 75.72% average classification accuracy compared to individual band and fusion of bands performance, as seen from Figures 8.4 and 8.5, respectively.

### 8.5.3    EVALUATION 2: WITHOUT-GLASS V/S WITH-GLASS

This section of the chapter details the experimental results based on evaluation protocol discussed in Section 8.5.1, i.e., training the model with multi-spectral ocular instances "Without-Glass" data and tested against the data "With-Glass" for gender classification. The purpose of conducting this set of evaluation is to analyse the effect of eyeglasses on the performance accuracy of the algorithms for ocular gender classification. This evaluation further demonstrates the real-life scenario, when the training and testing datasets are acquired under different environmental conditions to truly signify the robustness of the classification model.

To present our results at the same time to provide reasonable comparison with our proposed approach, we again present the performance analysis of individual eight bands and fusion of bands (based on three different fusion methods, as discussed earlier) independently using five different state-of-the-art feature extraction methods along with SVM classifier for two class gender prediction. Hence, we systematically present the performance of individual bands (Section 8.5.3.1) and fusion of bands performance (Section 8.5.3.2) in the following section, followed by their comparison with our proposed approach.

### 8.5.3.1   Individual Band Comparison

Table 8.7 summarises the average gender classification accuracy obtained after 10-fold cross-validation, and Figure 8.6 illustrates the mean and variance plot describing the classification accuracy. The results have shown a decrease in the overall classification accuracy of individual bands compared to the previous evaluation results indicating the effect of wearing eyeglasses. Based on the classification results obtained, we summarise our specific observation for this category of evaluation as below:

- The maximum average gender classification accuracy obtained across the individual band is 71.04% for 710 nm spectrum band using BSIF feature extractor with SVM classifier. On the other hand, the poor average gender classification accuracy is obtained across the individual band is 52.88% for 530 nm spectrum band using LBP feature extractor with SVM classifier.
- As it can be seen from the evaluation results, poor classification accuracy is obtained between 55%–65% across most of the individual spectrum bands using state-of-the-art feature descriptor and classification technique. This drastic degradation in the performance is due to the presence of eyeglasses in the ocular regions indicating the vulnerability of these algorithm towards the variation in the data.

### 8.5.3.2   Fused Band Comparison

In this section, we combine the individual eight spectral bands into single composite image using the three different fusion methods employed in the previous evaluation. The idea is to present the combined effect of individual bands against the variations in the probe ocular data such as wearing of eyeglasses. Table 8.8 tabulates the average gender classification accuracy, and Figure 8.5 illustrates the mean and variance plot describing the classification accuracy of three fusion methods across five different feature extraction methods. Based on the results obtained, we present major observations in this section as below:

### TABLE 8.7
### Average Gender Classification Accuracy (in %) across Individual Bands and Proposed Method, When Training Ocular Samples Images Belongs to *Without-Glass* and Testing Ocular Sample Images Belongs to *With-Glass* Category

| Algorithm | Spectral Bands | | | | | | | |
|---|---|---|---|---|---|---|---|---|
| | 530 nm | 590 nm | 650 nm | 710 nm | 770 nm | 890 nm | 950 nm | 1,000 nm |
| LBP-SVM | 52.88 | 54.00 | 56.44 | 58.47 | 59.42 | 61.97 | 58.47 | 58.85 |
| •LPQ-SVM | 57.10 | 56.52 | 64.54 | 64.30 | 66.43 | 66.01 | 59.72 | 66.35 |
| •HOG-SVM | 53.27 | 55.03 | 70.26 | 67.17 | 63.51 | 55.80 | 57.60 | 59.70 |
| •GIST-SVM | 63.48 | 66.32 | 66.88 | 66.73 | 64.77 | 63.27 | 64.89 | 64.08 |
| •BSIF-SVM | 60.34 | 64.15 | 69.79 | 71.04 | 70.25 | 68.53 | 67.38 | 67.00 |
| •Proposed approach | | | | 72.50 | | | | |

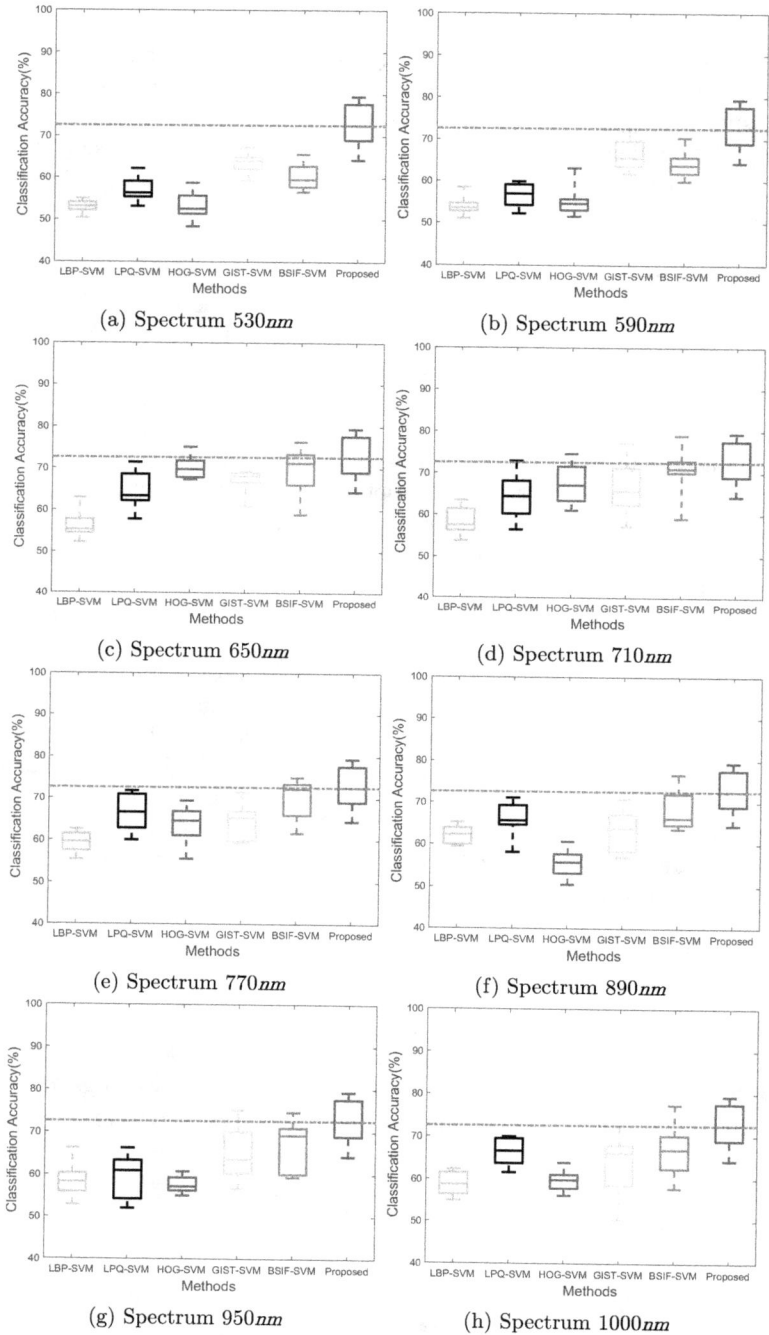

(a) Spectrum 530$nm$

(b) Spectrum 590$nm$

(c) Spectrum 650$nm$

(d) Spectrum 710$nm$

(e) Spectrum 770$nm$

(f) Spectrum 890$nm$

(g) Spectrum 950$nm$

(h) Spectrum 1000$nm$

**FIGURE 8.6**   Average classification accuracy (%) illustrated in terms of mean and variance plot for Without-Glass *v/s* With-Glass evaluation for gender prediction.

**TABLE 8.8**
**Average Gender Classification Accuracy (%) across Fusion of Bands and Proposed Method When Training Ocular Sample Images Belongs to *Without-Glass* and Testing Ocular Sample Images Belongs to *With-Glass* Category**

| | Algorithm | | | | | |
|---|---|---|---|---|---|---|
| Fusion Method | LBP-SVM | LPQ-SVM | HOG-SVM | GIST-SVM | BSIF-SVM | Proposed Approach |
| IMF | 54.23 | 57.25 | 56.75 | 63.4 | 59.79 | |
| GFF | 54.88 | 55.79 | 55.87 | 64.96 | 60.82 | 72.50 |
| 2-DWT | 59.65 | 62.74 | 68.98 | 62.71 | 67.11 | |

- The highest average classification accuracy of 68.98% is obtained with the 2-DFT fusion method using the HOG-SVM algorithm, while the lowest average classification accuracy of 54.23% is obtained for the IMF fusion method using the LBP-SVM algorithm.
- On similar lines with above evaluation (Section 8.5.2.2) conducted for fusion methods, 2-DFT demonstrates the better performance across all the five feature extraction along with SVM classifier compared to GFF and IMF.

It can be clearly seen from both the evaluation (individual bands and the fusion approach), there is degradation in the overall classification accuracy when the ocular sample data employed in the testing set is covered with eyeglasses. However, it is observed that our proposed approach still performs better compared to individual bands and fusion of the bands for this set of evaluation. Specifically, the proposed approach computes 72.50% average classification accuracy, demonstrating its superiority over other methods under varying environmental factors.

To summarise, the gender classification based on multi-spectral imaging collected for the ocular region has demonstrated reasonable classification accuracy presenting the significance of employing the inherent properties of multi-spectral imaging sensors. But performance becomes slightly poor when the ocular instances are covered with eyeglasses, which presents the vulnerability of multi-spectral imaging against the eyeglasses. Further, our proposed approach has shown its significance across both the evaluations (i.e., Without-Glass *v/s* Without-Glass and Without-Glass *v/s* With-Glass).

## 8.6   CONCLUSIONS

Gender prediction plays an important role as a soft label in biometrics. Gender classification based on the ocular region has been investigated recently in this direction. In this chapter, we have investigated a new challenge faced by the ocular biometrics for predicting gender in the presence of eyeglasses. We study the influence of eyeglasses by analysing the data with and without the presence of eyeglasses by capturing the data in multi-spectral sensors. Specifically, we have performed gender classification based on ocular data using 104 unique ocular images captured using multi-spectral sensor data across eight narrow spectrum bands. In order to present the robustness of gender classification under the presence of glasses, we have proposed an approach in which we have selected four discriminative spectral band images based on the highest entropy value to process independently using the bank of Gabor filters to extract the local and global features which we further concatenate to learn the model using ProCRC for gender prediction. Two sets of experimental evaluation are conducted – Without-Glass *v/s* Without-Glass and Without-Glass *v/s* With-Glass to demonstrate the significance of our approach. The obtained results have demonstrated the reasonable performance accuracy with multi-spectral imaging data under the presence of eyeglasses, as compared to the other state-of-the-art methods. Despite the promising results, number of factors such as degraded, unconstrained data, and longitudinal variability of gender classification are not carried out in this work, leaving certain challenges open in this direction.

## ACKNOWLEDGEMENT

This work is supported by the University Grant Commission (UGC), India (Grant No. 40–664/2012(SR) and Research Council of Norway (Grant No. IKTPLUSS 248030/O70).

## REFERENCES

[Computer software manual]. (2016, September).

Amolins, K., Zhang, Y., & Dare, P. (2007). Wavelet based image fusion techniques — an introduction, review and comparison. *{ISPRS} Journal of Photogrammetry and Remote Sensing, 62* (4), 249–263.

Bekios-Calfa, J., Buenaposada, J. M., & Baumela, L. (2011, April). Revisiting linear discriminant techniques in gender recognition. *IEEE Transactions on Pattern Analysis and Machine Intelligence, 33* (4), 858–864.

Bharadwaj, S., Bhatt, H. S., Vatsa, M., & Singh, R. (2010, September). Periocular biometrics: When iris recognition fails. In *2010 Fourth IEEE International Conference on Biometrics: Theory, Applications and Systems (BTAS)*, pp. 1–6.

BioCOP. (Database Available Online). Retrieved from http://biic.wvu.edu/

Bobeldyk, D., & Ross, A. (2016, September). Iris or periocular? exploring sex prediction from near infrared ocular images. In *2016 International Conference of the Biometrics Special Interest Group (BioSig)* pp. 1–7).

Bowyer, K. W., Hollingsworth, K., & Flynn, P. J. (2008). Image understanding for iris biometrics: A survey. *Computer Vision and Image Understanding, 110* (2), 281–307.

Burge, M. J., & Bowyer, K. W. (2013). *Handbook of Iris Recognition.* London: Springer-Verlag.

Cai, S., Zhang, L., Zuo, W., & Feng, X. (2016, June). A probabilistic collaborative representation based approach for pattern classification. In *2016 IEEE Conference on Computer Vision and Pattern Recognition (CVPR)*, pp. 2950–2959.

Castrillón-Santana, M., Lorenzo-Navarro, J., & Ramón-Balmaseda, E. (2016). On using periocular biometric for gender classification in the wild. *Pattern Recognition Letters, 82,* 181–189.

Dalal, N., & Triggs, B. (2005). Histograms of oriented gradients for human detection. In *IEEE Computer Society Conference on Computer Vision and Pattern Recognition, 2005,* Vol. 1, pp. 886–893.

Dame, N. (n.d.). Retrieved from https://cvrl.nd.edu/projects/data/.

Data on optometry and optics in Europe [Computer software manual]. (2017, May).

Dong, Y., & Woodard, D. (2011, 10). Eyebrow shape-based features for biometric recognition and gender classification: A feasibility study. *International Joint Conference on Biometrics*.

Drozdowski, P., Struck, F., Rathgeb, C., & Busch, C. (2018, February). Detection of glasses in near-infrared ocular images. In *2018 International Conference on Biometrics (ICB)*, pp. 202–208.

Gallagher, A. C., & Chen, T. (2009). Understanding images of groups of people. In *2009 IEEE Conference on Computer Vision and Pattern Recognition*, pp. 256–263.

Haghighat, M., Zonouz, S., & Abdel-Mottaleb, M. (2015). Cloudid: Trustworthy cloud-based and cross-enterprise biometric identification. *Expert Systems with Applications, 42* (21), 7905–7916.

Inc., Y. (2010, (Available Online)). *Flickr*. Retrieved from https://www.flickr.com/.

ISO/IEC JTC1 SC37 Biometrics. (2015). Iso/iec 29794-6:2015. information technology – biometric sample quality – part 6:iris image data [Computer software manual].

Jain, Anil K., Nandakumar, Kartik, & Dass, S. C. (2004). *Can Soft Biometric Traits Assist User Recognition?*, *Proc.Spie 5404*, Biometric Technology for Human Identification, (25 August 2004).

Kannala, J., & Rahtu, E. (2012). BSIF: Binarized statistical image features. In *Proceedings of the 21st International Conference on Pattern Recognition (ICPR2012)*, pp. 1363–1366.

Kuehlkamp, A., Becker, B., & Bowyer, K. W. (2017). Gender-from-iris or gender- from-mascara? *2017 IEEE Winter Conference on Applications of Computer Vision (WACV)*, pp. 1151–1159.

Kumari, S., Bakshi, S., & Majhi, B. (2012, 12). Periocular gender classification using global ICA features for poor quality images. *Procedia Engineering, 38,* 945–951.

Lee, K., Lim, S., Lee, K., Byeon, O., & Kim, T. (2001). Efficient iris recognition through improvement of feature vector and classifier. *ETRI Journal, 23,* 61–70.

Li, S., Kang, X., & Hu, J. (2013). Image fusion with guided filtering. *IEEE Transactions on Image Processing, 22* (7), 2864–2875.

Li, S., Kang, X., Hu, J., & Yang, B. (2013). Image matting for fusion of multi-focus images in dynamic scenes. *Information Fusion, 14* (2), 147–162.

Lyle, J. R., Miller, P. E., Pundlik, S. J., & Woodard, D. L. (2010, September). Soft biometric classification using periocular region features. In *2010 Fourth IEEE International Conference on Biometrics: Theory, Applications and Systems (BTAS)*, pp. 1–7.

Lyle, J. R., Miller, P. E., Pundlik, S. J., & Woodard, D. L. (2012). Soft biometric classification using local appearance periocular region features. *Pattern Recognition, 45* (11), 3877–3885.

Marsico, M. D., Nappi, M., Riccio, D., & Wechsler, H. (2015). Mobile Iris CHallenge evaluation (MICHE)-I, biometric iris dataset and protocols. *Pattern Recognition Letters, 57,* 17–23. (Mobile Iris CHallenge Evaluation part I (MICHE I)).

Merkow, J., Jou, B., & Savvides, M. (2010, September). An exploration of gender identification using only the periocular region. In *2010 Fourth IEEE International Conference on Biometrics: Theory, Applications and Systems (BTAS)*, pp. 1–5.

Moeini, H., & Mozaffari, S. (2017, January). Gender dictionary learning for gender classification. *Journal of Visual Communication and Image Representation, 42* (C), 1–13.

Ojala, T., Pietikainen, M., & Maenpaa, T. (2002). Multiresolution gray-scale and rotation invariant texture classification with local binary patterns. *IEEE Transactions on Pattern Analysis and Machine Intelligence, 24* (7), 971–987.

Ojansivu, V., & Heikkilä, J. (2008). *Blur Insensitive Texture Classification Using Local Phase Quantization* (Vol. 5099). In A. Elmoataz, O. Lezoray, F. Nouboud, D. Mammass (Eds.), *Image and Signal Processing*, pp. 236–243. ICISP 2008. Lecture Notes in Computer Science. Berlin, Heidelberg: Springer.

Oliva, A., & Torralba, A. (2001). Modeling the shape of the scene: A holistic representation of the spatial envelope. *International Journal of Computer Vision, 42* (3), 145–175.

Park, U., Ross, A., & Jain, A. K. (2009, September). Periocular biometrics in the visible spectrum: A feasibility study. In *2009 IEEE 3rd International Conference on Biometrics: Theory, Applications, and Systems*, pp. 1–6.

Phillips, P. J., Flynn, P. J., Scruggs, T., Bowyer, K. W., Jin Chang, Hoffman, K., & Worek, W. (2005). Overview of the face recognition grand challenge. In *2005 IEEE Computer Society Conference on Computer Vision and Pattern Recognition (CVPR'05)*, Vol. 1, pp. 947–954.

Phillips, P. J., Hyeonjoon M., Rizvi, S.A., & Rauss, P. J. (2000). The FERET evaluation methodology for face-recognition algorithms. *IEEE Transactions on Pattern Analysis and Machine Intelligence, 22* (10), 1090–1104.

Proença, H., & Alexandre, L. A. (2005). UBIRIS: A noisy iris image database. In *Proceedings of the 13th International Conference on Image Analysis and Processing*, pp. 970–977. Berlin, Heidelberg: Springer-Verlag.

Proença, H., & Alexandre, L. A. (2010). Iris recognition: Analysis of the error rates regarding the accuracy of the segmentation stage. *Image and Vision Computing, 28* (1), 202–206.

Raghavendra, R., Vetrekar, N., Raja, K. B., Gad, R., & Busch, C. (2018, January). Robust gender classification using extended multi-spectral imaging by exploring the spectral angle mapper. In *IEEE International Conference on Identity, Security and Behavior Analysis (ISBA)* pp. 1–8.

Raja, K. B., Raghavendra, R., & Busch, C. (2020). Fused spectral features in kernel weighted collaborative representation for gender classification using ocular images. In B. B. Chaudhuri, M. Nakagawa, P. Khanna, & S. Kumar (Eds.), *Proceedings of 3rd International Conference on Computer Vision and Image Processing*, 131–143. Singapore: Springer Singapore.

Raja, K. B., Raghavendra, R., Vemuri, V. K., & Busch, C. (2015). Smartphone based visible iris recognition using deep sparse filtering. *Pattern Recognition Letters, 57,* 33–42. (Mobile Iris CHallenge Evaluation part I (MICHE I)).

Rattani, A., Derakhshani, R., Saripalle, S. K., & Gottemukkula, V. (2016). ICIP 2016 competition on mobile ocular biometric recognition. In *2016 IEEE International Conference on Image Processing (ICIP)*, pp. 320–324.

Rattani, A., Reddy, N., & Derakhshani, R. (2017). Gender prediction from mobile ocular images: A feasibility study. In *2017 IEEE International Symposium on Technologies for Homeland Security (HST)*, pp. 1–6.

Rattani, A., Reddy, N., & Derakhshani, R. (2018). Convolutional neural networks for gender prediction from smartphone-based ocular images. *IET Biometrics, 7* (5), 423–430.

Santos, G., Grancho, E., Bernardo, M. V., & Fiadeiro, P. T. (2015). Fusing iris and periocular information for cross-sensor recognition. *Pattern Recognition Letters, 57,* 52–59. (Mobile Iris CHallenge Evaluation part I (MICHE I)).

Sequeira, A. F., Monteiro, J. C., Rebelo, A., & Oliveira, H. P. (2014). Mobbio: A multimodal database captured with a portable handheld device. In *2014 International Conference on Computer Vision Theory and Applications (VISAPP)*, Vol. 3, pp. 133–139.

Tapia, J., & Aravena, C. C. (2018, January). Gender classification from periocular NIR images using fusion of CNNS models. In *2018 IEEE 4th International Conference on Identity, Security, and Behavior Analysis (ISBA)*, pp. 1–6.

Tapia, J., Arellano, C., & Viedma, I. (2019a). Sex-classification from cellphones periocular iris images. In A. Rattani, R. Derakhshani, & A. Ross (Eds.), *Selfie Biometrics: Advances and Challenges* (pp. 227–242). Cham: Springer International Publishing.

Tapia, J., Rathgeb, C., & Busch, C. (2019b, May). Sex-prediction from periocular images across multiple sensors and spectra. *arXiv e-prints, arXiv:1905.00396*.

Tapia, J. E., Perez, C. A., & Bowyer, K. W. (2016). Gender classification from the same iris code used for recognition. *IEEE Transactions on Information Forensics and Security, 11* (8), 1760–1770.

Viedma, I., Tapia, J., Iturriaga, A., & Busch, C. (2019). Relevant features for gender classification in NIR periocular images. *IET Biometrics, 8* (5), 340–350.

Vetrekar, N. T., Raghavendra, R., & Gad, R. S. (2016). Low-cost multi-spectral face imaging for robust face recognition. In *2016 IEEE International Conference on Imaging Systems and Techniques (IST)*, pp. 324–329.

Vetrekar, N., Ramachandra, R., Raja, K., & Gad, R. (2018, November). Detecting glass in ocular region based on Grassmann manifold projection metric learning by exploring spectral imaging. In *2018 14th International Conference on Signal-Image Technology Internet-Based Systems (SITIS)*, pp. 106–113.

Vetrekar, N., Raghavendra, R., Raja, K. B., Gad, R. S., & Busch, C. (2017a). Extended multi-spectral imaging for gender classification based on image set. In *Proceedings of the 10th International Conference on Security of Information and Networks*, pp. 125–130.

Vetrekar, N., Ramachandra, R., Raja, K. B., Gad, R. S., & Busch, C. (2017b, December). Robust gender classification using multi-spectral imaging. In *2017 13th International Conference on Signal-Image Technology Internet-Based Systems (SITIS)*, pp. 222–228.

# 9 Investigation of the Fingernail Plate for Biometric Authentication using Deep Neural Networks

*Surabhi Hom Choudhury*
National Institute of Technology Silchar

*Amioy Kumar*
Intel Corporation Bangalore

*Shahedul Haque Laskar*
National Institute of Technology Silchar

## CONTENTS

9.1   Introduction ...........................................................................................206
    9.1.1   Motivation and Scope of Present Work ...........................................206
9.2   Related Work ........................................................................................207
9.3   Sample Acquisition and ROI Extraction..............................................209
    9.3.1   Sample Acquisition.........................................................................209
    9.3.2   ROI Extraction ............................................................................... 210
9.4   Feature Extraction ............................................................................... 211
    9.4.1   Transfer Learning using AlexNet..................................................... 211
    9.4.2   Transfer Learning using ResNet-18 ................................................ 212
    9.4.3   Transfer Learning using DenseNet-201 .......................................... 212
9.5   Multimodal System Design .................................................................. 213
    9.5.1   Score-Level Fusion ......................................................................... 213
    9.5.2   Rank-Level Fusion.......................................................................... 214
        9.5.2.1   Logistic Regression Method ............................................. 214
        9.5.2.2   Mixed Group Rank ........................................................... 214
        9.5.2.3   Inverse Rank Position ....................................................... 215
        9.5.2.4   Nonlinear Weighted Methods............................................ 215
9.6   Experiments, Results, and Analyses...................................................... 216
    9.6.1   Performance of Fingernail Plates in Verification Systems.............. 216

        9.6.1.1   Performance of Fingernail Plates in Unimodal
                  Verification Systems .......................................................... 216
        9.6.1.2   Performance of Fingernail Plates in Multimodal
                  Verification Systems .......................................................... 218
     9.6.2   Performance of Fingernail Plates in Identification Systems............220
        9.6.2.1   Performance of Fingernail Plates in Unimodal
                  Identification Systems ....................................................... 220
        9.6.2.2   Performance of Fingernail Plates in Multimodal
                  Identification Systems ....................................................... 222
9.7   Challenges and Scope of Fingernail Plates in Biometrics...........................234
9.8   Conclusions and Future Scope................................................................... 235
References.......................................................................................................... 235

## 9.1   INTRODUCTION

Biometric traits from the dorsal part of the hand have been subjected to limited investigation, as compared to traits from the palmer part. However, these limited studies have established the dorsal traits as biometrics of immense potential. Significant biometric traits from the dorsal part of the hand that have proved their worth in biometric authentication include geometry of the fingers [1], knuckles [2–5], and hand vein thermograms [6,7]. The fingernail plate, another biometric trait belonging to the dorsal part of the hand, has only been explored very recently for personal authentication [8]. The extremely minuscular research, that the fingernail plate has been subjected to, has provided reasonably appreciable results.

### 9.1.1   MOTIVATION AND SCOPE OF PRESENT WORK

While the nail plate regenerates with the creation of new cells, the spacing amidst the grooves found in the nail bed (as shown in Figure 9.1) has been found to remain proportionally constant throughout the life of a person [9]. Also, nail-ridge patterns illustrate a high extent of individuality, even in the case of identical twins [10].

It is of importance to note that potential trace evidences found in crime investigation sites often include photographs (as shown in Figure 9.2.) and videos of the hand dorsum. These type of evidences which encompass one or more fingernail plate(s) can be suitably processed to match with samples acquired from the person suspected of being the criminal.

A major disadvantage of popular biometrics like fingerprints and palmprints is that people leave these marks on whatever they touch. This renders a biometric system involving one or both of these two traits very susceptible to impersonation. This is not the case with nail plates, and as such these cannot be impersonated very easily. As such, authors have chosen this trait and subjected it to exhaustive investigation in this work, so as to examine the adequacy of the trait in biometric authentication.

Owing to the better performance of multibiometric systems [11,12], and aiming towards enhanced performance, the current work has been largely dedicated to a multimodal design, where one fingernail plate has been subjected to fusion with one or more fingernail plates. Even in case of criminal investigations, trace evidences

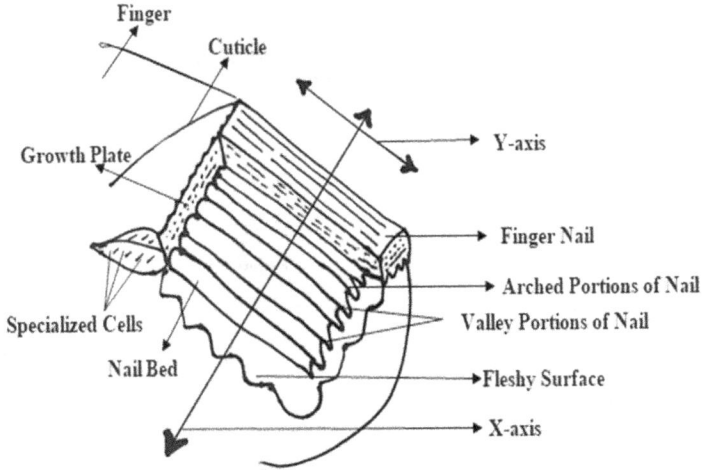

**FIGURE 9.1**   Anatomy of Human Fingernail.

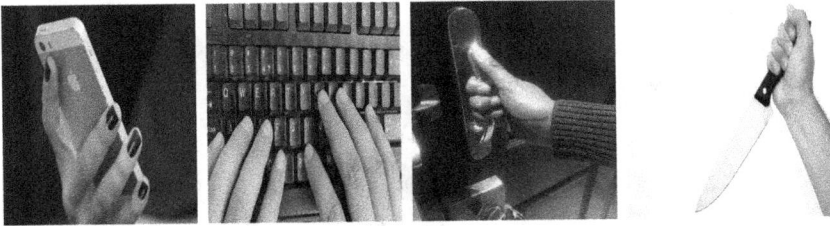

**FIGURE 9.2**   Samples of Trace Evidences encompassing the Fingernail Plate.

are more likely to comprise more than one fingernail plate, as seen in Figure 9.2. Also, nail damage caused by infection [8,13] are likely to reduce the achievable accuracy. Multimodal systems shall largely make up for such adverse results. All possibilities have been considered, and rigorous experiments have been performed to address all such situations and to investigate the nail plate in a multimodal biometric setup. A framework of the overall processing scheme of the proposed work has been provided in Figure 9.3.

## 9.2   RELATED WORK

The preliminary and pioneering works carried out using the fingernail involved the usage of very heavy sensors [14,15]. A variety of equipment like transmitted light comparison microscope, cross polarising filters equipped for polarised light, and acrylic resin were required for the flawless acquisition of nail samples [14]. For sampling, the study carried out in Ref. [15] required a whole range of apparatus including an acousto-optic 2D beam deflector, thermoelectric coolers, master oscillator, a pair of highly monochromatic light sources, and photodiode array or charge-coupled device (CCD) sensor.

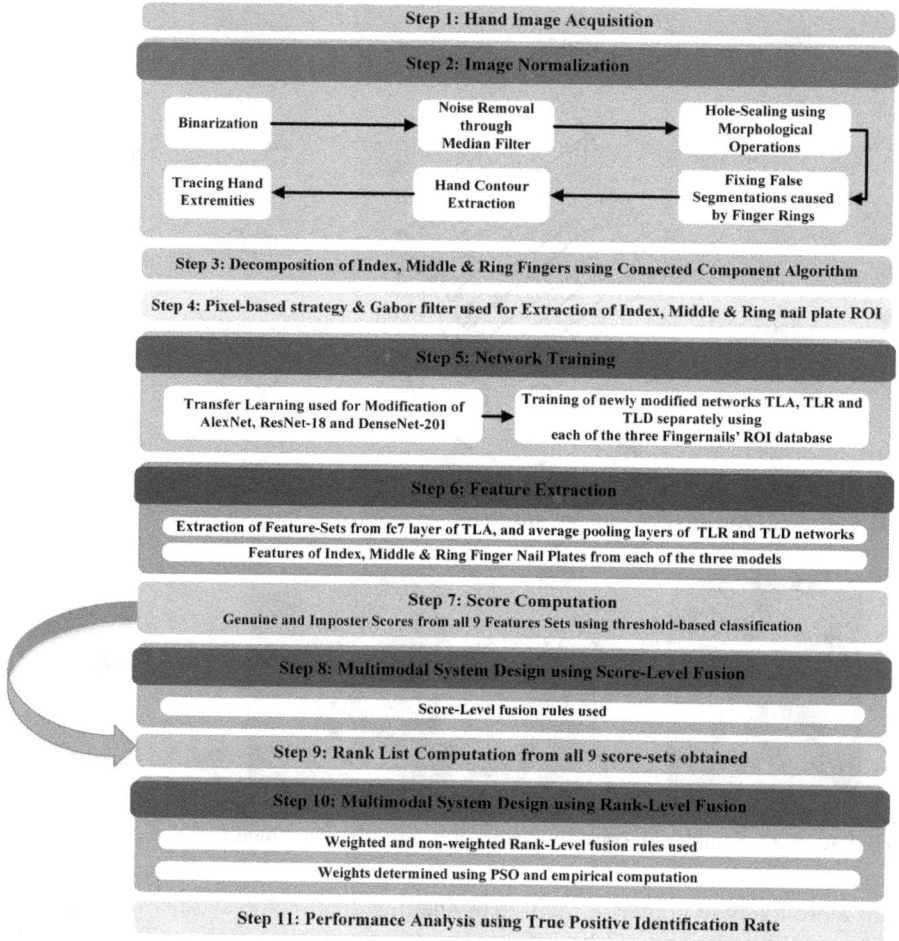

**FIGURE 9.3**  Framework of the Overall Processing Scheme of the Proposed System.

As per the best of the authors' knowledge, the very first work that threw light on the fingernail plate as a biometric trait [16] introduced the Region of Interest (ROI) extraction technique used in this work. However, the reported technique did not remove the grown nail region, a part that provides nothing but redundant information. The mentioned work used low resolution white light for the sample acquisition of fingernail images, Haar Wavelet as the feature extraction technique, and it investigated verification systems. The mentioned work did not explore identification systems. The next work in this domain [17] reported a method to remove the grown nail part. It used Haar Wavelet and Independent Component Analysis (ICA) as the feature extraction techniques. The said work mainly investigated the performance of the nail plate in verification systems; and only explored the same in identification systems at a very preliminary level. Also, the designing of the multibiometric systems adopted in this work was not robust enough. The subsequent work that checked the performance of nail plates in personal authentication of individuals [8] also used Haar Wavelet and ICA as the feature extraction algorithms. It reported designing a multibiometric setup as well, and exploited only a couple of preliminary fusion techniques. However, the limited experimentation gave appreciable results. The first work [5] that has carried out exhaustive experiments under a deep learning framework using a multibiometric fusion of the nail plate and the knuckle reported notable results. This study has used deep learning features exploiting the AlexNet model, and has established the nail plate as a potential biometric identifier in both verification and identification modes. The very promising results reported in the multimodal identification mode opened up avenues to further investigate the nail plate in the light of personal authentication.

The current study explores the fingernail plate further as a biometric trait in personal authentication, and also serves as a counter-narrative to the situation when the framework of Ref. [5] cannot be adopted, which may be caused due to the non-availability of usable and/or acceptable finger knuckle images. The primary objective of this work is to design efficient multimodal systems using the nail plate to check for better efficacy in personal authentication; and also to explore the nail plate under a broader spectrum of deep learning models. A major advantage that any system using the fingernail plate enjoys is its limited chance of impersonation, primarily because the traces of nail plate are not left by any person on anything that she/he touches during day-to-day activities.

## 9.3 SAMPLE ACQUISITION AND ROI EXTRACTION

### 9.3.1 SAMPLE ACQUISITION

A contact-free and peg-free image acquisition setup, as shown in Figure 9.4, has been used in the current work. The imaging setup uses (i) low-cost camera, (ii) wooden box with an aperture at the top to make room for the camera lens, and (iii) fluorescent light fixed below the roof of the box to provide illumination. It is to be noted that the setup

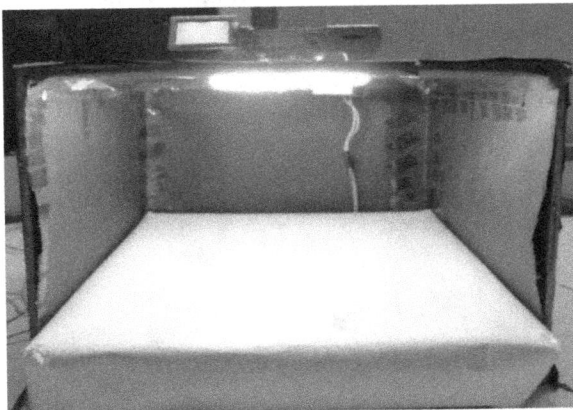

FIGURE 9.4  Image Acquisition Setup used for Constructing Hand Dorsal Image Database.

does not use any peg. For image acquisition, a volunteer needs to place his/her hand in the box without any constraints on the position or orientation of the palm. Thus, the sample acquisition setup can be asserted to be a contactless, unconstrained one.

For this work, a database of 890 images had been formed by collecting five hand dorsal images each from 178 volunteers. Each of the five images was acquired at a time-interval of seven days. Age of these volunteers span between 17 and 65 years. The volunteers included males and females. There were no restrictions imposed on them regarding the use of nail paint or finger-rings.

## 9.3.2 ROI Extraction

In order to obtain reliable ROIs of the nail plates, the images were subjected to a sequence of pre-processing steps, finger normalisation procedure, and ROI extraction techniques. The series of aforesaid steps were followed as is described in Ref. [5], and these steps have been portrayed in the block diagram in Figure 9.3. A sample of the initially captured hand image, and the decomposed and extracted fingers are shown in Figure 9.5 [5]. Figure 9.6 [5] shows the extracted ROIs of the fingernail plate of index, and the middle and ring fingers of the hand sample shown in Figure 9.5.

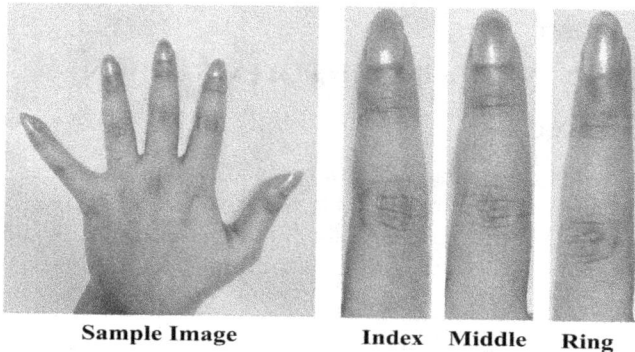

Sample Image          Index   Middle   Ring

**FIGURE 9.5**  Decomposed and Normalised Fingers from Sample Image.

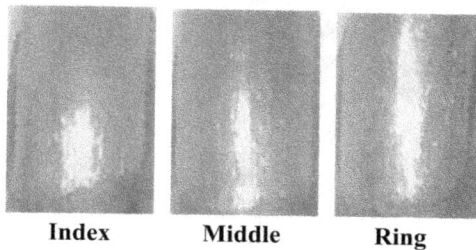

Index          Middle          Ring

**FIGURE 9.6**  ROIs of Fingernail Plates of Index, Middle, and Ring Fingers of the Sample Image in Figure 9.5.

## 9.4   FEATURE EXTRACTION

Deep learning, a distinctive and promising headliner from the domain of machine learning, has validated itself to be immensely advantageous for the purview of computer vision [18,19]. This is owed highly to the fact that this subset of Machine Learning has empowered computers to execute intricate and elaborate perception tasks like image classification, object detection, etc. [20] very effectively.

A colossal volume of labelled data is explored by the deep learning networks to learn about which specific features differentiate the various groups of data, and to consequently form a framework for feature extraction and classification [21]. One of the most phenomenal and favourable characteristics of the pre-trained deep learning models is that such models can be fine-tuned to serve purposes for which they were not trained in the first place. Such a method of fine-tuning a pretrained model to use the knowledge earned and stored while solving one problem, and applying the same to another problem is called Transfer Learning [22]. For such cases, the fine-tuned model itself is capable of serving as the feature extractor. Many such models have carried out different computer vision tasks [23,24] for which the original model was not trained. To address such problems, the last layer of the pretrained model is substituted with a classifier that agrees with the space dimensions of the newly assigned task. If the fine-tuning is carried out well, these models perform efficiently when applied to a new sphere of task.

The current work makes use of three different pre-trained deep learning models, viz. – AlexNet, ResNet-18, and DenseNet-201.

### 9.4.1   Transfer Learning using AlexNet

AlexNet [25] has been originally trained on a subset of the ImageNet database [26], which originally contained more than 15 million annotated images segregated into more than 22,000 categories. AlexNet consists of eight weighted layers; specifically five convolutional layers followed by three fully connected layers (*fc6, fc7, fc8*). The weighted layers are followed by one or more layers like Rectified Linear Units (ReLU) activation function, maxpooling function, Local Response Normalisation (LRN) function, etc. The output of the *fc8* layer is provided to a softmax layer. This capacitates the network to predict what probability the test subject has, of belonging to the different trained classes. Due to the reasonably smaller size of the current database, building a new deep learning network would prove to be unproductive. Thus, Transfer Learning has been opted for.

AlexNet has been suitably fine-tuned, and the newly modified network has been named as Transfer Learning using AlexNet (TLA). All layers of AlexNet, except the last one (namely, *fc8*), have been retained for TLA. A new fully connected (FC) layer and a softmax layer are added to the retained set of layers. The new FC layer has been taken to be of size equal to the number of classes (users) in the concerned database. In case of the fingernail plates' database used in this work, the number of classes is 178. Transfer learning requires slow learning over the layers that are retained and fast learning over the new layers. In order to warrant fast training over the newly added layers, the bias and weight learning rates are multiplied with a high value of 20 in the new FC layer. Authors have zeroed down on this value of '20' through empirical computation and found it to deliver optimum results. To ensure that the learning process is slow over the retained layers, the initial learning rate has been kept low (0.0001). The newly formed TLA network has been trained over the images of the fingernail plate database. Three TLA feature-sets pertaining to the index, middle, and ring fingernail plates have been extracted from the fully connected '*fc7*' layer of the trained TLA network.

### 9.4.2 Transfer Learning using ResNet-18

Since the advent of the AlexNet [25], a number of deeper Convolutional Neural Networks have been introduced. However, it was seen that increasing the network depth just by stacking layers often saturates or degrades performance accuracy. This was because of the vanishing gradient problem. Residual Neural Networks, or ResNets [27], demonstrated that the vanishing gradient problem can be tackled by splitting a very deep network into smaller blocks, which were inter-connected through skip connections.

ResNet-18 is a network trained on a section of the ImageNet database. This network is composed of five convolutional layers, which are superseded by an average pooling layer and a FC layer. The weighted layers are followed by layers like ReLU function, maxpooling function, etc.

Transfer Learning using ResNet-18 (TLR) has been used as one of the feature extraction techniques in this work. Except for the last FC layer, all layers of ResNet-18 are retained for TLR. A fresh FC layer is added, which has a size equal to the number of classes (178) of the current database. Transfer learning demands slow learning over the detained layers and fast learning over the fresh layers. With a view to ensure faster training over the new layer, the learning rate factors of the newly added layer are set at 20. Feature-sets of the index, middle and ring fingernail plates have been extracted from the average pooling layer of the modified deep learning model.

### 9.4.3 Transfer Learning using DenseNet-201

Densely Connected Convolutional Networks, or DenseNets, are popular as a logical extension of the ResNets. DenseNets concatenate outputs from previous layers, whereas ResNets sum them up. Major advantages of the DenseNets are that they diminish the vanishing gradient problem stated in Section 9.4.2, reinforce propagation of feature, promote reuse of feature, and reduce the number of parameters noticeably.

The DenseNet-201 [28] is formed of five dense blocks, where each dense block consists of a $1 \times 1$ convolutional layer for downsampling, followed by a $3 \times 3$ convolutional layer. The dense blocks are followed by an average pooling layer and a FC layer. There is one transition layer between every two dense blocks, which is made up of a $1 \times 1$ convolutional layer followed by a $2 \times 2$ average pooling layer. Every dense block follows the sequence: Batch Normalisation-ReLU-Convolution.

Transfer Learning using DenseNet-201 (TLD) is the third feature extraction technique used in this work. All layers of the DenseNet-201 have been retained,

**TABLE 9.1**
**Hyperparameters used in the Implemented Deep Learning Models**

| Model | Hyperparameters | | |
|-------|-----------------|---|---|
|       | Momentum | Initial Learning Rate | Mini-Batch Size |
| TLA | 0.9 | 0.0001 | 5 |
| TLR | 0.9 | 0.0003 | 10 |
| TLD | 0.9 | 0.0003 | 10 |

except the final FC layer. This FC layer has been replaced in the TLD with a new FC layer which has 178 number of outputs in tune with the current database. Similar to TLR, in TLD too, the learning rate factors of the new FC layer have been fixed to 20. The required feature-sets from the TLD model have been extracted from its average pooling layer.

Table 9.1 enlists the important hyperparameter settings of the above mentioned models.

## 9.5 MULTIMODAL SYSTEM DESIGN

Multimodal biometric systems designed to manage access to sheltered assets and information have been found to offer a better deal of security and user-convenience. Multimodal biometric setups handle frameworks which pursue the organisation or synchronisation of the usage of different biometric traits in a way that advances the method of information fusion [29]. In addition, such systems also cater to situations where one or more of the nail plates might not be accessible, rendering the investigation feasible only if it can be carried out using the available traits.

### 9.5.1 SCORE-LEVEL FUSION

For the work reported in this brief, four well-accepted score-level fusion rules have been used to check the efficacy of the proposed system in the verification mode, namely the sum rule, the product rule, the max rule, and the min rule.

If $S_1$ and $S_2$ are the scores from two different biometric traits, then their fused scores using the above mentioned rules are given as follows:

$$\text{Sum Rule: Score Sum} = \text{sum}(S_1, S_2) \tag{9.1}$$

$$\text{Product Rule: Score Product} = \text{product}(S_1, S_2) \tag{9.2}$$

$$\text{Max Rule: Score Max} = \text{max}(S_1, S_2) \tag{9.3}$$

$$\text{Min Rule: Score Min} = \text{min}(S_1, S_2) \tag{9.4}$$

## 9.5.2 RANK-LEVEL FUSION

Rank-level fusion amalgamates ranking lists procured from different biometric traits for deriving a final ranking list, in order to aid in the process of arriving at the final decision [30]. Also, certain systems provide scores or features, inappropriate for fusion [31]. In those cases as well, rank-level fusion is a very feasible choice for building multimodal systems [32]. Optimal performance accuracy in a multimodal system is achieved when the different traits under fusion are given appropriate weightage or importance. In this work, the following linear and nonlinear fusion rules have been put to use for rank-level fusion:

### 9.5.2.1 Logistic Regression Method

The Logistic Regression method [33] may be considered as an important tool for combining classifiers having non-uniform performances. The Borda count method [30] evaluates the fused score as the sum of rank scores of all the considered classifiers. The final ranking list for this method is achieved by sorting the fused scores. The method works under the assumption that all classifiers perform equally well. Such an assumption makes the system extremely vulnerable to weak classifiers. The Borda count method requires substantial modification when a combination of classifiers having varying performances is considered. Such modification demands the assignment of weights to the rank scores of each classifier, where the assigned weights reflect the relative importance of each classifier from the perspective of their rank-level fusion. Let the probability of getting the true class be $P(Y = 1 \mid x)$, and let it be denoted by $\pi(x)$ where $x = (x_1, x_2, \ldots, x_m)$ corresponds to the rank scores assigned to that class by classifiers $C_1, C_2, \ldots, C_m$. If it is assumed that $x_i$ has the largest value and if the class is ranked at the top by $C$, then

$$log \frac{\pi(x)}{1 - \pi(x)} = \frac{exp(\alpha + \beta_1 x_1 + \beta_2 x_2 + \ldots + \beta_m x_m)}{1 + exp(\alpha + \beta_1 x_1 + \beta_2 x_2 + \ldots + \beta_m x_m)} \qquad (9.5)$$

Here, $\alpha$, $\beta = (\beta_1, \beta_2, \ldots, \beta_m)$ are constant parameters. $log \frac{\pi(x)}{1 - \pi(x)}$ is called the logit, and it is linearly related to $x$. These constants can be used as weights for the rank scores of each classifier. This is because the relative magnitudes of these constants signify the marginal contribution of the classifiers to the logit. As such, the fused rank $R_{LR}$ is given by Equation (9.6) as follows:

$$R_{LR} = \beta_1 x_1 + \beta_2 x_2 + \ldots + \beta_m x_m \qquad (9.6)$$

In this work, the aforementioned constants have been computed by two methods, viz. one by empirical computation and the other by Particle Swarm Optimisation (PSO) [34] technique.

### 9.5.2.2 Mixed Group Rank

The Mixed Group Rank method [35] makes use of the classical Highest Rank (minimum value amongst all rank scores) and the Logistic Regression method.

The Mixed Group Rank method makes a linear weighted combination of the minimum ranks of all the possible subgroups in the considered group of matchers. The final fused rank is given by

$$R_{\text{MGR}} = \sum_{G \subseteq \{1,\dots,N\}} \left[ -\omega_G \min \{ r_j : j \in G \} \right] \tag{9.7}$$

Here, $G$ is the subgroup of matchers belonging to the entire group of matchers, $r_j$ is the rank assigned to user $j$, $\omega_G$ represents the weight assigned to the concerned subgroup of matchers, $N$ is the total matchers used. For this study where three matchers have been used, $i.e.$, the index, middle, and ring fingernail plates, Equation (9.7) shall be represented as follows:

$$R_{\text{MGR}} = -\omega_1 r_1 - \omega_2 r_2 - \omega_3 r_3 - \omega_{12} \min (r_1, r_2) - \omega_{23} \min (r_2, r_3)$$

$$- \omega_{31} \min (r_3, r_1) - \omega_{123} \min (r_1, r_2, r_3) \tag{9.8}$$

For the Mixed Group Rank method, all concerned weights have been evaluated through PSO and empirical computation.

### 9.5.2.3  Inverse Rank Position

The Inverse Rank Position algorithm [36] uses the inverse of the sum of inverses of all rank scores for every matcher, and the final fused rank is given by

$$R_{\text{IRP}} = \frac{1}{\displaystyle\sum_{x=1}^{N} r_x (f)} \tag{9.9}$$

Here, $r_x (f)$ is the rank assigned to user $f$ by the $x^{\text{th}}$ matcher, and $N$ is the total number of matchers used.

### 9.5.2.4  Nonlinear Weighted Methods

Nonlinear methods [30] for rank-level fusion have been used in this work. The ranked list of user identities are nonlinearly weighted and consolidated as follows:

$$\text{Hyperbolic Tangent Method:} C_p = \sum_{i=1}^{N} \tanh \left( \omega_i \, r_i(p) \right) \tag{9.10}$$

$$\text{Weighted Exponential Method:} C_p = \sum_{i=1}^{N} \omega_i \, \exp \left( \omega_i \, r_i(p) \right) \tag{9.11}$$

Here, $r_i(p)$ is the rank assigned to candidate $p$ by the $i^{\text{th}}$ matcher, $\omega_i$ represents the weights assigned to the $i^{\text{th}}$ matcher, and $C_p$ is the fused rank. For these non-linear matchers too, computation of weights has been performed via PSO, in addition to empirical computation [30].

## 9.6  EXPERIMENTS, RESULTS, AND ANALYSES

The experimentation in this work requires the nail plate ROI extraction from the index, middle, and ring fingers of all the hand images in the database of 890 hand dorsal images. Thus, 2,670 (178 volunteers × 5 images each × 3 fingers) ROIs have been extracted in total.

The TLA, TLR, and TLD feature-sets have been extracted from the all the three fingernail plate ROI databases. It is to be noted that for any kind of processing, the images are required to be sized as per the pre-defined image input size of the respective deep learning models. In accordance with the same, the images have been resized to $227 \times 227 \times 3$ for the TLA network and to $224 \times 224 \times 3$ for both the TLR and the TLD networks.

Choosing Euclidean distance as the similarity measure, and the training to test ratio as 4:1, the genuine and imposter scores have been computed from the features, which are then used to check the performance of the traits in verification systems. For analysing the performance of fingernail plate in identification systems, the corresponding ranking lists for all the three nail plates are obtained by sorting the scores.

### 9.6.1  PERFORMANCE OF FINGERNAIL PLATES IN VERIFICATION SYSTEMS

#### 9.6.1.1  Performance of Fingernail Plates in Unimodal Verification Systems

In this section, performances of the index, middle, and ring fingernail plates in unimodal verification systems have been comparatively analysed for each of the three deep learning models considered. Figure 9.7 compares the verification performance in the form of Receiver Operating Characteristic (ROC) curves of the three fingernail plates using TLA as the feature extraction technique. Similar comparative depiction of the nail plates for the TLR and TLD feature-sets are given in Figures 9.8 and 9.9, respectively.

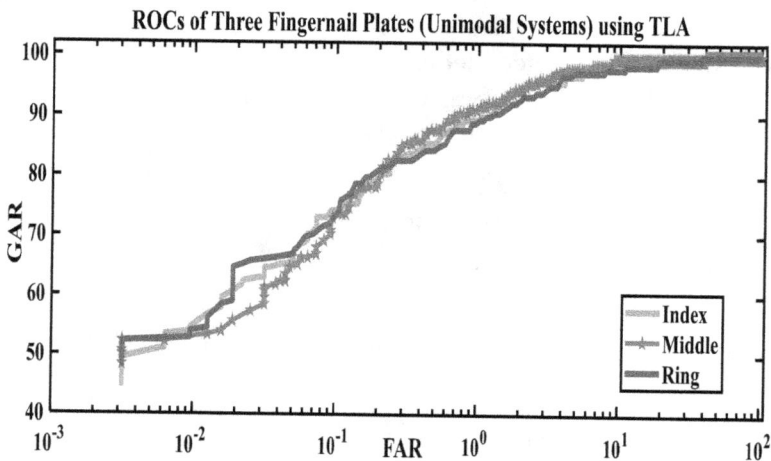

**FIGURE 9.7**  ROCs comparing the performance of Index, Middle, and Ring Fingernail Plates in Unimodal Verification Systems using TLA as the feature extraction technique.

Figure 9.7 demonstrates that verification systems built from either of the three nail plates shall give similar and appreciable results at all levels of security, when the TLA feature-sets are being used.

Figure 9.8 portrays that the index and middle fingernail plates outperform the ring fingernail plate in terms of verification performance when TLR feature-sets are being used.

Figure 9.9 shows that the verification systems comprising either the index or the ring fingernail plate shall give very good results when the TLD feature-sets are being used. It is evident from the figure that the middle fingernail plate also gives appreciable results, but lags behind the other two nail plates.

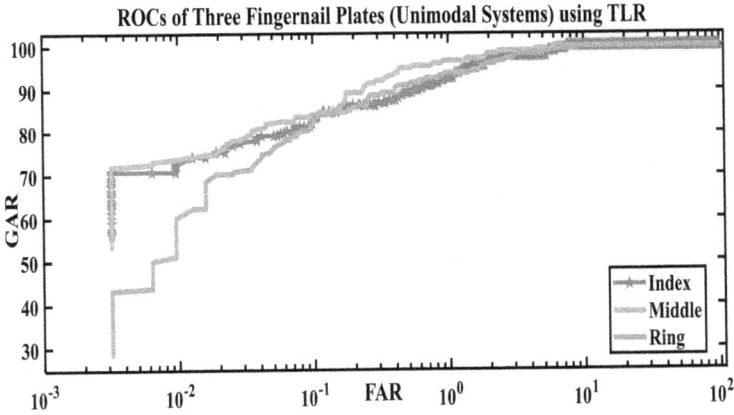

**FIGURE 9.8** ROCs comparing the performance of Index, Middle, and Ring Fingernail Plates in Unimodal Verification Systems using TLR as the feature extraction technique.

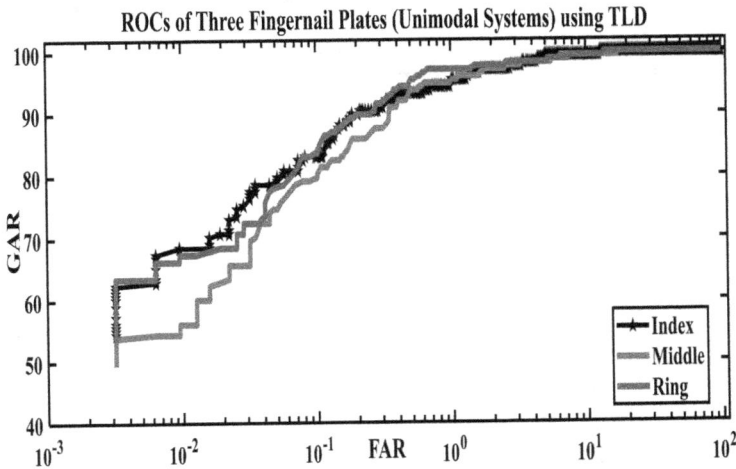

**FIGURE 9.9** ROCs comparing the performance of Index, Middle, and Ring Fingernail Plates in Unimodal Verification Systems using TLD as the feature extraction technique.

**TABLE 9.2**

**GAR (in %) at FAR = 0.012% for Various Unimodal Verification Biometric Systems built using Index, Middle, and Ring Fingernail Plates[a]**

| Model | Trait | | |
|-------|-------|--------|------|
|       | Index | Middle | Ring |
| TLA   | *56.74* | 53.37 | 56.18 |
| TLR   | **74.16** | ***74.16*** | 62.36 |
| TLD   | *68.53* | 60.11 | 67.41 |

[a] Numbers in bold and italics signify the best performance across one or the other parameter.

The values of Genuine Acceptance Rate (GAR) at False Acceptance Rate (FAR) = 0.012%, obtained from all the three nail plates from all the three deep learning models are tabulated in Table 9.2.

Analyses of the particular results tabulated in Table 9.2 show that amongst all the three nail plates considered, the best verification performance is provided by the index fingernail plate, for each of the three feature-sets. Table 9.2 also shows that TLR gives the best performance followed by TLD and TLA in the mentioned sequence at FAR = 0.012%.

### 9.6.1.2 Performance of Fingernail Plates in Multimodal Verification Systems

In this section, fusion of the scores obtained from all the three fingernail plates has been carried out using the four fusion rules – sum, product, min, and max – as detailed in the Section 9.5.1.

At the very outset, the fusion of TLA scores obtained from all three nail plates has been implemented. Figure 9.10 depicts the verification performance of the same

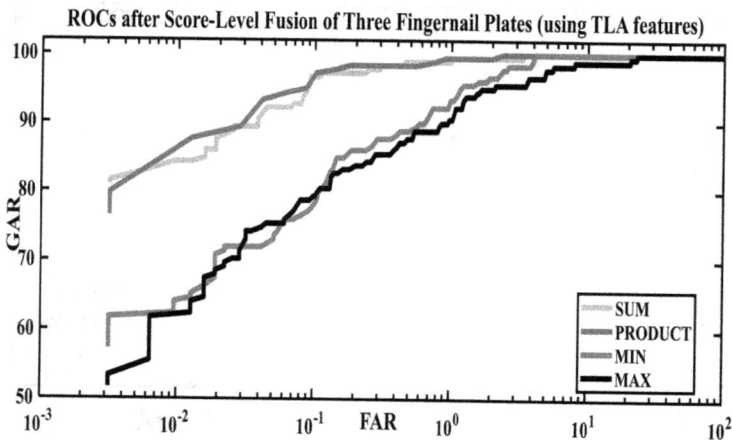

**FIGURE 9.10** ROCs after Score-Level Fusion of scores from Index, Middle, and Ring Fingernail Plates using TLA feature-sets and four different score-level fusion rules.

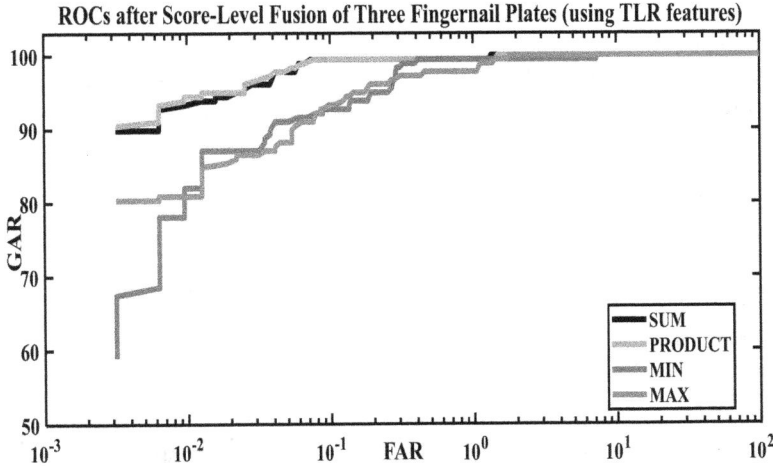

**FIGURE 9.11**  ROCs after Score-Level Fusion of scores from Index, Middle, and Ring Fingernail Plates using TLR feature-sets and four different score-level fusion rules.

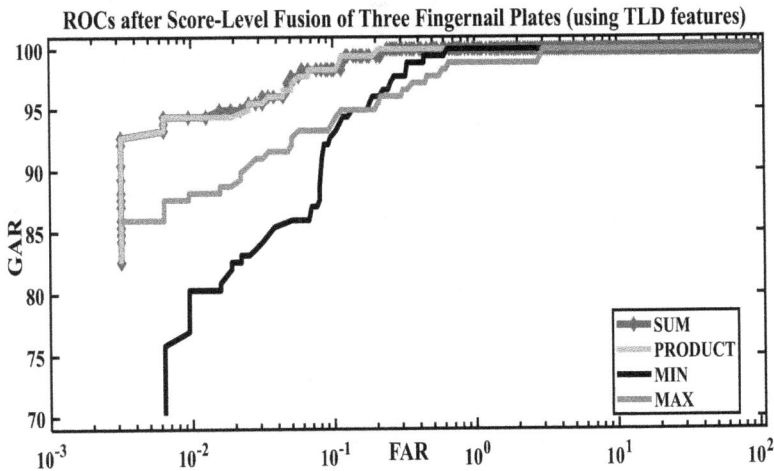

**FIGURE 9.12**  ROCs after Score-Level Fusion of scores from Index, Middle, and Ring Fingernail Plates using TLD feature-sets and four different score-level fusion rules.

in the form of ROCs. Similar performances after score-level fusion of TLR and TLD based scores are given in Figures 9.11 and 9.12, respectively.

Analysing the results in Figures 9.10–9.12 and comparing them with that of unimodal verification systems reported in Section 9.6.1.1 shows that the multimodal systems outperform their unimodal counterparts with all each of the four rules used. These results also show that for the chosen traits, the 'product rule' fusion method performs better than the other three score-level fusion methods for almost all operating points.

**TABLE 9.3**

**Comparison of GAR (in %) at FAR = 0.012% of Various Multimodal Verification Biometric Systems with that of Unimodal Verification Systems**[a]

| Model | Rule/Trait | | | | | | |
|-------|-----|---------|-----|-----|-------|--------|------|
|       | Sum | Product | Min | Max | Index | Middle | Ring |
| TLA | 84.27 | *87.64* | 65.17 | 64.04 | 56.74 | 53.37 | 56.18 |
| TLR | 93.82 | ***94.94*** | 82.02 | 84.83 | 74.16 | 74.16 | 62.36 |
| TLD | *94.38* | *94.38* | 80.34 | 88.20 | 68.53 | 60.11 | 67.41 |

[a] Numbers in bold and italics signify the best performance across one or the other parameter.

Table 9.3 makes a comparison of the values of GAR obtained at FAR = 0.012% after score-level fusion with that of the unimodal fingernail plate systems.

The results tabulated in Table 9.3 show that the best verification performance is obtained from the fusion of TLR features-based scores, followed by TLD and TLA in the mentioned sequence. This is the same sequence of performance that is observed in the unimodal verification systems. Table 9.3 also shows that the highest GAR at FAR = 0.012% is given by the product rule. However, the sum rule also performs well and even equals the performance given by the product rule in the case where the TLD match scores are fused.

Thus, it may be said that appreciable verification performance can be achieved by the combination of index, middle, and ring fingernail plates.

However, there are situations where score-level fusion is not feasible or not practiceable. Moreover, certain circumstances like criminal investigation demand for the authentication to be carried out in identification mode. Keeping the same in mind, the next section analyses the performance of fingernail plates under rank-level fusion.

### 9.6.2 Performance of Fingernail Plates in Identification Systems

For identification systems, the performance may be evaluated using the parameter *True Positive Identification Rate (TPIR)*. TPIR is the proportion of times the identity determined by the system is actually the true identity of the person who is providing the biometric sample [37]. If the biometric system provides the identities of the top $x$ matches, the Rank – $x$ TPIR, $R_x$, is the proportion of times the true identity of the individual is contained in the top $x$ matching identities. For performance analysis of an identification system, the TPIR at various ranks may be depicted in the *Cumulative Match Characteristics (CMC)* curve where TPIR, $R_x$, are plotted against Rank $x = 1, 2, \ldots N$ where $N$ is the total number of people enrolled in the database.

### 9.6.2.1 Performance of Fingernail Plates in Unimodal Identification Systems

In this section, performances of the three fingernail plates in unimodal identification systems have been analysed for each of the three deep learning models – TLA, TLR, and TLD. Figure 9.13 makes a comparison of the performance of the index, middle, and ring fingernail plate using the TLA feature-set. Similar comparative performance portrayal of the three fingernail plates are depicted in Figures 9.14 and 9.15 for the TLR and TLD feature-sets, respectively.

It is observed from Figures 9.13–9.15 that identification systems built from either the index or the middle fingernail plate shall provide good performance. The ring fingernail plate also fares well, but lags in performance behind the other two nail plates. Analysing the results depicted in these three figures also shows that for all the three nail plates, the best identification accuracy is given by TLD, followed by TLR and TLA, in that sequence. This, of course, is because of the depth of the respective models. Also, for all the three nail plates, TLD provides a high identification accuracy of above 93%. The other two models, TLA and TLD, also fare reasonably well.

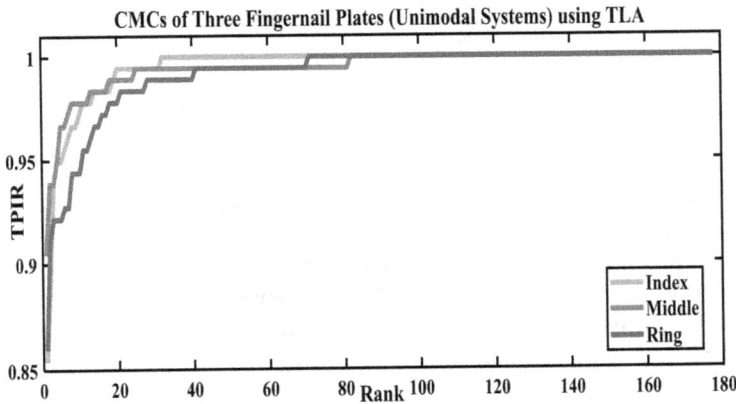

FIGURE 9.13  CMCs comparing the Performance of Index, Middle, and Ring Fingernail Plates in Unimodal Identification Systems using TLA as the feature extraction technique.

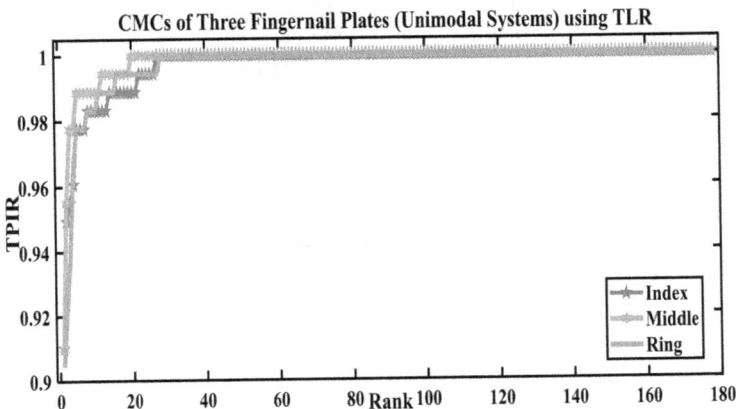

FIGURE 9.14  CMCs comparing the Performance of Index, Middle, and Ring Fingernail Plates in Unimodal Identification Systems using TLR as the feature extraction technique.

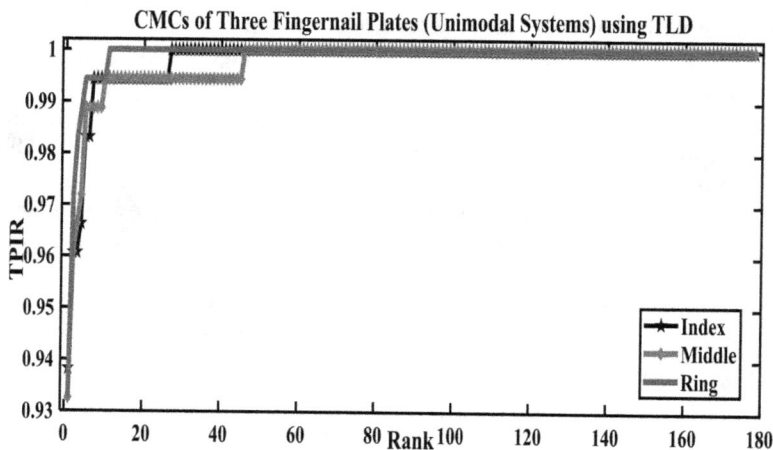

**FIGURE 9.15**  CMCs comparing the Performance of Index, Middle, and Ring Fingernail Plates in Unimodal Identification Systems using TLD as the feature extraction technique.

Table 9.4 tabulates the values of Rank-1 TPIRs obtained from all the three nail plates from the three deep learning models considered.

The lowest Rank-1 TPIR is obtained from the index nail plate when the TLA feature-set is used; and that is as high as 85.39%. Table 9.4 also demonstrates that the best Rank-1 identification performance is provided by the middle fingernail plate for TLA, by the index and the middle fingernail plates for TLR, and by index fingernail plate for TLD.

### 9.6.2.2   Performance of Fingernail Plates in Multimodal Identification Systems

In this section, the index, middle, and ring fingernail plates have been subjected to different frameworks of rank-level fusion. For all the experiments conducted, the performance of rank-level fusion of these three nail plates has been analysed for all the three deep learning models: TLA, TLR, and TLD.

**TABLE 9.4**

**Rank-1 TPIR (in %) for Various Unimodal Identification Biometric Systems built using Index, Middle, and Ring Fingernail Plates[a]**

| Model | Trait | | |
|---|---|---|---|
| | Index | Middle | Ring |
| TLA | 85.39 | *90.45* | 85.95 |
| TLR | *91.01* | *91.01* | 90.45 |
| TLD | ***93.82*** | 93.26 | 93.26 |

[a]  Numbers in bold and italics signify the best performance across one or the other parameter.

### 9.6.2.2.1 Experiment A

For the first set of experiments under this section, the three ranking lists of all the three fingernail plates have been fused separately for the three models: TLA, TLR, and TLD. In this experiment, fusion has been performed using two different linear, weighted fusion methods, namely the Logistic Regression and the Mixed Group Rank. Here, the weights have been chosen via empirical computation. Figure 9.16 compares the performance of the nail plates after rank-level fusion through Logistic Regression and Mixed Group Rank, where the TLA feature-set is used. Similar comparative performances of rank-level fusion for the TLR and TLD models are shown in Figures 9.17 and 9.18, respectively.

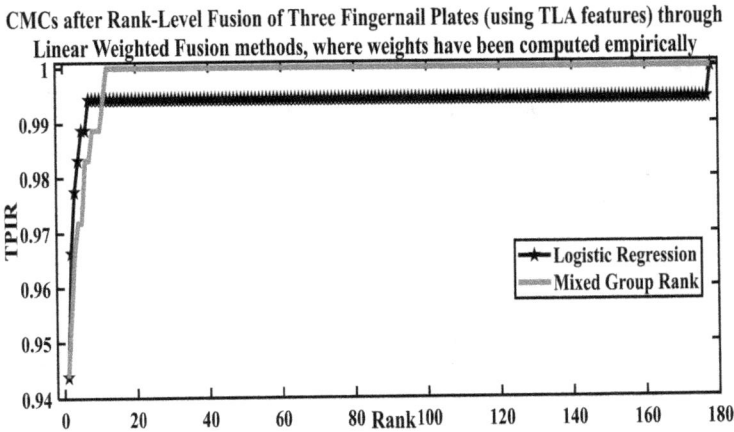

**FIGURE 9.16** CMCs after Rank-Level Fusion of all three Fingernail Plates for TLA feature-set using Logistic Regression and Mixed Group Rank, where all weights have been computed empirically.

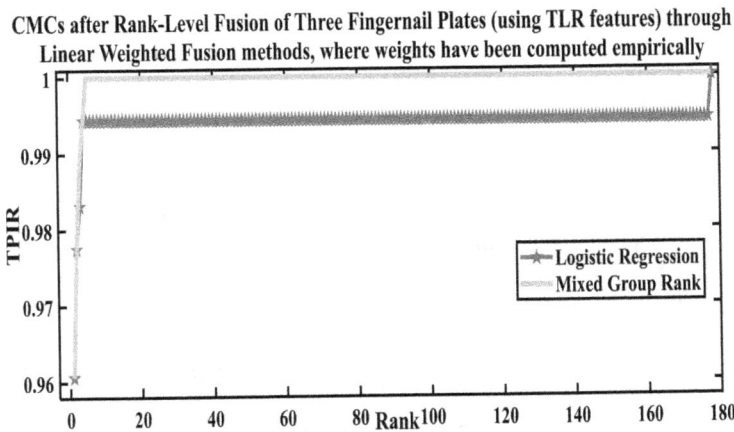

**FIGURE 9.17** CMCs after Rank-Level Fusion of all three Fingernail Plates for TLR feature-set using Logistic Regression and Mixed Group Rank, where all weights have been computed empirically.

**CMCs after Rank-Level Fusion of Three Fingernail Plates (using TLD features) through Linear Weighted Fusion methods, where weights have been computed empirically**

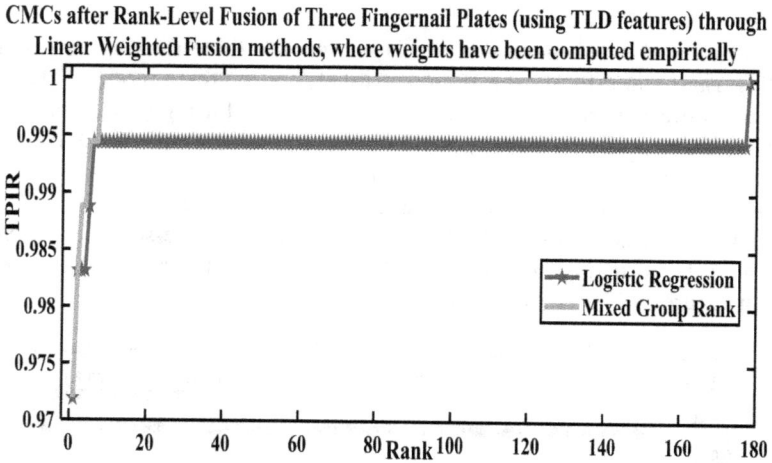

**FIGURE 9.18** CMCs after Rank-Level Fusion of all three Fingernail Plates for TLD feature-set using Logistic Regression and Mixed Group Rank, where all weights have been computed empirically.

Figures 9.16–9.18 establish that an identification system built from the index, middle, and ring fingernail plates gives good performance accuracy, where the TLD-based results are better than those based on TLR and TLA. Analyses of the results depicted in Figures 9.16–9.18 also bring out an interesting point. The corresponding Rank-1 TPIR values are the same for all three models, when Logistic Regression and Mixed Group Rank are the chosen fusion rules. However, with increasing ranks, the TPIR values increase in two different trends for the two fusion methods. While for all three deep learning models, 100% TPIR is achieved at Rank-12 for the Mixed Group Rank rule, the same is achieved by using the Logistic Regression method only at the final rank, *i.e.*, Rank 178.

A tabular depiction of the TPIR values at a few selected ranks is given in Table 9.5 to illustrate the aforementioned trend of results.

*9.6.2.2.2 Experiment B*

The weights for the three different fingernail plates in Experiment A have been assigned through empirical computation. With an aim to further improve accuracy, the exact experiments under Experiment A have been repeated by computing weights using PSO. This has been done to ensure optimal weight assignment to the three nail plates, and thus to provide better results. In Figure 9.19 the performance of the nail plates is depicted when their corresponding TLA feature-sets are subjected to rank-level fusion using Logistic Regression and Mixed Group Rank methods. Similar performances are depicted in Figures 9.20 and 9.21 when TLR and TLD feature-sets are used. All the three aforementioned figures show that when PSO is used for the computation of weights, the results improve for each of the cases, with the best Rank-1 accuracy being provided by the TLD based ranking lists for the Mixed Group Rank rule.

**TABLE 9.5**
**True Positive Identification Rates (in %) of the Multimodal Systems where Nail Plates of All Three Fingers are fused using Two Linear Weighted Fusion Methods (Weights Computed Empirically)**

### Logistic Regression Method

| Rank →<br>Model ↓ | 1 | 2 | 3 | 8 | 178 |
|---|---|---|---|---|---|
| TLA | 94.38 | 96.63 | 98.88 | 99.44 | 100 |
| TLR | 96.07 | 97.75 | 99.44 | 99.44 | 100 |
| TLD | 97.19 | 98.32 | 99.44 | 99.44 | 100 |

### Mixed Group Rank Method

| Rank →<br>Model ↓ | 1 | 2 | 3 | 8 | 12 |
|---|---|---|---|---|---|
| TLA | 94.38 | 95.51 | 96.63 | 98.88 | 100 |
| TLR | 96.07 | 97.75 | 98.88 | 100 | 100 |
| TLD | 97.19 | 98.32 | 98.88 | 100 | 100 |

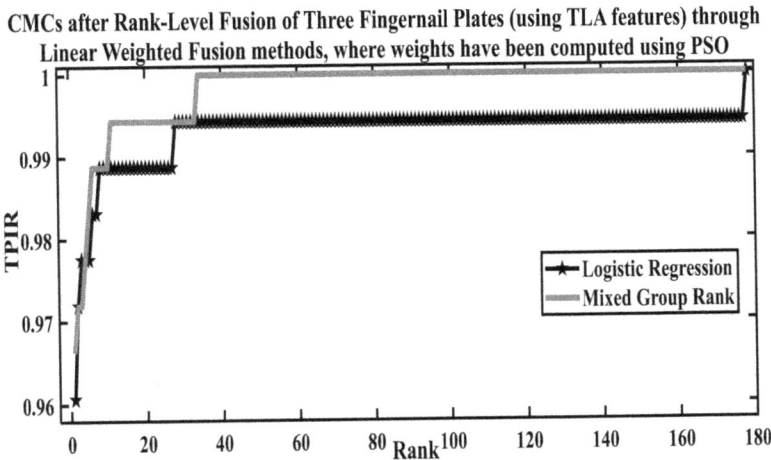

CMCs after Rank-Level Fusion of Three Fingernail Plates (using TLA features) through Linear Weighted Fusion methods, where weights have been computed using PSO

**FIGURE 9.19** CMCs after Rank-Level Fusion of all three Fingernail Plates for TLA feature-set using Logistic Regression and Mixed Group Rank, where all weights have been computed using PSO.

Figures 9.19–9.21 show that the three fingernail plates can be combined to build a reliable identification system. Systems where the fusion is performed using Logistic Regression give appreciable performance. However, the systems which employ Mixed Group Rank outdo the former.

Table 9.6 portrays the improvement in Rank-1 TPIR values obtained in this set of experiments when the weights are optimised using PSO.

CMCs after Rank-Level Fusion of Three Fingernail Plates (using TLR features) through
Linear Weighted Fusion methods, where weights have been computed using PSO

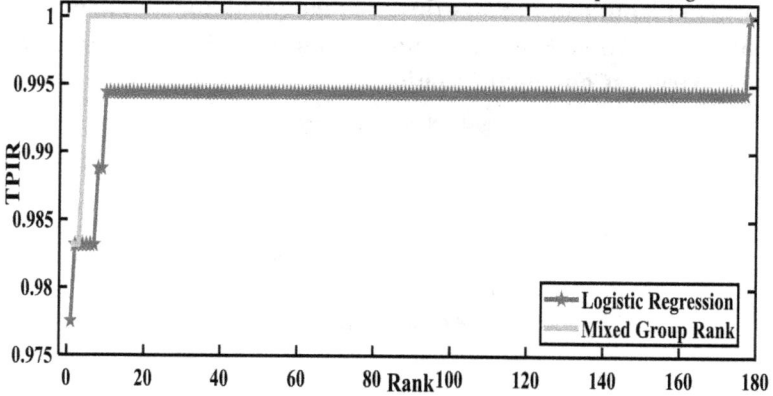

**FIGURE 9.20** CMCs after Rank-Level Fusion of all three Fingernail Plates for TLR feature-set using Logistic Regression and Mixed Group Rank, where all weights have been computed using PSO.

CMCs after Rank-Level Fusion of Three Fingernail Plates (using TLD features) through
Linear Weighted Fusion methods, where weights have been computed using PSO

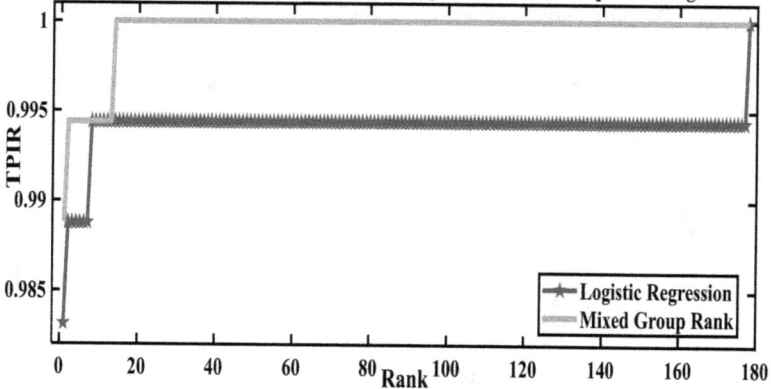

**FIGURE 9.21** CMCs after Rank-Level Fusion of all three Fingernail Plates for TLD feature-set using Logistic Regression and Mixed Group Rank, where all weights have been computed using PSO.

### 9.6.2.2.3 Experiment C

To check the efficacy of the proposed system further, the next experiment has implemented rank-level fusion of the three nail plates using another fusion rule: the Inverse Rank Position method. Figure 9.22 shows the results obtained after carrying out this experiment. It is seen that fusion of the TLR feature-based scores through this method gives the highest Rank-1 identification accuracy (98.88%). Table 9.7 illustrates the Rank-1 TPIRs for the three models.

**TABLE 9.6**

**Rank-1 Identification Rates (in %) of the Multimodal Systems where Nail Plates of All Three Fingers are fused using Two Different Linear Weighted Fusion Methods[a]**

| Model | Fusion Method | |
|---|---|---|
| | Logistic Regression | Mixed Group Rank |
| | **Weights computed Empirically** | |
| TLA | 94.38 | 94.38 |
| TLR | 96.07 | 96.07 |
| TLD | 97.19 | 97.19 |
| | **Weights determined using PSO** | |
| TLA | 96.07 | 96.63 |
| TLR | 97.75 | 98.32 |
| TLD | 98.32 | *98.88* |

[a] Number in bold and italics signify the best performance obtained.

### 9.6.2.2.4 Experiment D

Appreciable identification accuracy obtained from experiments A-C performed under the current section motivated authors to explore the fingernail plate multimodal identification system further. With the same intent, multimodal systems have been designed where all three fingernail plates have been fused using two different nonlinear weighted fusion rules.

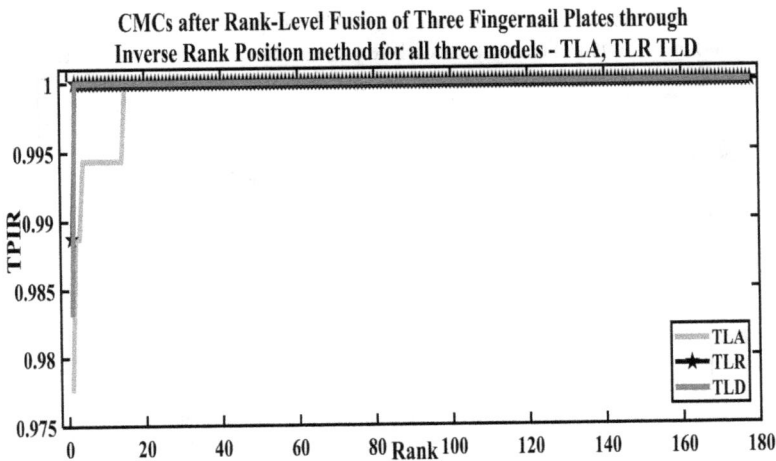

**FIGURE 9.22** CMCs after Rank-Level Fusion of all three Fingernail Plates for TLA, TLR, and TLR using Inverse Rank method.

**TABLE 9.7**

**Rank-1 Identification Rates (in %) of the Multimodal Systems where Nail Plates of All Three Fingers are fused using the Inverse Rank Position method[a]**

| TLA | TLR | TLD |
|---|---|---|
| 97.75 | *98.88* | 98.32 |

[a] Number in bold and italics signify the best performance obtained.

The experiment carried out under this section builds a multimodal system where the index, middle, and ring fingernail plates have been fused using Weighted Exponential and Hyperbolic Tangent: two different nonlinear, weighted fusion rules. Figure 9.23 gives the comparative depiction of the performance of the index, middle, and ring fingernail plates when their corresponding TLA based scores are fused at the rank-level using Weighted Exponential and Hyperbolic Tangent, while the same finding for TLR feature-set is shown in Figure 9.24, and that for TLD feature-set is given in Figure 9.25. For this experiment, all weights have been determined via empirical computation.

Figures 9.23–9.25 demonstrate that while the identification performance might be considered to be satisfactory, the results obtained in this experiment, especially the Rank-1 TPIRs are lower than that obtained through the Inverse Rank Position method, or those obtained through the Linear Weighted methods even when respective weights are computed empirically. Also, the Rank-1 TPIRs obtained using TLD model is less than that obtained using both TLR and TLA when Hyperbolic Tangent is used as the fusion rule. This is highly unlikely as TLD is much deeper than TLA, and this may have been caused because of possible inappropriate weight attribution to the three nail plates considered. Table 9.8 enlists the Rank-1 TPIRs obtained under this experiment.

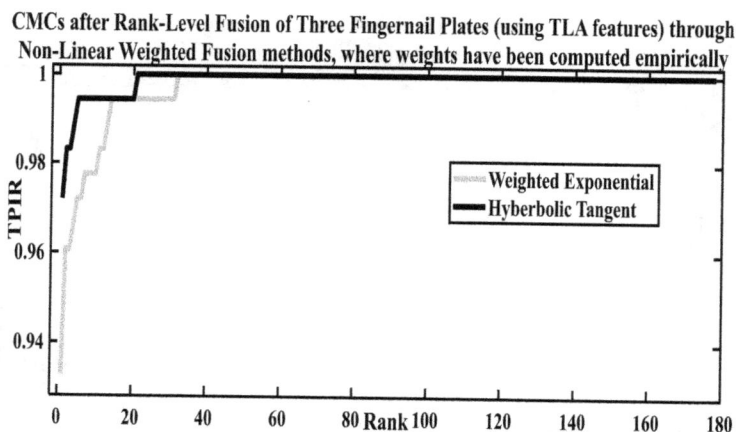

**FIGURE 9.23** CMCs after Rank-Level Fusion of all three Fingernail Plates for TLA feature-set using Weighted Exponential and Hyperbolic Tangent, where all weights have been computed empirically.

CMCs after Rank-Level Fusion of Three Fingernail Plates (using TLR features) through
Non-Linear Weighted Fusion methods, where weights have been computed empirically

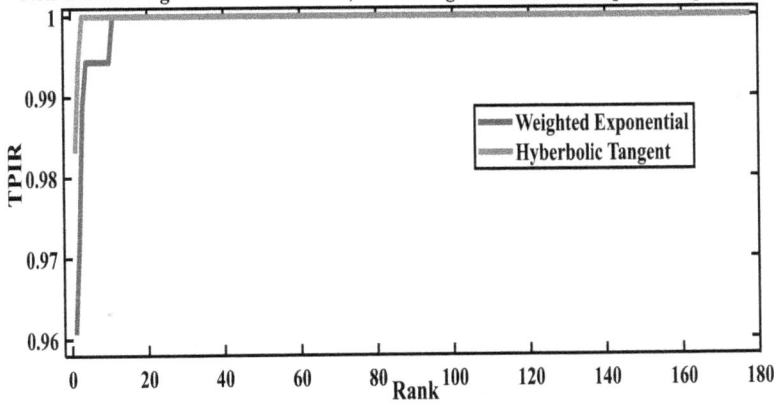

**FIGURE 9.24** CMCs after Rank-Level Fusion of all three Fingernail Plates for TLR feature-set using Weighted Exponential and Hyperbolic Tangent, where all weights have been computed empirically.

CMCs after Rank-Level Fusion of Three Fingernail Plates (using TLD features) through
Non-Linear Weighted Fusion methods, where weights have been computed empirically

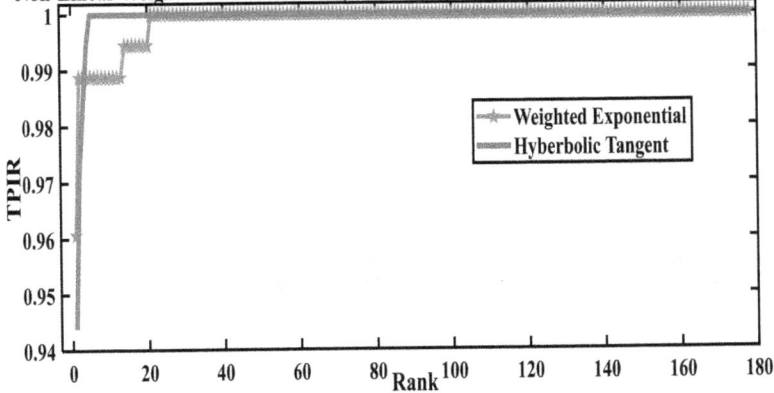

**FIGURE 9.25** CMCs after Rank-Level Fusion of all three Fingernail Plates for TLD feature-set using Weighted Exponential and Hyperbolic Tangent, where all weights have been computed empirically.

### 9.6.2.2.5 Experiment E

With an aim to achieve better identification accuracy, and also to check the sort of unlikely results obtained under the previous set of experiments, the same have been repeated after computing weights using PSO. Also, it is important to note that results obtained from Experiment B confirms that optimisation of weights improves identification accuracy considerably. Thus under this experiment, the three nail

**TABLE 9.8**

**Rank-1 Identification Rates (in %) of the Multimodal Systems where Nail Plates of All Three Fingers are fused using Two Different Nonlinear Weighted Fusion methods (Weights computed Empirically)**

| Model | Fusion Method | |
|---|---|---|
| | Weighted Exponential | Hyperbolic Tangent |
| TLA | 93.26 | 97.19 |
| TLR | 96.07 | 98.32 |
| TLD | 96.07 | 94.38 |

plates have been fused using Weighted Exponential and Hyperbolic Tangent rules, where weights have been calculated using PSO.

Figure 9.26 gives the comparative depiction of the performance of fused nail plates using the TLA feature-set when the fusion is performed using Weighted Exponential and Hyperbolic Tangent. The same findings for TLR and TLD feature-sets are given in Figures 9.27 and 9.28, respectively.

Analysing the CMCs in Figures 9.26–9.28 reveals that in this experiment, the highest Rank-1 TPIR of 99.44% has been provided by both TLR and TLD, when fusion is performed through the Hyperbolic Tangent rule. Comparing these results with those obtained in the previous experiment shows that the performance accuracy has improved significantly when the weight attribution has been performed using PSO.

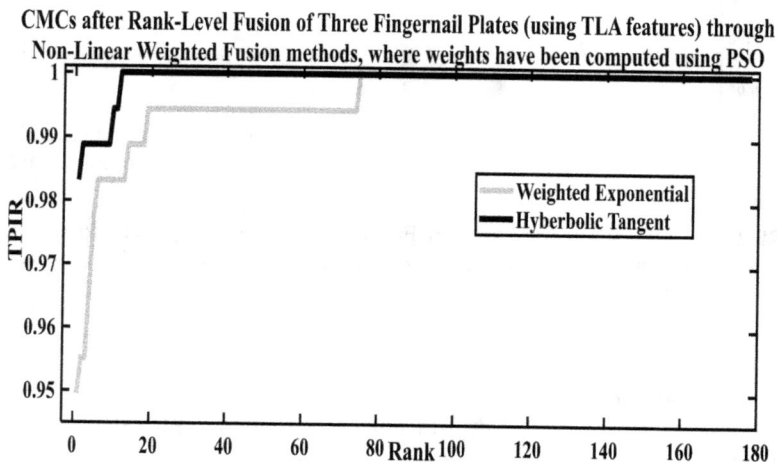

**FIGURE 9.26** CMCs after Rank-Level Fusion of all three Fingernail Plates for TLA feature-set using Weighted Exponential and Hyperbolic Tangent, where all weights have been computed using PSO.

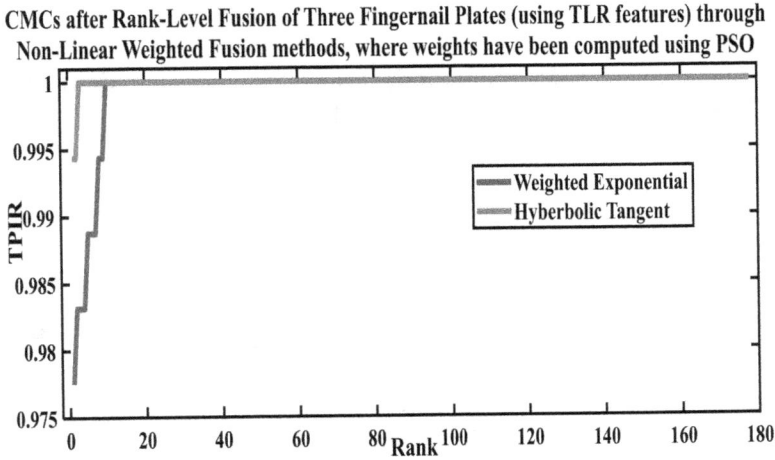

CMCs after Rank-Level Fusion of Three Fingernail Plates (using TLR features) through Non-Linear Weighted Fusion methods, where weights have been computed using PSO

**FIGURE 9.27** CMCs after Rank-Level Fusion of all three Fingernail Plates for TLR feature-set using Weighted Exponential and Hyperbolic Tangent, where all weights have been computed using PSO.

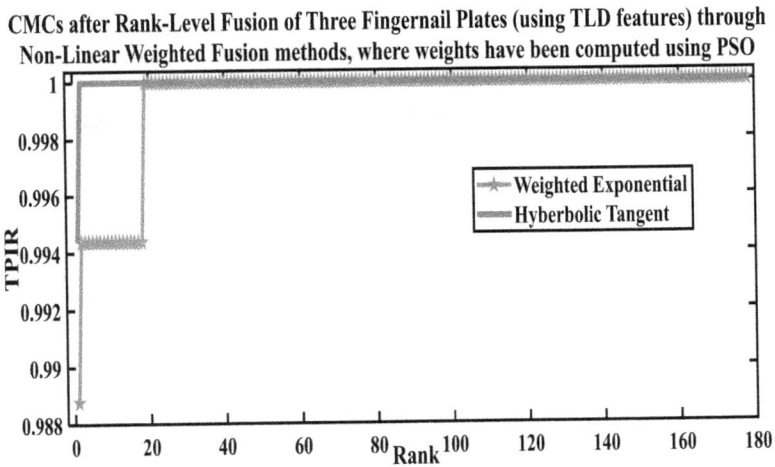

CMCs after Rank-Level Fusion of Three Fingernail Plates (using TLD features) through Non-Linear Weighted Fusion methods, where weights have been computed using PSO

**FIGURE 9.28** CMCs after Rank-Level Fusion of all three Fingernail Plates for TLD feature-set using Weighted Exponential and Hyperbolic Tangent, where all weights have been computed using PSO.

The Rank-1 TPIRs of the different systems in this experiment have been given in Table 9.9 to alleviate their comparative representation.

To make a comparison of the methodologies adopted and results obtained in the current work with that of some of the previously reported works which investigated the fingernail, a tabular depiction has been made in Table 9.10.

**TABLE 9.9**

**Rank-1 Identification Rates (in %) of the Multimodal Systems where Nail Plates of Three Fingers are fused using Nonlinear Weighted Fusion Methods (Weights computed using PSO)[a]**

| Model | Fusion Method | |
|---|---|---|
| | Weighted Exponential | Hyperbolic Tangent |
| TLA | 94.94 | 98.32 |
| TLR | 97.75 | *99.44* |
| TLD | 98.88 | *99.44* |

[a] Numbers in bold and italics signify the best performance obtained.

**TABLE 9.10**

**Comparison of Proposed Work with Significant Existing Works on Fingernail[a]**

| Ref. No. | Part of Finger Explored | Feature Extraction Technique | Database | Results of Unimodal System | Results of Fusion |
|---|---|---|---|---|---|
| [15] | Index Fingernail Bed | None | Not reported | Binary representation of the relative positions of capillary loops is obtained, which is unique to every individual. | Not explored |
| [14] | Nail-Ridges | None | Not reported | Bands of colour obtained. Each represents a single ridge or valley of nail surface. | Not explored |
| [16] | Nail Surface of Index, Middle, and Ring Fingers | Hand-crafted Approach: Haar Wavelet | Database of 5 images/180 users = 900 images per modality | Highest accuracy reported: Verification: GAR = 50% (at FAR = 0.01%) *Identification not investigated.* | Fusion of three nail surfaces done. Verification results as high as GAR = 72% (at FAR = 0.01%) for Product Rule. *Identification not investigated.* |

*(Continued)*

## TABLE 9.10 (*Continued*)
## Comparison of Proposed Work with Significant Existing Works on Fingernail[a]

| Ref. No. | Part of Finger Explored | Feature Extraction Technique | Database | Results of Unimodal System | Results of Fusion |
|---|---|---|---|---|---|
| [17] | Nail Plate of Index, Middle, and Ring Fingers | Hand-crafted Approaches: Haar Wavelet and ICA | Database of 5 images/180 users = 900 images per modality | Highest accuracy reported: Verification: GAR = 55% (at FAR = 0.01%) Identification: 81% Rank-1 TPIR | Fusion of three nail plates done. Verification and Identification accuracy as high as GAR = 85% (at FAR = 0.01%) and 96.5% Rank-1 TPIR respectively reported |
| [8] | Nail Plate of Index, Middle, and Ring Fingers | Hand-crafted Approaches: Haar Wavelet and ICA | Database of 5 images/180 users = 900 images per modality | Highest accuracy reported: Verification: GAR = 60% (at FAR = 0.01%) Identification: 89% Rank-1 TPIR | Fusion of three nail plates done. Verification and Identification accuracy as high as GAR = 80% (at FAR = 0.01%) and 96.5% Rank-1 TPIR respectively reported |
| [5] | Knuckle and Nail Plate of Index, Middle and Ring Fingers | Deep Learning Approach: AlexNet | Database of 5 images/178 users = 890 images per modality | Highest accuracy reported from Nail Plates: Verification: GAR = 56.74% (at FAR = 0.01%) Identification: 90.45% Rank-1 TPIR | Fusion of a) three nail plates, b) three knuckles, c) each nail plate with corresponding knuckle done. For fusion of nail plate and knuckle, Verification and Identification accuracy as high as GAR = 96.63% (at FAR = 0.01%) and 98.31% Rank-1 TPIR, respectively, reported For fusion of three nail plates, Verification and Identification accuracy as high as GAR = 87.64% (at FAR = 0.01%) and 98.31% Rank-1 °TPIR, respectively, reported |

*(Continued)*

**TABLE 9.10 (*Continued*)**

**Comparison of Proposed Work with Significant Existing Works on Fingernail[a]**

| Ref. No. | Part of Finger Explored | Feature Extraction Technique | Database | Results of Unimodal System | Results of Fusion |
|---|---|---|---|---|---|
| *This Work* | Nail Plate of Index, Middle, and Ring Fingers | Deep Learning Approaches: AlexNet, ResNet and DenseNet | Same database as in Ref. [5] | Highest accuracy Verification: GAR = *74.16%* (at FAR = 0.01%) Identification: *93.82%* Rank-1 TPIR | Verification and Identification accuracy as high as GAR = *94.94%* (at FAR = 0.01%) and *99.44%* Rank-1 TPIR respectively |

[a]  Numbers in bold and italics signify the best performance obtained in the current work.

## 9.7   CHALLENGES AND SCOPE OF FINGERNAIL PLATES IN BIOMETRICS

- **Fusion**: Fingernail plates have been seen to provide appreciable biometric authentication results, when these are fused with finger knuckles [5]. Also, the performance of the fusion of multiple fingernail plates may be a prospective scope, something which has been explored to a certain extent in this work. The fingernail plate may also be investigated in combination with other biometric traits which do not belong to the dorsal part of the human hand like fingerprint, iris, face, and voice. However, such systems would require sample acquisition in multiple steps.

- **Application in Forensics**: Drug analyses in nails have received significant attention because of the ability of nails to amass drugs, when subjected to long-term exposure [38]. Nail plates may be subjected to exhaustive experimentation in order to observe the changes or deformities caused by long- or short-term drug abuse. Additionally, trace evidences obtained from crime scenes encapsulating fingernail plates (as shown in Figure 9.2) shall be able to provide considerable help in forensic investigation.

- **Acceptability**: The sample collection of fingernail plate images can be done using a very low-cost camera, without any constraints being imposed on the user. This has been implemented in the current work. As such, sample collection of this trait may be considered to be considerably user-friendly as it does not require active user cooperation.

- **Spoofing**: The nail plate enjoys a certain degree of advantage over popular hand biometrics like fingerprints. This is because people inadvertently leave behind fingerprints on whatever they touch, which is not possible in the case of nail plates. However, it would be significant to work towards designing anti-spoofing techniques particularly aimed towards nail plates.

## 9.8   CONCLUSIONS AND FUTURE SCOPE

This work has tried to further explore the fingernail plate as a biometric trait, considering three different deep learning methods for feature extraction. The nail plate has anatomically distinctive features which are less prone to impersonation.

This report analyses the performance of the nail plate in unimodal verification and identification systems. The results show that the nail plates under all the three considered feature extraction techniques provide substantially significant authentication performance.

Multimodal systems are known to compensate the drawbacks of a trait and to provide improved performance accuracy. In view of this, the fingernail plate has been explored in various multimodal systems. Score-level fusion has been implemented through four different rules for all three deep learning models. The results show that the multimodal verification systems perform much better than their unimodal counterparts. Various multimodal identification systems have been designed using different weighted and non-weighted fusion rules. The weights of the three nail plates for the weighted fusion methods have been attributed through empirical computation, and by using PSO in order to ensure optimal weight attribution. Comprehensive experiments have been carried out, and results depict that identification accuracy as high as 99.44% can be achieved when right weightage is given to the traits. The results also confirm the fact that more the depth of the model, better the results.

The significant results given by the experiments performed in this work give inspiration to further probe the scope of the nail plate in biometric authentication. It shall be interesting to design the implemented system in an adaptive framework [29], which shall lessen the computational time and cost. This is because such a system shall provide the results after deciding on the optimal fusion rule for the selected biometric traits. The future scope of this proposed work includes checking its efficacy in feature-level multibiometric systems. A larger database also may be prepared so as to scale up research in this field.

## REFERENCES

1. Kumar, A., Zhou, Y., "Human Identification using Finger Images." *IEEE Transactions on Image Processing*, Vol. 21, No. 4, 2012, pp. 2228–2244.
2. Kumar, A., Ravikanth, C., "Personal Authentication using Finger Knuckle Surface." *IEEE Transactions on Information Forensics and Security*, Vol. 4, No. 1, 2009, pp. 98–110.
3. Kumar, A., Hu, Z., "Personal Identification using Minor Knuckle Patterns from Palm Dorsal Surface." *IEEE Transactions on Information Forensics & Security*, Vol. 11, No. 10, 2016, pp. 2338–2348.
4. Kumar, A., Prathyusha, K. V., "Personal Authentication using Hand Vein Triangulation and Knuckle Shape." *IEEE Transactions on Image Processing*, Vol. 18, No. 9, 2009, pp. 2127–2136.
5. Hom Choudhury, S., Kumar, A., Laskar, S. H., "Biometric Authentication through Unification of Finger Dorsal Biometric Traits." *Information Sciences*, Vol. 497, 2019, pp. 202–218.
6. Huang, D., Tang, Y., Wang, Y., Chen, M., Wang, Y., "Hand-Dorsa Vein Recognition by Matching Local Features of Multisource Keypoints." *IEEE Transactions on Cybernetics*, Vol. 45, No. 9, 2015, pp. 1823–1837.

7. Qin, H., El-Yacoubi, M. A., "Deep Representation-based Feature Extraction and Recovering for Finger-Vein Verification." *IEEE Transactions on Information Forensics and Security*, Vol. 12, No. 8, 2017, pp. 1816–1829.

8. Kumar, A., Garg, S., Hanmandlu, M., "Biometric Authentication using Finger Nail Plates." *Expert Systems with Applications*, Vol.41, No. 2, 2014, pp. 373–386.

9. Diaz, A. A., Boehm, A. F., Rowe, W. F., "Comparison of Fingernail Ridge Patterns of Monozygotic Twins." *Journal of Forensic Sciences*, Vol. 35, No. 1, 1990, pp. 97–102.

10. Daniel III, C. R., Piraccini, B. M., Tosti, A., "The Nail and Hair in Forensic Science." *Journal of the American Academy of Dermatology*, Vol. 50, No. 2, 2004, pp. 258–261.

11. Hong, L., Jain, A., "Integrating Faces and Fingerprints for Personal Identification." *IEEE Transactions on Pattern Analysis and Machine Intelligence*, Vol. 20, No. 12, 1998, pp. 1295–1307.

12. Kumar, A., Kumar, A., "Adaptive Management of Multimodal Biometrics Fusion Using Ant Colony Optimization." *Information Fusion*, Vol. 32, No. B, 2016, pp. 49–63.

13. Roberts, D. T., Taylor, W. D., Boyle, J., "Guidelines for Treatment of Onychomycosis." *The British Journal of Dermatology*, Vol. 148, No. 3, 2002, pp. 402–410.

14. Apolinar, E., Rowe, W. F., "Examination of Human Fingernail Ridges by Means of Polarized Light." *Journal of Forensic Sciences*, Vol. 25, No. 1, 1980, pp. 154–161.

15. Topping, A., Kuperschmidt, V., Gormley, A., "Method and Apparatus for the Automated Identification of Individuals by the Nail Beds of their Fingernails." U.S. Patent, 5751835A, 1998.

16. Garg, S., Kumar, A., Hanmandlu, M., "Biometric Authentication using Finger Nail Surface." *12th International Conference on Intelligent Systems Design and Applications (ISDA)*, IEEE, Kochi, 2012, pp. 497–502.

17. Garg, S., Kumar, A., Hanmandlu, M., "Finger Nail Plate: A New Biometric Identifier." *International Journal of Computer Information Systems and Industrial Management Applications*, Vol. 6, 2014, pp. 126–138.

18. LeCun, Y., Bottou, L., Bengio, Y., Haffner, P., "Gradient-Based Learning Applied to Document Recognition." *Proceedings of the IEEE*, Vol. 86, No.11, 1998, pp. 2278–2324.

19. LeCun, Y., Bengio, Y., Hinton, G., "Deep Learning." *Nature*, Vol. 521, No. 7553, 2015, pp. 436–444.

20. Bengio, Y., "Learning Deep Architectures for AI." *Foundations and Trends® in Machine Learning*, Vol. 2, No. 1, 2009, pp. 1–127.

21. Amato, G., Carrara, F., Falchi, F., Gennaro, C., Meghini, C., Vairo, C., "Deep Learning for Decentralized Parking Lot Occupancy Detection." *Expert Systems with Applications*, Vol. 72, 2017, pp. 327–334.

22. Goodfellow, I., Bengio, Y., Courville, A., *Deep Learning*, Cambridge, MA: The MIT Press, 2016.

23. Razavian, A. S., Azizpour, H., Sullivan, J., Carlsson, S., "CNN Features off-the-shelf: An Astounding Baseline for Recognition." *Proceedings of the CVPR*, 2014, pp. 512–519.

24. Azizpour, H., Razavian, A. S., Sullivan, J., Maki, A., Carlsson, S., "From Generic to Specific Deep Representations for Visual Recognition." *Proceedings of the CVPR*, 2015, pp. 36–45.

25. Krizhevsky, A., Sutskever, I., Hinton, G. E., "ImageNet Classification with Deep Convolutional Neural Networks." *Proceedings of the Advances in Neural Information Processing Systems*, 2012, pp. 1097–1105.

26. ImageNet. http://www.image-net.org, Jun. 2015.

27. He, K., Zhang, X., Ren, S., Sun, J., "Deep Residual Learning for Image Recognition." *Proceedings of the CVPR*, 2016, pp. 1–9.

28. Huang, G., Liu, Z., van der Maaten, L., Weinberger, K. Q., "Densely Connected Convolutional Networks." *Proceedings of the CVPR*, 2017, pp. 2261–2269.

29. Kumar, A., Kanhangad, V., Zhang, D., "A New Framework for Adaptive Multimodal Biometrics Management." *IEEE Transactions on Information Forensics and Security*, Vol. 5, No. 1, 2010, pp. 92–102.

30. Ross, A., Nandakumar, K., Jain, A.K., *"Handbook of Multibiometrics."* Springer-Verlag, New York, 2006.

31. Monwar, M.M. "A Multimodal Biometric System based on Rank Level Fusion," Ph.D. Thesis, *University of Calgary*, Dec. 2012.

32. Kumar, A., Shekhar, S., "Personal Identification using Multibiometrics Rank-Level Fusion." *IEEE Transactions on Systems, Man, and Cybernetics, Part C (Appl. and Rev.)*, Vol. 41, No. 5, 2011, pp. 743–752.

33. Ho, T.K., Hull, J.J., Srihari, S.N., "Decision Combination in Multiple Classifier Systems." *IEEE Transactions on Pattern Analysis and Machine Intelligence*, Vol. 16, No. 1, 1994, pp. 66–75.

34. Kennedy, J., Eberhart, R., "Particle Swarm Optimization." *1995 IEEE International Conference on Neural Networks*, Vol. 4, 2002, pp. 1942–1948.

35. Melnik, O., Vardi, Y., Zhang, C-H., "Mixed Group Ranks: Preference and Confidence in Classifier Combination." *IEEE Transactions on Pattern Analysis and Machine Intelligence*, Vol. 26, No. 8, 2004, pp. 973–981.

36. Jovic, M., Hatakeyama, Y., Dong, F., Hirota, K., "Image Retrieval based on Similarity Score Fusion from Feature Similarity Ranking Lists." *Fuzzy Systems and Knowledge Discovery, ser. Lecture Notes in Computer Science*, Springer Berlin Heidelberg, Vol. 4223, 2006, pp. 461–470.

37. Nandakumar, K. "Multibiometric Systems: Fusion Strategies and Template Security," Ph.D. Thesis, *Michigan State University*, 2008.

38. Cappelle, D., Yegles, M., Neels, H., van Nuijs, A. L. N., Doncker, M.D., Maudens, K., Covaci, A., Crunelle, C. L., "Nail Analysis for the Detection of Drugs of Abuse and Pharmaceuticals: A Review." *Forensic Toxicology*, Vol. 33, 2015, pp. 12–36.

# 10 Fraud Attack Detection in Remote Verification Systems for Non-enrolled Users

*Ignacio Viedma, Sebastian Gonzalez,*
*Ricardo Navarro, and Juan Tapia*
TOC Biometrics Labs

## CONTENTS

10.1 Introduction .......................................................................................... 239
10.2 Related Work ......................................................................................... 241
    10.2.1 Remote Authentication Framework Using Biometrics .................... 241
    10.2.2 Image Manipulation and Deep Learning Techniques ..................... 242
10.3 Fake ID Card Detection for Non-enrolled Users ....................................... 243
    10.3.1 Databases ..................................................................................... 244
    10.3.2 Hand-Crafted Feature Extraction (BSIF, uLBP, and HED) ........... 244
    10.3.3 Automatic Feature Extraction (CNN) ............................................ 245
10.4 Experiments and Results ........................................................................ 246
    10.4.1 Feature Extraction Classification .................................................. 246
    10.4.2 Classification Using CNN Algorithms .......................................... 247
        10.4.2.1 Small-VGG Trained from Scratch ................................... 248
        10.4.2.2 Pre-trained VGG16 Model and Bottleneck ...................... 250
        10.4.2.3 Pre-trained VGG16 Model and Fine-Tuning .................... 253
10.5 Conclusions ........................................................................................... 254
Acknowledgement ........................................................................................ 254
References .................................................................................................... 254

## 10.1 INTRODUCTION

Identity verification systems are widely used in daily life. Most of these systems rely on official documents containing identifying information about a person (i.e. passport, ID card, driving licence, membership cards, and social services card, among others). These documents usually include a face image of the person which is used to validate identity.

Recent advances in computer vision techniques have explored the use of biometric information for identity verification [6]. The face [27], fingerprints, and iris [18] are amongst the most used and reliable features for automatic identity verification. However, most systems usually require the physical presence of the individual in order to capture the information from face, fingerprint, or iris images.

The recent massive increase in the use of mobile phones has opened a new form of remote authentication. In these situations, authentication is mainly based on comparing the input data of the user (i.e. selfie or fingerprint) with the information previously registered from the same individual (database).

These systems have been applied in several industries including banking. However, most of the methods require all the users to be registered in a database. The information captured by the mobile device is then matched with the existing information of the user previously saved in a database (Figure 10.1).

The enrolling requirement for all users of the system limits the use of this kind of authentication. Activities such as opening a new account in a bank, for instance, would always require the presence of the user in the bank to be enrolled.

In order to overcome this limitation and build a fully remote system, an authentication system based on the verification of users against the information provided by an official identity document (i.e. ID card, passport, or driver's license amongst others) is proposed.[1] This system uses the biometric information obtained from a self-taken photo (selfie) and compares it against the photo of their official identity document. This process is known as a biometric match [17]. Such systems do not require a database to be consulted as all the data needed are given by the official ID card and selfie image. This eliminates vulnerability to hacking and theft of private information.

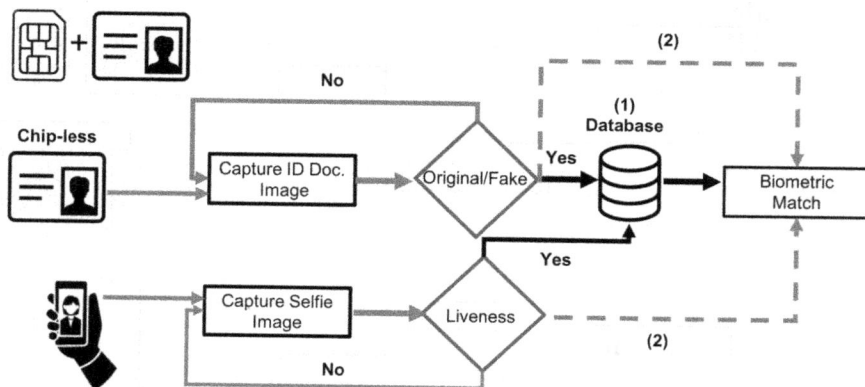

**FIGURE 10.1** Graphical representation of remote authentication (1) and remote verification (2) using selfie face images.

---

[1] https://tocbiometrics.com/

The key challenge of this authentication system is to ensure that neither the ID card nor the selfie has been manipulated by the user.

This work studies several algorithms used to detect whether ID cards have been altered. Two typical scenarios of image manipulation were studied: (i) physical and (ii) digital (see Figure 10.2). There are other cases of possible manipulation of ID cards, including more extreme scenarios, such as 'fake' identities with soft-biometric features generated using algorithms such as Generative Adversarial Network (GAN) [3]. However, this ongoing research only focuses on the two scenarios mentioned above.

The remainder of the chapter is organised as follows. Related work is reviewed in Section 10.2. The proposed method to detect ID card manipulation is described in Section 10.3. Experiments and results are reported in Section 10.4. Finally, the conclusions of this work are shown in Section 10.5.

## 10.2   RELATED WORK

### 10.2.1   REMOTE AUTHENTICATION FRAMEWORK USING BIOMETRICS

As technologies progress, the financial, government, and other sectors have increasingly opted for online services due to their lower cost compared to in-office services. As a result, remote authentication system has become critical in order to ensure user identity during transactions. The accelerated evolution in consumer smartphone cameras has brought with it an increased interest from the industry for mobile biometric verification systems. The capacity to reach the customer remotely for services such as e-commerce, digital banking, and general fintech requires robust systems for automatic identity verification. Remote biometric authentication system based on fingerprints [2,16,29] and the face [11,19,20,24] are amongst the most popular authentication systems.

In the case of authentication systems based on faces, there are two main categories: (i) remote authentication for enrolled users and for (ii) non-enrolled users. Most of the literature addresses the first scenario where biometric data from individuals (users of the system) are previously captured and stored in a database [9,20,24]. The main goal of the system is to ensure that the input data from the user match the biometric information previously stored. Stokkenes et al. [24], for instance, proposed online banking authentication based on features extracted from faces using bloom filters. This information is encoded and used as a key for opening banking services. Similar work, which involves fusing biometric information, has also been explored by Czyzewski et al. [9].

This scenario requires an enrolling process that sometimes can limit the application of such systems. Storing sensitive information from users such as biometric data can also be risky for companies due to regulations concerning personal data. Several approaches have been proposed to enforce security in such systems. Perera et al. [20], for instance, proposed an Active Authentication system that attempts to continuously monitor user identity after access has been initially granted. A similar approach has recently been reported by Oza and Patel [19]. Those approaches are a step towards security but do not solve other problems such as spoofing attacks.

The second category, (2) remote authentication systems for non-enrolled users, uses two inputs: a selfie face and an additional proof of identity. The most common proofs of identity are national ID cards and driving licences, amongst others. In this approach, the data contained in the embedded chip in an ID card can be read remotely by a Near Field Communication (NFC)-enabled mobile device and then matched with a frontal face photograph (selfie) of the person in question. Unfortunately, this approach is limited since only a few countries provide national ID cards that include embedded chips with user identity information. In countries such as Brazil, for instance, with a population of over 210 million people, the national ID card does not contain such an embedded chip. Furthermore, the ID card may vary from state to state.

In such cases, an additional challenge added to the remote authentication system is to validate the presented document as a proof of identity.

Remote authentication systems using non-enrolled users are computationally less expensive as they just match the information between the two inputs to the system. They do not require previous enrolment of the users and store any private information.

## 10.2.2   IMAGE MANIPULATION AND DEEP LEARNING TECHNIQUES

As discussed in previous sections, most 2D face-based biometric authentication systems use an image (selfie) as input information. In the case of non-enrolled users, the picture of an identification document is also required by the system. Altering a face photo or an ID document to trick an authentication system is a threat that needs to be detected in order to protect such systems and people's identity. Spoofing can directly attack biometric systems affecting people's security by creating fake biometric data [6,17,18]. Existing antispoofing methods generally move in the following directions: analysing the texture image captured by the sensor, detecting any evidence of liveness on the image [4], or combining both approaches together [12,18].

Image manipulation, on the other hand, has been a widely studied topic in the image processing and computer vision fields. Algorithms for tampering, in-painting, texture, and colour transformation amongst others have all been reported in the literature [30]. There are several algorithms to detect attacks on image-based biometric systems. The state-of-the-art technique for image analysis is Convolutional Neural Network (CNN). This is based on the use of algorithms that allow representations of the best features to be found in a hierarchical method [5,12,26].

One of the first applications of CNN was perhaps the LeNet-5 network described by Ref. [15] for optical character recognition. Compared to modern deep CNN, their network was relatively modest due to the limited computational resources of the era and the algorithmic challenges of training bigger networks. Although much potential has been laid in deeper CNN architectures (networks with more layers), only recently have they became prevalent, following the dramatic increase in both computational power, due to the availability of Graphical Processing

Units (GPU); the amount of training data readily available on the Internet; and the development of more effective methods for training such complex models. One recent and notable example is the use of deep CNN for image classification on the challenging ImageNet benchmark [10]. Deep CNN has additionally been successfully applied to human pose estimation, facial key-point detection, speech recognition, and action classification, amongst others [13,25]. However, there are smaller networks, such as small-VGG [23], that represent a trade-off between a shallow and a deeper CNN.

## 10.3 FAKE ID CARD DETECTION FOR NON-ENROLLED USERS

As reviewed previously, remote authentication systems aim to match an input data (i.e. face) with data from the same individual stored in a database. In these cases, the key is to detect whether the input data are fake or real in order to ensure a secure authentication system. This work, on the contrary, focuses on authentication systems where no data from the user are previously available. Such remote authentication systems have two inputs: a selfie and an official ID card. This work concentrates on the first step of the authentication system and aims to detect whether an ID card is real or fake. In other words, the goal is to detect if an ID card has been deliberately manipulated by replacing the face photo. This implies localising in the images for any ridges, edges, spots, or other forms of information that do not belong to an unaltered (original) ID card. There are several ways of manipulating ID cards. However, only two scenarios are considered in this work: (i) when the face image is manipulated manually and (ii) when the face photo is altered digitally. A graphical example is shown in Figure 10.2.

To detect and classify if the ID card is fake or real, traditional methods based on texture features and CNNs were studied [8]. The database and the algorithms used are described as follows.

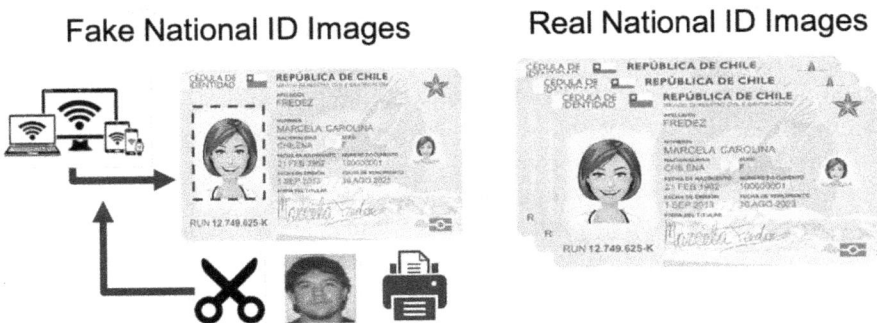

**FIGURE 10.2** Graphical representation of two scenarios to create fake images. (a) The face image coming from a digital device and from manual manipulation (i.e. a printed photo). (b) Real ID images.

## 10.3.1 Databases

Two databases were used in this work. The first one corresponds to a database of Chilean national IDs. This is a private database which contains 1,525 images of real Chilean ID cards from 316 different people. The second database was made by manipulating these Chilean ID cards using two techniques: manual and digital.

**Fake ID Card Database (Manually Manipulated)**: This database contains 762 Chilean ID cards where the face image has been replaced with face images of other people printed and stuck on the ID card. This is an easy and cheap technique used to fake ID cards and can be used without any knowledge of digital photo processing for normal or traditional users. This kind of attack is very common within remote verification ID systems.

**Fake ID Card Database (Digitally Manipulated)**: A total of 762 ID card images were manipulated by automatically detecting the face and replacing it with random face images. This technique allows a large quantity of fake ID images to be created in a short period of time. Alternatively, the face can be replaced manually using Photoshop or similar software to retouch images. This allows the fake face image to be better merged with the rest of the ID card making it difficult to detect. However, this technique is time-consuming which limits the feasibility of creating larger databases for training and testing algorithms.

## 10.3.2 Hand-Crafted Feature Extraction (BSIF, uLBP, and HED)

The first approach proposed in this work is based on machine learning techniques. Texture features are extracted from the 2D image of the ID card using three different algorithms: Uniform Local Binary Patterns (uLBP), Binary Statistical Image Feature filter (BSIF), and Holistically Nested-Edge Detection (HED).

BSIF [14] is a local descriptor constructed by binarising the responses to linear filters. The code value of pixels is considered as a local descriptor of the image intensity pattern in the pixels' surroundings. The value of each element (i.e. bit) in the binary code string is computed by binarising the response of a linear filter with a zero threshold. Each bit is associated with a different filter, and the length of the bit string determines the number of filters used.

uLBP [1] is a grey-scale texture operator which characterises the spatial structure of the local image texture. Given a central pixel in the image, a binary pattern number is computed by comparing its value with those of its neighbours.

The edge-detection algorithm (HED) [28] was developed to address two important issues in the vision problem: (i) holistic image training and prediction, and (ii) multi-scale and multi-level feature learning. The HED performs image-to-image prediction by means of a deep learning model that leverages fully CNNs and deeply supervised nets. HED automatically learns rich hierarchical representations (guided by deep supervision on side responses) that are important in order to resolve the challenging ambiguity in edge and object boundary detection.

These algorithms have been shown to outperform the state-of-the-art in texture methods. Image features are then classified in two classes (fake and real) using a Random Forest Classifier [7].

### 10.3.3 Automatic Feature Extraction (CNN)

As a second approach, two deep learning algorithms were tested. Deep learning techniques have been shown to be very effective in localising ridges, edges, and spots in the images [22], making them suitable for this problem.

First, a small-VGG [15] network (CNN-1) was used to classify fake and real ID cards. The small-VGG network comprises only three convolutional blocks and a fully connected layer with a small number of neurons. The choice of a smaller network design was motivated both from the desire to reduce the risk of overfitting as well as the nature of the problem, which attempts to solve a two-class classification task (fake and real). Figure 10.3 shows a scheme of the algorithm architecture. The three channels are processed directly by the network.

In order to find the best implementation, different parameters such as sparse connectivity, shared weight, pooling techniques, and hyper-parameters were defined. In this work, sparse connectivity was used by default, while shared weight, pooling techniques, and hyper-parameters (batch size, epochs, learning rate, and momentum) were explored in-depth in the experimental section. They were all tuned while fitting the network.

The batch size in the iterative gradient descent is the number of patterns shown to the network before the weights are updated. There is also an optimisation in the training of the network, defining how many patterns to read at a time and keep in memory. The number of epochs is the number of times that the entire training data set is shown to the network during training. The learning rate parameter controls how much to update the weight at the end of each batch and the momentum controls how much to let the previous update influence the current weight update.

Second, a pre-trained VGG16 [23] model with bottleneck and fine-tuning techniques was also tested. This model has been pre-trained on a large data set called

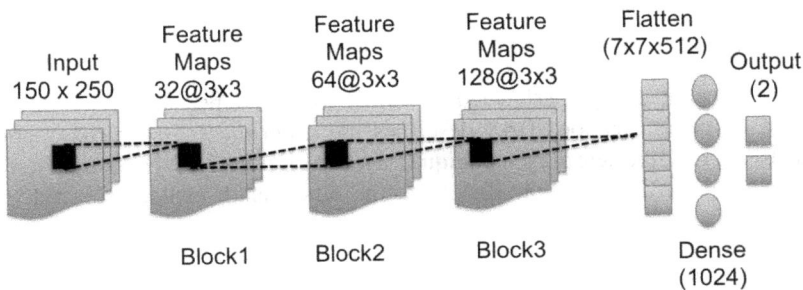

**FIGURE 10.3**   Architecture of the small-VGG network.

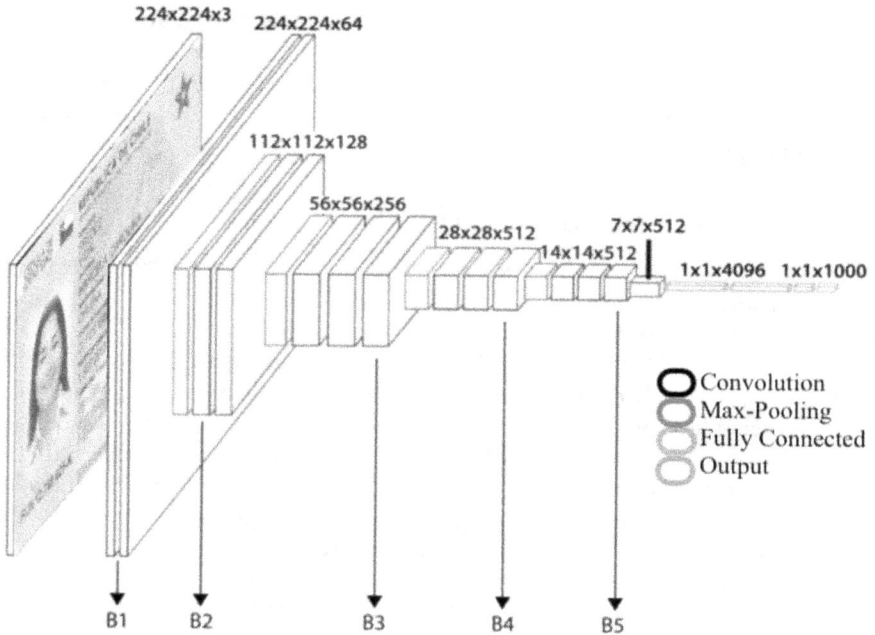

**FIGURE 10.4** VGG-16 Architecture. B1 up to B5 represent the convolutional blocks.

ImageNet. This data set contains a total of 1,000 classes, none of them including ID card images. This model had already learned features that are useful for most computer vision problems such as a ridge, lines, spots, and others. Leveraging such features allows better accuracy results to be reached than any method that would only rely on the available data. The architecture of the VGG16 model is shown in Figure 10.4.

The following section described the experiments and results obtained for classifying real and fake national ID cards when using machine learning and deep learning techniques.

## 10.4   EXPERIMENTS AND RESULTS

A set of experiments were performed in order to assess the best algorithm for detecting fake ID cards. Section 10.4.1 describes the experiments and results obtained when using machine learning techniques such as BSIF-uLBP-HED and Random Forest (RF) Classifier. Section 10.4.2, on the other hand, explores several CNN-based algorithms.

### 10.4.1   FEATURE EXTRACTION CLASSIFICATION

The descriptor BSIF has two parameters: the filter size and the number of features extracted. In this work, all the filters were used to compute the best window size and the best number of bits for the left and right eyes. Thus, the following window sizes, $5 \times 5$, $7 \times 7$, $9 \times 9$, $11 \times 11$, $13 \times 13$, $15 \times 15$, and $17 \times 17$, from 5 bits up to 12 bits were calculated for each filter. The number of bits represents the number of filters used in the convolution.

The uLBP transformation also has two parameters: The radius and the number of neighbours. The original operator used a $3 \times 3$ window size containing 9 values. In this work, several radii values from radius 1 up to 8 using grey-scale images were explored.

The HED automatically learns rich hierarchical representations (guided by deep supervision on side responses) that are important in order to resolve the challenging ambiguity in edge and object boundary detection. The Gaussian filter size (to highlight the borders) and the scale factor are two of the main parameters to be set (see Tables 10.5 and 10.6).

All the features extracted were classified using a RF approach. The RF was set up with the following parameters found after a grid search: 'N Trees': 600, 'Min samples split': 5, 'Min samples leaf ': 1, 'Max features': 'Sqrt', 'Max depth': 20, 'Criterion': 'Entropy', and 'Bootstrap': True. Figure 10.5 shows an example of applying these algorithms to a real and a fake ID card image.

Two experiments to extract features from the images were used. The first one uses the whole image ($150 \times 250$) and the second one ($150 \times 125$) only uses the left part of the image ($150 \times 125$). This corresponds to the region where the face photo is located. Tables 10.1 and 10.2 show the parameters and results achieved when the BSIF algorithm was used. Tables 10.3 and 10.4 show results for uLBP, whereas Tables 10.5 and 10.6 report parameters and results for HED algorithm. TN, FP, FN, and TP represent True Negative, False Positive, False Negative, and True Positive, respectively.

## 10.4.2 CLASSIFICATION USING CNN ALGORITHMS

Intensity images of whole national ID cards were used to classify fake or real cards using a small-VGG and a VGG16 network. This section presents the results for three different experiments: (i) using the small-VGG trained from scratch, (ii) using

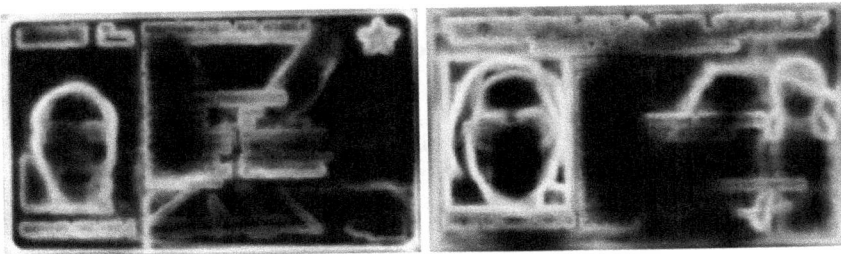

FIGURE 10.5 Graphical representation of two ID cards with HED feature extraction method applied. The border is detected and highlighted by the algorithm.

**TABLE 10.1**

**Parameters of the BSIF Algorithm (Filter Size and Bits) and the Classification Results (Acc.: Accuracy) Reached When Using 252 Full Images**

| Filter Sizes | Bits | TN | FP | FN | TP | Sensitivity (TPR) | Specificity (TNR) | Acc. |
|---|---|---|---|---|---|---|---|---|
| 3 × 3 | 6 | 81 | 45 | 14 | 112 | 0.89 | 0.64 | 0.77 |
| 5 × 5 | 12 | 81 | 45 | 20 | 106 | 0.84 | 0.64 | 0.74 |
| 7 × 7 | 10 | 85 | 41 | 27 | 99 | 0.79 | 0.67 | 0.73 |
| 9 × 9 | 8 | 83 | 43 | 34 | 92 | 0.73 | 0.66 | 0.69 |
| 11 × 11 | 8 | 87 | 39 | 27 | 99 | 0.79 | 0.69 | 0.74 |
| 13 × 13 | 5 | 89 | 37 | 37 | 89 | 0.71 | 0.71 | 0.71 |
| 15 × 15 | 7 | 89 | 37 | 25 | 101 | 0.80 | 0.71 | 0.75 |
| 17 × 17 | 7 | 88 | 38 | 23 | 103 | 0.82 | 0.70 | 0.76 |

TPR and TNR represent True Positive Rate and True Negative Rate, respectively.

**TABLE 10.2**

**Parameters of the BSIF Algorithm (Filter Size and Bits) and the Classification Results (Acc.: Accuracy) Reached When Using 255 Half Images**

| Filter Sizes | Bits | TN | FP | FN | TP | Sensitivity (TPR) | Specificity (TNR) | Acc. |
|---|---|---|---|---|---|---|---|---|
| 3 × 3 | 6 | 93 | 33 | 34 | 92 | 0.73 | 0.74 | 0.73 |
| 5 × 5 | 12 | 91 | 35 | 36 | 90 | 0.71 | 0.72 | 0.72 |
| 7 × 7 | 10 | 80 | 46 | 28 | 98 | 0.78 | 0.63 | 0.71 |
| 9 × 9 | 8 | 84 | 42 | 34 | 92 | 0.73 | 0.67 | 0.70 |
| 11 × 11 | 8 | 89 | 37 | 26 | 100 | 0.79 | 0.71 | 0.75 |
| 13 × 13 | 5 | 89 | 37 | 29 | 97 | 0.77 | 0.71 | 0.74 |
| 15 × 15 | 7 | 87 | 39 | 21 | 105 | 0.83 | 0.69 | 0.76 |
| 17 × 17 | 7 | 93 | 33 | 21 | 105 | 0.83 | 0.74 | 0.79 |

TPR and TNR represent True Positive Rate and True Negative Rate, respectively.

a pre-trained model (VGG16) with bottleneck approach, and (iii) the pre-trained VGG16 model with fine-tuning approach.

The experiments were performed using a NVIDIA 1080-TI GPU with 11 GB of RAM, defining a batch size of 64 images in each round. For all the experiments, the frameworks of Keras and Tensorflow were used (Figure 10.6).

## 10.4.2.1 Small-VGG Trained from Scratch

A grid search was used to find the best hyper-parameters for the small-VGG CNN. A suite of different mini batch sizes from $n = 16$ to $n = 1,024$ in steps of $2^n$ were evaluated. To set the learning rate, a small set of standard values ranging from 0.1 to 0.9

**TABLE 10.3**

**Parameters of the uLBP Algorithm (Neighbours and Radii) and the Classification Results (Acc.: Accuracy) Reached When Using 252 Full Images**

| Radii | TN | FP | FN | TP | Sensitivity (TPR) | Specificity (TNR) | Acc. |
|---|---|---|---|---|---|---|---|
| 8,2 | 92 | 34 | 37 | 89 | 0.71 | 0.73 | 0.72 |
| 8,3 | 85 | 41 | 47 | 79 | 0.63 | 0.67 | 0.65 |
| 8,4 | 80 | 46 | 51 | 75 | 0.60 | 0.63 | 0.62 |
| 8,5 | 78 | 48 | 55 | 71 | 0.56 | 0.62 | 0.59 |
| 8,6 | 71 | 55 | 49 | 77 | 0.61 | 0.56 | 0.59 |
| 8,7 | 65 | 61 | 51 | 75 | 0.60 | 0.52 | 0.56 |
| 8,8 | 60 | 66 | 51 | 75 | 0.60 | 0.48 | 0.54 |
| 8,2 to 8,8 | 99 | 27 | 44 | 82 | 0.65 | 0.79 | 0.72 |

TPR and TNR represent True Positive Rate and True Negative Rate, respectively.

**TABLE 10.4**

**Parameters of the uLBP Algorithm (Neighbours and Radii) and the Classification Results (Acc.: Accuracy) Reached When Using 252 Half Images**

| Radii | TN | FP | FN | TP | Sensitivity (TPR) | Specificity (TNR) | Acc. |
|---|---|---|---|---|---|---|---|
| 8,2 | 89 | 37 | 31 | 95 | 0.75 | 0.71 | 0.73 |
| 8,3 | 85 | 41 | 34 | 92 | 0.73 | 0.67 | 0.70 |
| 8,4 | 76 | 50 | 38 | 88 | 0.70 | 0.60 | 0.65 |
| 8,5 | 78 | 48 | 52 | 74 | 0.59 | 0.62 | 0.60 |
| 8,6 | 65 | 61 | 48 | 78 | 0.62 | 0.52 | 0.57 |
| 8,7 | 70 | 56 | 45 | 81 | 0.64 | 0.56 | 0.60 |
| 8,8 | 67 | 59 | 43 | 83 | 0.66 | 0.53 | 0.60 |
| 8,2–8,8 | 93 | 33 | 26 | 100 | 0.79 | 0.74 | 0.77 |

TPR and TNR represent True Positive Rate and True Negative Rate, respectively.

in steps of 0.1 were tested. For the momentum, values in the ranges of $1e-1$ to $1e-5$ were considered.

A database of 3,050 images plus data augmentation was used for training the algorithm. The data were divided into 70/30 for training and testing the classifier, respectively. This number of images (people) is larger than other databases used in literature. All images were re-sized to $150 \times 250$ pixels.

Figure 10.7 shows a graph of the training process when using 100 and 300 epochs. The loss and accuracy curves were noisy achieving a low classification rate. This instability persists when increasing the number of epochs and reducing the learning rate. In Figure 10.7a and b, the blue line shows the low error rate reached by the

**TABLE 10.5**

**Parameters of the HED Algorithm (Gaussian Filter Size and Scale factor) and the Classification Results (Acc.: Accuracy) Reached When Using 252 Full Images**

| Gaussian Filter | Scale Factor | TN | FP | FN | TP | Sensitivity (TPR) | Specificity (TNR) | Acc. |
|---|---|---|---|---|---|---|---|---|
| 3 × 3 | 0.5 | 92 | 53 | 33 | 74 | 0.69 | 0.63 | 0.66 |
| 3 × 3 | 0.7 | 99 | 40 | 30 | 83 | 0.73 | 0.71 | 0.72 |
| 3 × 3 | 1.0 | 82 | 50 | 44 | 76 | 0.63 | 0.62 | 0.63 |
| 5 × 5 | 0.5 | 84 | 47 | 44 | 77 | 0.64 | 0.64 | 0.64 |
| 5 × 5 | 0.7 | 85 | 49 | 41 | 77 | 0.65 | 0.63 | 0.64 |
| 5 × 5 | 1.0 | 89 | 45 | 39 | 79 | 0.67 | 0.66 | 0.67 |
| 7 × 7 | 0.5 | 71 | 61 | 40 | 80 | 0.67 | 0.54 | 0.60 |
| 7 × 7 | 0.7 | 88 | 43 | 40 | 81 | 0.67 | 0.67 | 0.67 |
| 7 × 7 | 1.0 | 86 | 50 | 43 | 73 | 0.63 | 0.63 | 0.63 |

**TABLE 10.6**

**Parameters of the HED Algorithm (Gaussian Filter Size and Scale factor) and the Classification Results (Acc.: Accuracy) Reached When Using 252 Half Images**

| Gaussian Filter | Scale Factor | TN | FP | FN | TP | Sensitivity (TNR) | Specificity (TPR) | Acc. |
|---|---|---|---|---|---|---|---|---|
| 3 × 3 | 0.5 | 92 | 60 | 33 | 67 | 0.67 | 0.61 | 0.63 |
| 3 × 3 | 0.7 | 99 | 46 | 30 | 77 | 0.72 | 0.68 | 0.70 |
| 3 × 3 | 1.0 | 82 | 56 | 44 | 70 | 0.61 | 0.59 | 0.60 |
| 5 × 5 | 0.5 | 84 | 63 | 44 | 61 | 0.58 | 0.57 | 0.58 |
| 5 × 5 | 0.7 | 85 | 55 | 41 | 71 | 0.63 | 0.61 | 0.62 |
| 5 × 5 | 1.0 | 89 | 61 | 39 | 63 | 0.62 | 0.59 | 0.60 |
| 7 × 7 | 0.5 | 71 | 66 | 40 | 75 | 0.65 | 0.52 | 0.58 |
| 7 × 7 | 0.7 | 88 | 49 | 40 | 75 | 0.65 | 0.64 | 0.65 |
| 7 × 7 | 1.0 | 86 | 56 | 43 | 67 | 0.61 | 0.61 | 0.61 |

model in the validation set. The grey line, on the other hand, shows the classification accuracy reached for the CNN in the validation set.

### 10.4.2.2   Pre-trained VGG16 Model and Bottleneck

In order to improve the classification results from previous experiments, a pre-trained VGG16 model was used to extract features using the bottleneck technique.

Figure 10.8a shows graphical training process results for bottleneck. Table 10.7 shows the summary of the results using different setting parameters. The best results were reached by the learning rate of $1e-5$, 300 epochs and batch size of 64.

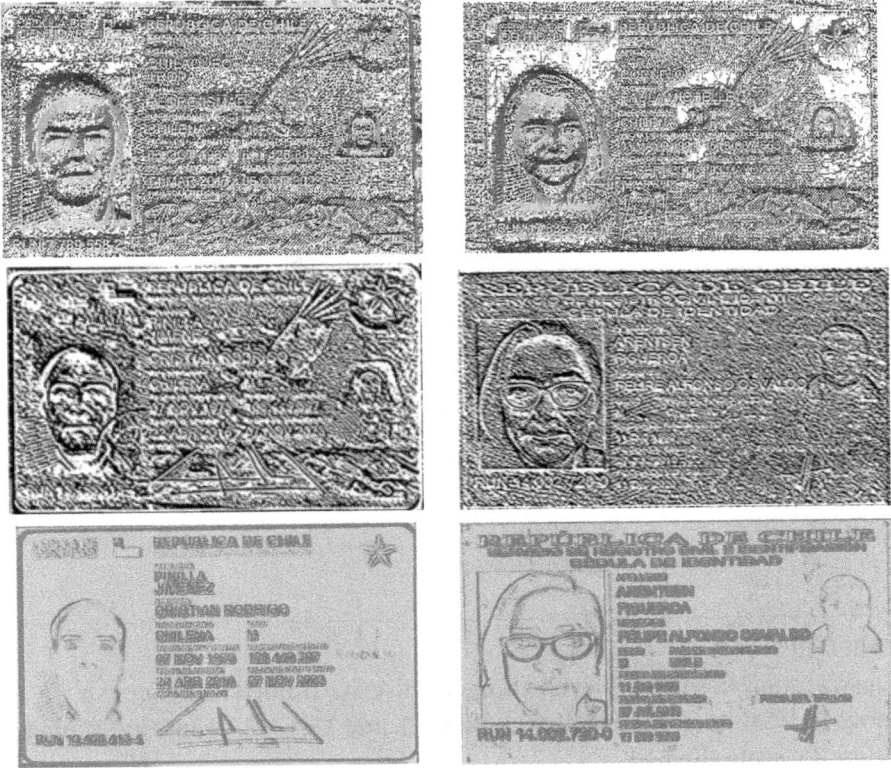

**FIGURE 10.6** Image feature texture analysis. Top images were computed using uLBP. Left corresponds to an original ID card and right to a fake ID card. The middle row shows original (left) and fake (right) ID card when using BSIF 7 × 7. The bottom rows are the same BSIF images but using a colour representation.

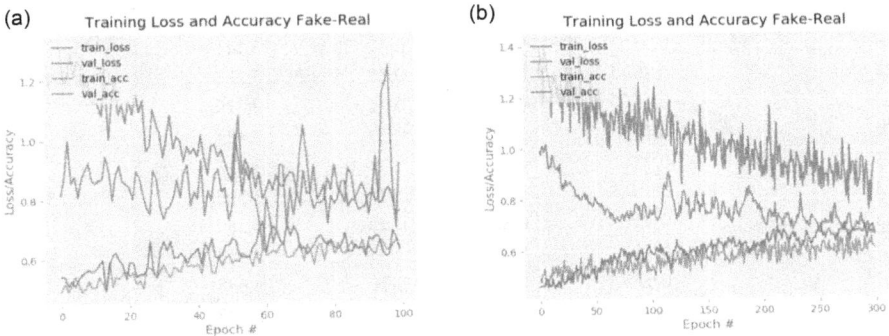

**FIGURE 10.7** Analysis of the training loss and accuracy between fake and real images when training a small-VGG from scratch. In (a), 100 epochs were used, while (b) shows the results for 300 epochs.

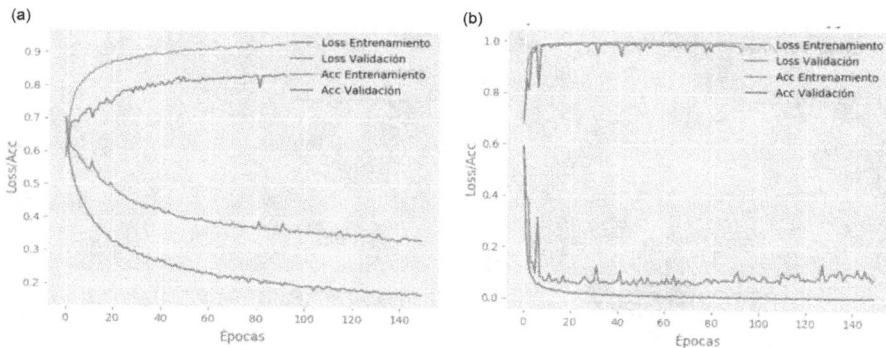

**FIGURE 10.8** Analysis of the bottleneck and fine-tuning techniques applied to a pre-trained VGG16 model. (a) The training loss and accuracy between fake and real images when using only bottleneck. (b) The training loss and accuracy between fake and real images when using only fine-tuning. The x-axis represents the number of epochs and the y-axis, the accuracy.

**TABLE 10.7**

**Classification Results Achieved Using Small-VGG Trained from Scratch (Row 1), VGG-16 Using Bottleneck (Row 2), and VGG-16 Using Fine-Tuning (from Row 3 to the End)**

| Trained Interval | TPR AUC | TNR AUC | ACC | Model Size MB | Conv. Block |
|---|---|---|---|---|---|
| Scratch | 0.60 | 0.61 | 0.60 | 500 | All |
| Bottleneck | 0.79 | 0.72 | 0.76 | 2,857 | -.- |
| L0 | 0.94 | 0.94 | 0.94 | 3,658 | All Trainable |
| L1 | 0.78 | 0.68 | 0.73 | 2,857 | Block 5 |
| L2 | 0.86 | 0.79 | 0.83 | 2,952 | |
| L3 | 0.85 | 0.81 | 0.83 | 3,046 | |
| L4 | 0.93 | 0.88 | 0.91 | 3,141 | |
| L6 | 0.86 | 0.77 | 0.81 | 3,235 | Block 4 |
| L7 | 0.91 | 0.90 | 0.91 | 3,329 | |
| L8 | 0.77 | 0.81 | 0.79 | 3,424 | |
| L9 | 0.95 | 0.91 | 0.93 | 3,518 | |
| L11 | 0.96 | 0.93 | 0.95 | 3,565 | Block 3 |
| L12 | 0.93 | 0.86 | 0.90 | 3,589 | |
| L13 | 0.94 | 0.92 | 0.93 | 3,613 | |
| L15 | 0.94 | 0.87 | 0.91 | 3,648 | |
| L16 | 0.93 | 0.93 | 0.93 | 3,648 | Block 2 |
| L17 | 0.93 | 0.95 | 0.94 | 3,654 | |
| L19 | 0.96 | 0.90 | 0.93 | 3,657 | Block 1 |
| L20 | 0.88 | 0.83 | 0.86 | 3,658 | |

The classification accuracy achieved was 84.00% in the B5 block. This result outperforms the traditional small-VGG trained from scratch (Section 10.4.2.1).

Table 10.7 shows the different results obtained when adjusting each convolutional block of the model and when only using the features extracted by the bottleneck technique.

An example of the activation map using grad-cam algorithm [21] of the national ID card is shown in Figure 10.9. This heat map represents the most relevant areas that are considered during classification.

The colours of the heat map represent the relevance of the features, where the warm colours (red, orange, yellow, and purple) are the most relevant features and cold colours (blue) are the less relevant features. For colour images description check the online version.

### 10.4.2.3 Pre-trained VGG16 Model and Fine-Tuning

In order to better improve the tampering detection on national ID cards, a pre-trained model to extract features using the fine-tuning technique was used.

The fine-tuning technique refers to initialising a CNN with pre-trained parameters instead of random parameters and then re-training it on a new dataset using very small weight updates. This allows the process of network learning to be accelerated and the generalisation skill to be improved, thanks to the initial information that is delivered to the network [5]. This process was completed in three steps:

1. instantiate the convolutional base of VGG16 and load its weights
2. add the previously defined fully connected model on top and load its weights
3. freeze the layers of the VGG-16 models up to the last convolutional block

Figure 10.8b shows a graph of the training process for the best fine-tuning result. The re-trained network from the blocks B1 and B2 reached the best classification accuracy of 95.65% (see the results in Table 10.7). Indeed, if the network is re-trained in a deeper layer, the classification results decreased. However, when using the fine-tuning approach, all the results outperform those obtained when using a small-VGG trained from scratch or when using VGG-16 with bottleneck.

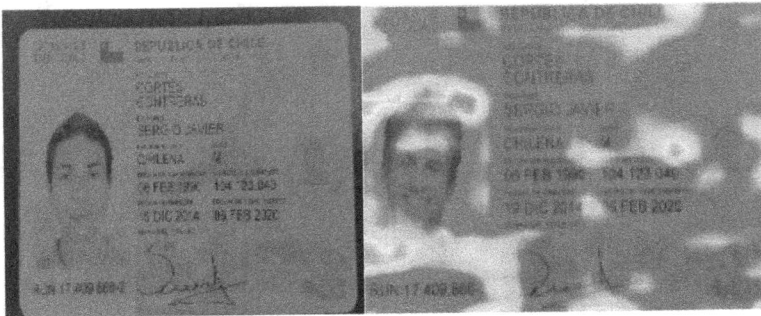

**FIGURE 10.9** Heat map of the most relevant features from the different blocks (stages) belonging to VGG-16.

## 10.5  CONCLUSIONS

This work studied the problem of detecting fake ID cards in remote identification systems. The situation where the ID card is altered by replacing the face photo either manually or digitally was analysed. Several algorithms were tested in order to classify whether or not the ID card has been manipulated. Machine learning algorithms such as the feature extractors BSIF, uLBP, and HED were proposed, and the classification between fake and real was performed using an RF algorithm. The best classification results achieved by those algorithms were only 79%. Texture extraction methods were able to identify borders, lines, and spots coming from the manipulation of the face photo (as shown in Figure 10.5). However, the classification results were not competitive. In order to improve the results, two CNNs were tested. First, a small-VGG trained from scratch, and then, a pre-trained VGG16 model for which bottleneck and fine-tuning techniques were applied. The best classification results were achieved when applying fine-tuning techniques (95.0%) Block 3. Fine-tuning was shown to be the most adequate approach as fake images present simple features such as lines and edges. The first layers from pre-trained models were able to capture these textures. Although this research is preliminary and only included a limited case of ID card manipulation, they show promising results.

## ACKNOWLEDGEMENT

This work was supported by the company TOC Biometric, R+D Centre SR-226.

## REFERENCES

1. Timo Ahonen, Student Member, Abdenour Hadid, Matti Pietikainen, and Senior Member. Face description with local binary patterns: Application to face recognition. *IEEE Transactions on Pattern Analysis and Machine Intelligence*, 28:2037–2041, 2006.
2. H. Alshehri, M. Hussain, H. A. Aboalsamh, and M. A. Al Zuair. Cross-sensor fingerprint matching method based on orientation, gradient, and gabor-hog descriptors with score level fusion. *IEEE Access*, 6:28951–28968, 2018.
3. Claudia Arellano, Juan Tapia. Soft-biometrics encoding conditional GAN for synthesis of NIR periocular images. *Future Generation Computer Systems*, 97:503–511, 2019.
4. Geetika Arora, Kamlesh Tiwari, and Phalguni Gupta. *Selfie Biometrics. Advances in Computer Vision and Pattern Recognition. Liveness and Threat Aware Selfie Face Recognition.* Springer, pp. 197–210, 09 2019.
5. Yoshua Bengio, Pascal Lamblin, Dan Popovici, and Hugo Larochelle. *Greedy Layerwise Training of Deep Networks.* MIT Press, Cambridge, MA, pp. 153–160, 2007.
6. Bir Bhanu and Ajay Kumar. *Deep Learning for Biometrics.* Springer Publishing Company, Incorporated, 1st edition, 2017.
7. Leo Breiman. Random forests. *Machine Learning*, 45(1):5–32, October 2001.
8. I. Chingovska, A. Anjos, and S. Marcel. On the effectiveness of local binary patterns in face anti-spoofing. In *2012 BIOSIG - Proceedings of the International Conference of Biometrics Special Interest Group (BIOSIG)*, pp. 1–7, September 2012.
9. A. Czyzewski, P. Hoffmann, P. Szczuko, A. Kurowski, M. Lech, and M. Szczodrak. Analysis of results of large-scale multimodal biometric identity verification experiment. *IET Biometrics*, 8(1):92–100, 2019.

10. J. Deng, W. Dong, R. Socher, L.-J. Li, K. Li, and L. Fei-Fei. ImageNet: A Large-Scale Hierarchical Image Database. In *CVPR09*, 2009.
11. M. E. Fathy, V. M. Patel, and R. Chellappa. Face-based active authentication on mobile devices. In 2015 *IEEE International Conference on Acoustics, Speech and Signal Processing (ICASSP)*, pp. 1687–1691, April 2015.
12. J. Galbally, S. Marcel, and J. Fierrez. Biometric antispoofing methods: A survey in face recognition. *IEEE Access*, 2:1530–1552, 2014.
13. Felix Juefei-Xu, Eshan Verma, Parag Goel, Anisha Cherodian, and Marios Savvides. Deepgender: Occlusion and low resolution robust facial gender classification via progressively trained convolutional neural networks with attention. In *The IEEE Conference on Computer Vision and Pattern Recognition (CVPR) Workshops*, June 2016.
14. Juho Kannala and Esa Rahtu. BSIF: Binarized statistical image feature. *21st International Conference on Pattern Recognition (ICPR 2012)*, Tusuka, Japan, pp. 1363–1366, 2012.
15. Y. LeCun, B. Boser, J. S. Denker, D. Henderson, R. E. Howard, W. Hubbard, and L. D. Jackel. Backpropagation applied to handwritten zip code recognition. *Neural Computation*, 1(4):541–551, December 1989.
16. J. K. Lee, S. R. Ryu, and K. Y. Yoo. Fingerprint-based remote user authentication scheme using smart cards. *Electronics Letters*, 38(12):554–555, June 2002.
17. Sebastien Marcel, Mark S. Nixon, and Stan Z. Li. *Handbook of Biometric Anti-Spoofing: Trusted Biometrics Under Spoofing Attacks*. Springer Publishing Company, Incorporated, 2014.
18. Sebastien Marcel, Mark S. Nixon, Julian. Fierrez, and Nicholas Evans. *Handbook of Biometric Anti-Spoofing: Presentation Attack Detection*. Editors: Marcel, S., Nixon, M.S., Fierrez, J., Evans, N. (Eds.); Springer International Publishing, 2018, 2nd ed. ISBN: 978–3319926261, 09 2018.
19. P. Oza and V. M. Patel. Active authentication using an autoencoder regularized CNN-based one-class classifier. In 2019 *14th IEEE International Conference on Automatic Face Gesture Recognition (FG 2019)*, pp. 1–8, May 2019.
20. P. Perera and V. M. Patel. Face-based multiple user active authentication on mobile devices. *IEEE Transactions on Information Forensics and Security*, 14(5):1240–1250, May 2019.
21. Ramprasaath R. Selvaraju, Abhishek Das, Ramakrishna Vedantam, Michael Cogswell, Devi Parikh, and Dhruv Batra. Grad-cam: Why did you say that? Visual explanations from deep networks via gradient-based localization. *CoRR*, abs/1610.02391, 2016.
22. Yichun Shi and Anil K. Jain. Docface: Matching ID document photos to selfies. *CoRR*, abs/1805.02283, 2018.
23. K. Simonyan and A. Zisserman. Very deep convolutional networks for large-scale image recognition. *CoRR*, abs/1409.1556, 2014.
24. M. Stokkenes, R. Ramachandra, and C. Busch. Biometric transaction authentication using smartphones. In *2018 International Conference of the Biometrics Special Interest Group (BIOSIG)*, pp. 1–5, September 2018.
25. Yi Sun, Xiaogang Wang, and Xiaoou Tang. Hybrid deep learning for face verification. In *The IEEE International Conference on Computer Vision (ICCV)*, December 2013.
26. Yaniv Taigman, Ming Yang, Marc'Aurelio Ranzato, and Lior Wolf. Deepface: Closing the gap to human-level performance in face verification. In *Proceedings of the 2014 IEEE Conference on Computer Vision and Pattern Recognition*, CVPR '14, pp. 1701–1708, 2014.
27. J. E. Tapia and C. A. Perez. Clusters of features using complementary information applied to gender classification from face images. *IEEE Access*, 7:79374–79387, 2019.
28. Saining Xie and Zhuowen Tu. Holistically-nested edge detection. *International Journal of Computer Vision*, 125(1–3):3–18, December 2017.

29. F. J. Zareen and S. Jabin. Authentic mobile-biometric signature verification system. *IET Biometrics*, 5(1):13–19, 2016.

30. Peng Zhou, Xintong Han, Vlad I. Morariu, and Larry S. Davis. Learning rich features for image manipulation detection. In *The IEEE Conference on Computer Vision and Pattern Recognition (CVPR),* June 2018, pp. 1053–1061.

# 11 Indexing on Biometric Databases

*Geetika Arora, Jagdiah C. Joshi,*
*Karunesh K. Gupta, and Kamlesh Tiwari*
Birla Institute of Technology and Science Pilani

## CONTENTS

11.1 Introduction .......................................................................257
11.2 Indexing Facial Images.........................................................260
    11.2.1 Predictive Hash Code ................................................261
    11.2.2 Results.......................................................................262
11.3 Indexing Fingerprint Images ................................................264
    11.3.1 Coaxial Gaussian Track Code ....................................268
    11.3.2 Results.......................................................................270
11.4 Indexing Finger-Knuckle Print Database .............................270
    11.4.1 Boosted Geometric Hashing.......................................270
    11.4.2 Results.......................................................................272
11.5 Indexing Iris Images.............................................................273
    11.5.1 Indexing of Iris Database Based on Local Features.......273
    11.5.2 Results.......................................................................275
11.6 Indexing Signature Images ..................................................276
    11.6.1 KD-Tree-Based Signature Database Indexing..............276
    11.6.2 Results.......................................................................276
11.7 Conclusion ..........................................................................277
References......................................................................................278

## 11.1 INTRODUCTION

Authentication is essential to enforce access control. A natural way to recognise a person for authentication is through his physiological or behavioural characteristics. Physiological characteristics include face, iris, fingerprint, palm print, knuckle print, *etc.* that can be acquired from an individual's body. However, behavioural characteristics are those that can be observed when a person accomplishes a specific task, such as talk, walk, and signs. All these characteristics are called biometric traits and are supposed to be unique to a person. Biometric traits are binding to a person and, therefore, are always available with the user and are difficult to steal. Figure 11.1 shows images of some of the popular biometric traits. To build an automated authentication system for this recognition approach, one has to choose a suitable trait having specific properties such as universality, uniqueness, permanence,

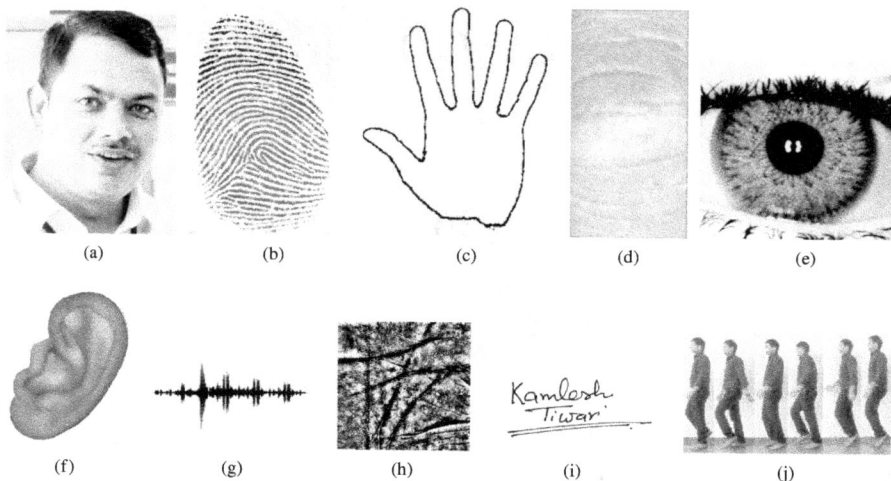

**FIGURE 11.1** Samples of some popular biometric traits such as (a) face, (b) fingerprint, (c) hand geometry, (d) knuckle, (e) iris, (f) ear, (g) voice, (h) palm print, (i) signature, and (j) gait.

collectability, acceptability, and circumvention [26]. Getting all these properties fully satisfied in a single trait is difficult because physiological characteristics depend on biological tissue that is prone to aging. Behavioural characteristics can be influenced by the environment and emotions. Decision on the choice of suitable biometric trait depends on the application for which the recognition system is to be deployed. If the authentication is to be deployed for a high-security area such as opening a currency chest, then highly accurate biometric trait, such as iris, could be deployed even when the trait has lesser user acceptance. It should be noted that providing iris sample is inconvenient to the user so its acceptance is low. When the system is to be deployed to less-security area such as entrance to a shopping mall then we can use traits that have high social acceptance such as face even when its accuracy is a bit low.

Authentication has two aspects. Either we wish to determine whether the given two samples of a biometric trait are acquired from the same person or we wish to determine the identity of an acquired sample. The former is called as verification while the latter is known as identification. Of these two, identification is more challenging and it is based on the biometric template database collected during the time of registration or enrolment. It can be observed that if the size of the gallery database is $n$ then, to establish the identity of any acquired sample, it has to be compared with all the existing $n$ samples. Due to this reason, identification takes more time when the size of the database increases. But, the increase in the size of database is bound to happen with time because a working system would keep registering more and more users in the database over time. Consider an example, assume a CCTV camera has recorded a face image of some suspicious person roaming around. A natural question would be to identify that person. To answer this, one would need to refer a facial database and compare the face in question with all faces present in the database. The time needed to perform this investigation would be proportional to the number of images in the database, and this time would increase if more face images are included in the database. This process of identity discovery is called as identification.

To automate the identification process, one has to determine suitable feature that represents the sample and then develop a matching strategy. However, to reduce the time taken for comparison, there is a need to develop a technique that can efficiently produce a small candidate list that would have high similarity score for the query image. This will result in filtering out highly dissimilar samples from being compared, thus speeding up the identification process and saving time. Such a technique is called as indexing. Indexing has two coupled phases. In the first phase, a suitable index structure is created, *i.e.*, every sample is associated with a feature vector which conforms to an index of the database. The second phase, called as *retrieval*, produces a candidate list by fetching the samples from the database that are contained in the most similar index as that of the query sample. Both the methods are related and complement each other. Figure 11.2 depicts the process of indexing on a biometric database. The process starts with extracting features from the biometric sample and constructing a feature vector that best represents that sample. The generated feature vector is used for index generation and the index is stored in an index table. Whenever a query biometric sample is given to an identification system, its corresponding features are also extracted using the same feature extraction module and is mapped to the most similar index or bin in the index table. The templates stored in those bins are retrieved as a candidate list for comparison. The size of the candidate list is small as compared to the original database size. Developing an indexing scheme for a biometric database is challenging because of the inherent properties of the biometric traits and features.

One such challenge is due to high dimensionality of the biometric features as a result of which even multi-dimensional data structures like KD-tree and R-tree do not minimize the search space, and the efficiency is no better than exhaustive search. Another challenge is that features obtained from the two samples taken at different time instants of the same subject and trait are not guaranteed to be same. They are similar but not the same. Moreover, there could be the presence of some false

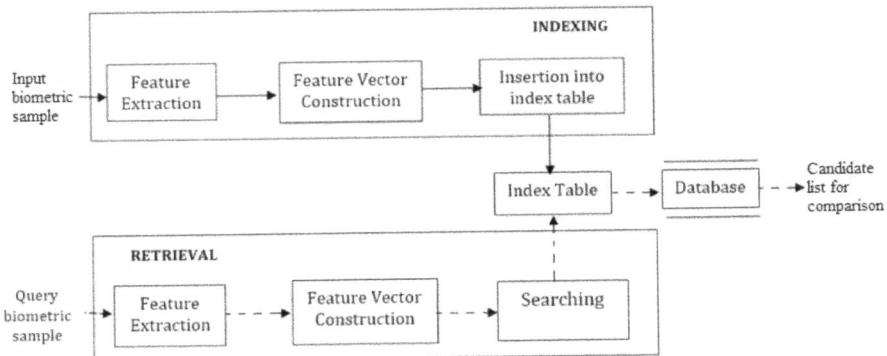

FIGURE 11.2   Typical block diagram of a biometric database indexing scheme.

features and some true features that may miss out. One more problem is related to the feature dimensions that it is not ordered. Of the available features, it is not easy to decide which one to consider first and which one has to be taken second. Understand this with the help of a point $(x, y)$ in 2D space; note that $(y, x)$ is an entirely different point and if you do not know which value of $x$ and which is $y$ then you would not be able to locate the actual point. Another challenge with biometric sample in general is the creative user behaviour and unique interaction with the sensor. This may lead to various issues such as pose, occlusion, illumination, rotation, and shear transformations. Interoperability could also be an issue that arises when we use different kind of sensors to acquire different samples.

It should be noted that the biometric indexing is not the same as the general database indexing. The difference lies in the size of the retrieved list. If the index of a query item is obtained in the general database setting, it retrieves a unique database item having the same hash value equal to the query item. However, in biometric indexing, it would return a list of similar items from the database, none of them may have the same hash value. The purpose here is to continue with the search by further matching with the retrieved items for identification.

Evaluation of an indexing scheme is done based on some parameters such as *hit rate*, *penetration rate*, and *bin miss rate*. Hit rate is the percentage of genuine matches that are successful at top $t$ matches from the total number of queries made. Penetration rate refers to the percentage of database that must be returned as the candidate list for a successful retrieval. *Bin miss rate* is another parameter that represents fraction of genuine biometric templates misplaced in a wrong class.

This chapter introduces biometric indexing and its challenges. It also provides details of an indexing technique for some popular biometric traits databases. Four physiological biometric traits are considered, *viz.*, face, fingerprint, finger-knuckle print, and iris along with a behavioural trait, *viz.* signature. The next section describes facial image indexing scheme by learning predictable binary code. Section 11.3 explains fingerprint indexing using the special code called Coaxial Gaussian Track Code (CGTC). A method for finger-knuckle print indexing is explained in the subsequent section using boosted geometric hashing. Section 11.5 focuses on Iris database indexing by making use of local features. A technique to index behavioural biometric trait, *viz.* signature, is explained in Section 11.6. Conclusions are presented at the end.

## 11.2 INDEXING FACIAL IMAGES

Human face is one of the most natural choices for biometric recognition. Authentication using facial data has been in place from the time people started using photographs [8]. Face is suitable for law enforcement and surveillance, such as CCTV control, suspect tracking, shoplifting, and investigation. The authentication process has been fairly manual till the early 2000s after which the automated face recognition using facial database has been started. Face recognition is a visual pattern recognition problem that takes a face as a three-dimensional input that may have pose, illumination, or emotion variation and identifies it on the basis of its two-dimensional representation. An automatic face recognition system broadly involves four modules, face detection, alignment, feature extraction, and lastly, matching and decision-making. The block

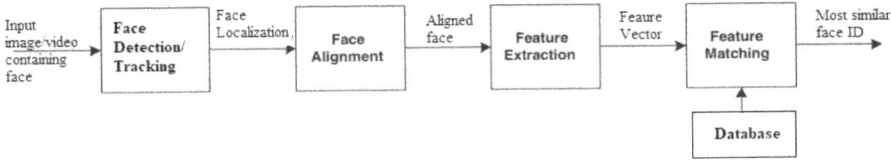

**FIGURE 11.3**   Block diagram of a face recognition system [27].

diagram of the face-recognition process is shown in Figure 11.3. Face detection aims at identifying face from an image and segmenting out the face portion from the background. However, in case of videos, face needs to be tracked down. After getting the face from the image/video, it is aligned in a uniform coordinate system. The face is then normalised geometrically and photometrically to account for pose and illumination variation, respectively. The pose variation is addressed by normalizing the face using the localization points of the face components, such as eyes, nose, mouth, and facial outline. After normalization, representative yet discriminating features are extracted from the face. Lastly, matching is performed using the extracted features, thus making the feature extraction process highly important [27]. Features can be referred to as shallow and deep features [53]. Shallow features are the ones that are extracted using handcrafted local image descriptors such as SIFT, LBP, and HOG and are then concatenated to form a representation that describes the face as a whole. Deep features, on the other hand, are extracted from a learned function, also referred to as a deep neural network, that takes a face image as input and outputs the salient features describing that image.

Face recognition can be done as either a verification or an identification process based on the type of application it is targeted for. Apart from identification of the facial images, there has been a demand to upscale the database and search ideal match in that large database. For example, social networking websites such as Facebook and Instagram have a large number of users who upload millions of images on daily basis. Now the task in hand is to auto-tag the people in the images. Also, in criminal investigations, it is required to find a match of a probe image from a database containing millions of images. These processes are very compute-intensive as they require the number of verifications proportional to the size of the database. The efficiency decreases with increase in the size of the database. To fix the efficiency deterioration, there is a need to develop a strategy that can perform pre-filtering on the database in constant time to produce a small fixed-length candidate set of fingerprints having probabilistic guarantees of hit rate. This is achieved by indexing the database.

## 11.2.1  PREDICTIVE HASH CODE

Hashing methods, which refer to learning binary code representations with Hamming distance calculation, have been used lately for retrieval of images in large-scale databases. These methods speed up the searching process, but due to variations of illumination, pose, and expression in facial images, the hashing codes tend to become unstable. Therefore, to apply hashing on facial features, feature should be

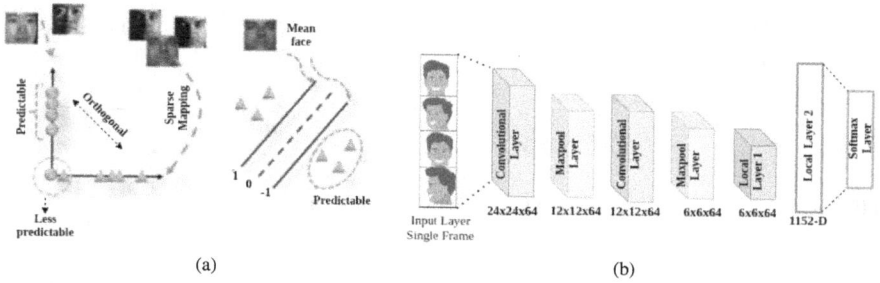

(a)                                                                      (b)

**FIGURE 11.4** Block diagram and the architecture. (a) Block diagram depicting learning of predictable binary codes [22]; (b) architecture of the CNN that is utilized to improve the predictability of the binary code [22].

predictable even with the presence of facial variations. To address this, a predictable hash code (PHC) that embeds facial features into Hamming space has been proposed in Ref. [22]. The code is learned in such a way that inter-class distance is maximised while minimizing the intra-class distance. To do so, mean face of each class is identified. But faces of the same class can also suffer from huge variations due to the presence of pose, illumination, and expression, thus making hamming distance a strong constraint. To address this, it has been enforced that the codes corresponding to facial images of the same person must be similar to the code of the mean face of the same subject. To account for maximizing inter-class distance between the codes, it has been ensured that the codes corresponding to the mean faces of different classes are orthogonal to each other. Expectation maximization has been utilised to find the linear mapping of the face image to a predictable binary code. The process is depicted in Figure 11.4a. This is implemented as two different models: one that uses L1-norm and the other uses L2-norm. L1-norm outputs a sparse mapping while L2-norm gives a dense mapping. A convolutional neural network (CNN)-based architecture has also been used that takes the non-preprocessed grey-scale face image as input and outputs its feature representation. It is employed to enhance the predictability of the binary codes and is trained using a softmax layer that has nodes equal to the number of classes. The architecture of the proposed network is given in Figure 11.4b (images taken from Ref. [22]).

## 11.2.2 RESULTS

The aforementioned technique has been tested on three publicly available standard data sets, FRGC [54], AR [57], and YouTube Celebrities [52] data set. A subset of facial images from FRGC data set, *i.e.* first 20 images of each subject, has been taken for experimentation. This makes it a total of 3,720 images collected from 186 subjects, which are cropped to a size of 32 × 32. Some cropped facial images of the same person are shown in Figure 11.5a. The experiment has been conducted in two phases, closed set and open set scenario. The first 10 images of 100 subjects have been taken as a training set. In the closed set scenario, the training set has been considered

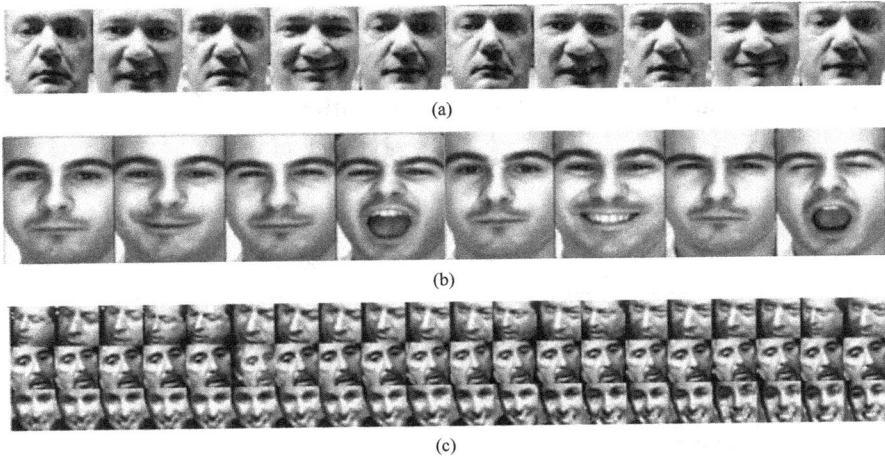

**FIGURE 11.5**   Facial images available in some of the popular open-source facial database. (a) Cropped facial images taken from FRGC dataset [54]; (b) face images of a subject from AR database [57]; (c) cropped facial images of three subjects taken from YouTube Celebrities dataset [52].

as gallery and the remaining 10 images of the same subjects have been taken for testing. However, in the open set scenario, the first 10 images of the remaining 86 subjects have been considered for gallery and the remaining 10 images have been used as query images. The proposed PHCs are denoted using PHC-L2 and PHC-L1 (L2-norm and L1-norm, respectively). The proposed technique has been compared using popular hashing methods, such as Locality Sensitive Hashing (LSH) [18], Spectral Hashing (SH) [62], Iterative Quantization (ITQ) [20], Linear Discriminative Analysis Hash (LDAH) [59], Binary Reconstruction Embedding (BRE) [35], Kernel-Based Supervised Hashing (KSH) [39], and Fast Supervised Hashing (FastH) [38]. For ITQ, its supervised version (CCA-ITQ) and unsupervised version (PCA-ITQ) are included. Recognition accuracy of the proposed technique PHC-L2 and PHC-L1 on FRGC data set for closed set and open set is shown in Ref. [22] that supersedes all other methods such as LSH, SH, ITQ, LDAH, BRE, KSH, FastH, CCA-ITQ, and PCA-ITQ.

The AR database [57] consists of 4,000 images collected from 126 individuals. It covers images with different illumination, facial expressions, and occlusion. But in the proposed technique, the experiment has been conducted by taking eight facial images of 100 subjects each. These images have been down-sampled to a size of $28 \times 23$. Some of the down-sampled images from AR database are shown in Figure 11.5b. Images of first 50 subjects have been used for training while the images from remaining 50 subjects have been used for testing. Of the testing images, the first four of each subject have been used as probe images and the remaining four have been used as gallery images. The recognition accuracy vs. the number of bits used for feature on the YouTube celebrities data set as shown in Ref. [22] depicts that in the initial phase, supervised hashing methods perform better than the unsupervised ones because of

**TABLE 11.1**

**Predictability Analysis (Average Recognition Accuracy (in %) ± Standard Deviation) for the Proposed Feature Representation [22]**

| | Feature Size (Bits) | | |
|---|---|---|---|
| | 32 | 64 | 128 | |
| Pixels | 21.13 ± 3.01 | 46.31 ± 1.43 | 54.82 ± 2.68 | 58.37 ± 4.36 |
| CNN | 59.13 ± 2.10 | 74.29 ± 3.20 | 80.94 ± 1.65 | 83.15 ± 0.68 |

the presence of illumination and pose variation in the frontal images of both the sessions. This was an open-set problem because the training and testing images are from different subjects. Hence, it can be said that for such problem, larger number of bits are required to attain good recognition rate.

The third data set, *i.e.* YouTube celebrities data set [52], consists of video clips instead of images. It has 1,910 clips from 47 individuals collected from YouTube. Forty-one images have been collected for every person by clipping three videos each, and then, these images have been cropped to a size of 30 × 30 as shown in Figure 11.5c. Six images for every person have been used for testing, thus making it a total of 44,172 and 239,997 images in the testing and training set, respectively. The training images have been used for training the CNN that outputs a 1,152-dimensional feature vector for every image. The database has also been augmented by flipping the images for better training of the network. The similarity is computed between every pair of test and train set using Hamming distance. The plot between recognition accuracy vs. number of bits for AR and YouTube Celebrities data set as shown in Ref. [22] shows that PHC-L2 has highest accuracy per bit as compared to all other methods such as LSH, SH, ITQ, LDAH, BRE, KSH, FastH, CCA-ITQ, and PCA-ITQ. The predictability (accuracy ± standard deviation) of the feature representation using the proposed methods (both pixels and CNN) has also been analysed for different feature length (number of bits). The result is shown in Table 11.1. A higher recognition rate with lower standard deviation indicates that the proposed features are predictable. It has also been observed that the code obtained from CNN features tends to attain better recognition rate (68.79%) than the code obtained by using the image pixels directly (58.37%).

## 11.3 INDEXING FINGERPRINT IMAGES

Fingerprint is an impression formed when the inner surface of the finger comes in contact with a surface. Earlier used in crime forensics, fingerprints have gained popularity and have become most widely used biometric trait for access control in day-to-day life applications [45]. A fingerprint has various properties that make it a capable and reliable trait to be used for authentication. Fingerprints are found to be unique with every individual and even in different fingers of the same individual. They do not undergo temporal changes and are acceptable by the society. Acquisition and identification of a fingerprint sample is easier and is inexpensive [21]. Fingerprints are considered as the most expressive biometric trait because sufficient amount of unique features can be found on it. It consists of a pattern of curves also called as ridges and the area between the ridges known as valley. The ridge-ending and bifurcation points are known as minutiae points. The point with the highest curvature is known as core point. Delta is a point from where the ridges spread in three directions. An island is a line that stands alone and does not touch any other line type. All the aforementioned characteristics are unique in every fingerprint and are shown in Figure 11.6.

An automatic fingerprint identification system (AFIS) aims to establish the identity of a human being based on the acquired fingerprint. But with the increase in the size of the database, the process of identification becomes compute-intensive as the number of comparisons increases. It may also result in an increase in the number of false positives. To tackle this problem, there is a need to reduce the search space. To achieve this, the pre-selection techniques have been applied on the fingerprint databases that can be broadly classified into two categories: (i) exclusive classification and (ii) continuous classification or indexing. In exclusive classification, a fingerprint pattern is classified into three major categories, namely, loops, whorls, and arches. It is found that 65% of fingerprints are loops, 30% are whorls, and the remaining 5% are arches [7]. The loops and arches have been further subclassified into radial and ulnar loop and tented and plain arches. This categorisation divides a given fingerprint database into five mutually exclusive classes, and therefore, it is termed as exclusive classification. During retrieval, the class of the query or the probe fingerprint is determined, and the fingerprint templates stored in the chosen class are retrieved for

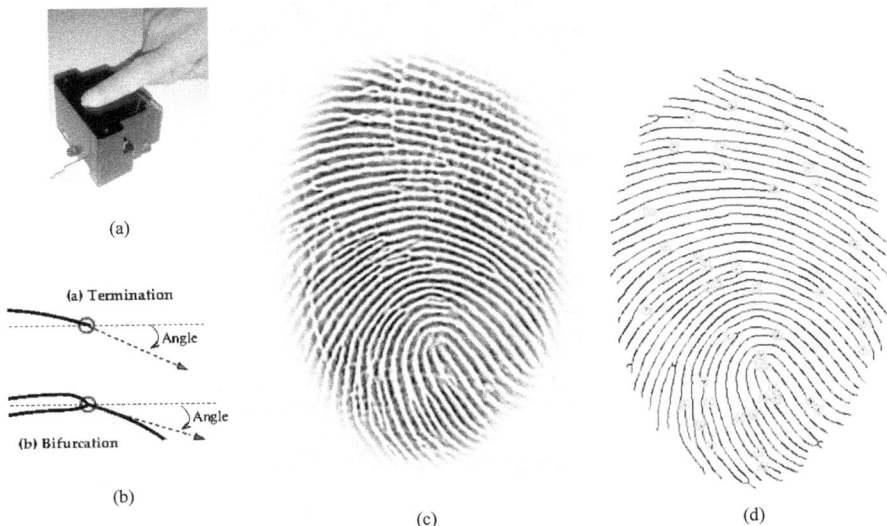

FIGURE 11.6 Fingerprint scanner, feature points, fingerprint images, and minutiae feature marked on skeleton fingerprint images. (a) Acquisition; (b) features; (c) fingerprint; (d) minutiae.

comparison. However, this technique suffers from a limitation due to non-uniform distribution of the fingerprints among the five classes. It has been seen that 93.4% of the fingerprints lie in three classes, namely, left loop, right loop, and whorl. Thus, there is no significant reduction in the search space for comparison. Also, it is difficult to determine the class of the probe fingerprint [63]. To address these limitations, indexing of the fingerprint database was introduced. During indexing, rather than putting a fingerprint in one class, it is represented by a feature vector that contains its most distinguishing yet important characteristics. Every feature vector is associated with an index that is stored in the index table. During retrieval, the feature vector of the probe fingerprint is extracted and the corresponding most-similar index is found out. The candidates lying against the chosen index are fetched for comparison with the probe fingerprint [34]. Fingerprint indexing techniques can be categorised into four classes based on the type of feature they utilize for indexing. These categories are as follows: (i) texture-based, (ii) minutiae-based, (iii) hybrid, and (iv) deep neural network-based approaches [21].

**Algorithm 11.1: Indexing-FP** *(hashTable, fp)*

**Input**: 2D lookup table *hashTable* and a fingerprint *fp*
**Output**: Updated *hashTable*

1. The core point $(cp)$ is detected $cp = (cp.x, cp.y, cp.\theta)$ of *fp*
2. The fingerprint *fp* is transformed by rotating the *fp* anti-clockwise by $cp.\theta$ degree with center at $(cp.x, cp.y)$
3. The fingerprint *fp* is divided into 72 sectors keeping core point as center and starting numbering from the horizontal.
4. **for** every minutia $m_j \in fp$ **do**
5. The minutiae points $m_j$ are extracted and the CGTC vector is constructed
6. The sector $m_j.s$ and distance $m_j.d$ in which $m_j$ lied, with respect to the core point, is obtained.
7. The fingerprint id $fp_{id}$ and its corresponding CGTC vector $m_{j(CGTC)}$ is inserted in *hashTable* at location $(m_j.s, m_j.d)$.
8. **return** hashTable

**Texture-based Indexing Techniques**: The texture-based indexing approaches use global features such as ridge orientation, ridge frequency, type of ridge pattern, and its flow structure, delta and core point(s). Cappelli et al. proposed a technique that utilizes scalar and vector features obtained from ridge orientation and frequency for indexing [10]. Liu et al. have used orientation fields of ridges and valleys for feature vector construction [40]. Polar complex moments (PCMs) have been employed to extract rotation invariant feature representation. It has been found that these features are not good at handling occlusion, translation, rotation, *etc.* They also require the fingerprint to be aligned with respect to the core point which is not possible in low-quality images.

**Minutiae-based Indexing Techniques**: As the name suggests, the minutiae-based indexing techniques consider properties of minutiae points for feature vector construction. These can be further sub-classified as minutia-singles, doubles, triplets, quadruplets, and minutia cylinder-based techniques on the basis of the number of minutiae that have been utilised for the feature vector. A single minutia-based indexing technique that utilizes geometric hashing has been proposed in Ref. [29]. It constructs a fixed length feature vector corresponding to each minutia that gets indexed into the hash table exactly once. An indexing approach that uses minutiae pair has been proposed in Ref. [5]. It uses Euclidean distance between minutiae points for feature vector generation. Minutiae triplets-based indexing techniques use descriptors derived from the triangles formed using the minutia points. A technique that defines triangle set based on extension of Delaunay triangulation [12] has been proposed in Ref. [17]. It uses a relative direction of each minutia with respect to the opposite side of the triangle along with the number of ridges that lie between minutiae pairs. Additionally, it also deals with the problem of missing and spurious minutiae. Minutia quadruplets have been considered to be more robust to distortions. An indexing approach that utilizes quadruplets of minutiae has been proposed in Ref. [25]. It considers seven geometric features such as difference of opposite internal angles, height, area for creating index and diagonals. $K$-means [42] clustering is then implemented on the obtained feature vectors to group similar feature vectors together. The indexing techniques based on quadruplets were found to be accurate, but with the increase in the consideration of more minutiae points together, the computational complexity also tends to increase. The minutiae cylinder code is a three-dimensional data structure that depicts spatial relationship between distance and orientation of neighbouring minutiae. An MCC-based indexing technique that encodes each minutia's neighbourhood into a fixed length vector has been proposed in Ref. [11]. Indexing of the generated feature vectors has been done using LSH [14], and retrieval is done by applying hash functions to the feature vectors.

**Algorithm 11.2: Retrieval-FP** *(hashTable, qf p)*

**Input**: *hashTable*: Index table, $q_f p$: query fingerprint

1. Minutiae and core point from the $q_f p$ are extracted $q_c p = (q_c p.x, q_c p.y, q_c p.\theta)$ of $q_f p$
2. Minutiae points are geometrically transformed, by rotating $q_f p$ with angle $q_c p.\theta$ anti-clockwise having centre at $(q_c p.x, q_c p.y)$
3. **for** all minutiae point $m_i \in q_f p$ **do**
4. The sector and distance $(m_i.s, m_i.d)$ of $m_i$ are determined and the CGTC vector $m_{i(CGTC)}$ of $m_i$ is constructed
5. $S_{m_i} = \phi$
6. **for** all $m_j \in \delta d \times \delta s$ neighbourhood of $(m_i.s, m_i.d)$ **do**
7. All the CGTC vectors $m_{j(CGTC)}$ of $m_j$ are retrieved from *hashTable*
8. **if** $dist(m_{i(CGTC)}, m_{j(CGTC)}) < Th$ **then**

9. **if** $id(m_j) \notin S_{m_i}$ **then**
10. $id(m_j)$ is inserted in $S_{m_i}$ with a score $e$
11. else
12. $S_{m_i}[id(m_j)] = S_{m_i}[id(m_j)] \times e$
13. $K = \phi$
14. **for** all minutiae point $m_i \in Q$ **do**
15. **for** *all id* $\in S_{m_i}$ **do**
16. if $id \notin K$ then
17. $id$ is inserted in $K$ with score $S_{m_i}[id]$
18. else
19. $K[id] = K[id] \times S_{m_i}[id]$
20. For each $id$, $K$ is arranged in decreasing order of score
21. **return** top-k id's from $K$ as candidate set

**Hybrid Indexing Techniques**: These techniques combine texture and minutiae features for indexing. Such a technique has been proposed in Ref. [65], which utilizes minutia triplets along with FOMFE coefficients [61] to generate two different feature vectors. Two different candidate lists for comparison are generated from these vectors which are then combined using fuzzy rules to output a single list of candidates.

**Deep Neural Network-Based Techniques**: These techniques employ deep learning framework for feature extraction instead of using handcrafted features for the purpose of indexing. A CNN [37]-based indexing technique has been proposed in Ref. [9]. It first aligns all the fingerprints in a unified coordinate system using the orientation field dictionary and then trains a CNN to learn a 2,048-d fixed length feature vector from the fingerprint image. This process is repeated for the query fingerprint, and it is compared with the gallery templates to fetch top-$k$ most similar candidates.

Let us now look closely a minutiae-based fingerprint indexing technique that constructs a special feature vector for the purpose termed as CGTC around every minutiae point.

### 11.3.1 COAXIAL GAUSSIAN TRACK CODE

A fingerprint indexing technique that uses directional and spatial information of minutiae points to construct a feature vector called as CGTC has been proposed in Ref. [2]. A minutiae point mi is represented as a three tuple $(x_i, y_i, \theta)$, where $x_i, y_i$ are the coordinates of the minutia point and $\theta$ represents the orientation. These three characteristics are required to construct the CGTC vector which is of fixed length. Figure 11.7 shows an example of CGTC vector for a specific minutia point. The technique is divided into three phases: (i) CGTC vector construction, (ii) indexing, and (iii) retrieval.

Indexing. Let there is a fingerprint $F_i$ with $F_{iM}$ number of minutiae points and a core point represented by the coordinates (c_xi, c_yi) and c $\theta$ orientation. For the *i*th minutia $m_i \in F_{iM}$, a l-bit CGTC vector is constructed using computing binary values from the minutiae points lying in its neighbourhood. CGTC is computed for every minutia point, thus, it is rotation and translation invariant. After constructing

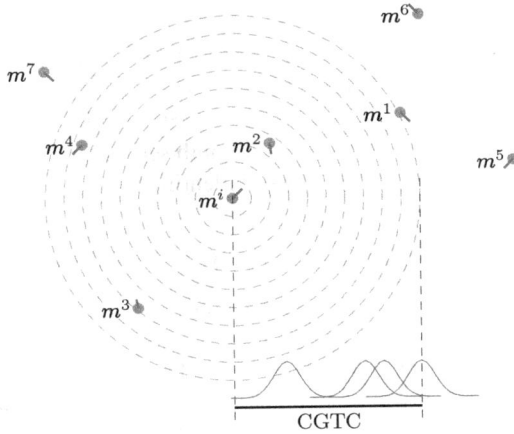

**FIGURE 11.7** Example of a CGTC vector for minutiae $m_i$.

CGTC vector corresponding to every minutia point of each fingerprint, an index table has to be generated that would contain all the feature vectors. The fingerprint $F_i$ is first rotated using the orientation angle of its core point c_$\theta$ in such a way that it becomes parallel to the positive $x$-axis. The fingerprint is then divided into 72 sectors of 5° starting from horizontal. Corresponding to each minutia point mi lying in $F_i$ determines its sector ($m_i.s$) and Euclidean distance ($m_i.d$) with respect to the core point. The CGTC vector along with the fingerprint ID, to which the minutia point belongs, is inserted into the index table at location ($m_i.s$, $m_i.d$).

Retrieval. During the retrieval stage, the minutiae points lying in the probe fingerprint $Q_i$ are extracted, and the CGTC vector corresponding to every minutia is constructed. Subsequently, the core point of the $Q_i$ is extracted and it is rotated in a manner such that the orientation of the core point becomes parallel to the horizontal. Now, following the similar procedure, for every minutia $m_{Q_i}$ lying in $Q_i$, distance and sector $m_{Q_i} \cdot d$ and $m_{Q_i} \cdot s$ with respect to the core point are computed. All the CGTC vectors along with the fingerprint IDs stored at location $m_{Q_i} \cdot s$, $m_{Q_i} \cdot d$ and $\delta_s \times \delta_d$ neighbourhood in the index table are fetched as the candidates for comparison. For all $m_{Q_i}$, candidate list is retrieved from the index table and CGTC vectors are compared using the hamming distance [50]. A score list is maintained which consists of fingerprint ID and its score. If the computed hamming distance is greater than the pre-defined threshold value, the fingerprint ID along with a score e, whose value is set to 2, is inserted into the score table. This score gets exponentially multiplied every time the same fingerprint ID is encountered. Therefore, for every minutia $m_{Q_i}$, a score list will be obtained and these lists are then concatenated to obtain a single score list for a fingerprint. It is then sorted in decreasing order of scores and the top-k candidates are retrieved as a candidate set for identification. The algorithms (taken from [2]) for indexing and retrieval are given as Algorithms 11.1 and 11.2, respectively.

## 11.3.2 RESULTS

As discussed, the technique has been tested on FVC 2004 DB1a database [44]. It consists of 800 fingerprints collected from 100 subjects. It has been reported that the technique achieves 95% hit rate at just 3.57% penetration rate, and to achieve 100% hit rate, 7.86% penetration rate is required. In both scenarios, the system reported 28 and 13 times speed-up than the non-indexed identification process.

## 11.4 INDEXING FINGER-KNUCKLE PRINT DATABASE

Finger-Knuckle Print (FKP) refers to the pattern obtained from the outer surface around the phalangeal joint of the finger. This image pattern has been found to contain rich lines and creases that contribute to the unique and permanent features that can be used for authentication. The hand-based biometrics such as fingerprint and knuckle print attract much attention due to high acceptability and convenience among the users. They have been found to be unique due to the presence of a large amount of distinctive information [64]. They also possess an advantage in terms of easy capturing using low-cost devices and with no requirement of an additional hardware system. Also, the acquired feature is small in size and, hence, can be used for applications involving a larger population [60]. But in countries, where a majority of the population is involved in agricultural or labourer activities, a serious damage happens to the inner part of the hand. This causes a drop in the quality of the acquired biometric sample, which further harms the feature extraction and identification process [1]. In such a scenario, features found on the outer part of the finger-knuckle print surface can be used for the purpose of authentication. The features in FKP also tend to survive longer as they lie on the outer area of the hand. The unique features of fingerprints such as minutiae and singular points have been observed to fade with time [49]. Hence, FKP can be used as a feasible biometric trait for authentication. The FKP identification system aims to find top-$k$ matches from the database for a given probe sample. To make an identification system efficient, indexing is applied in order to reduce the search space, thereby reducing the number of comparisons [21]. Indexing is accomplished by extracting features from the FKP image and then using that feature vector for indexing.

### 11.4.1 BOOSTED GEOMETRIC HASHING

A geometric hashing-based FKP indexing technique has been proposed in Ref. [28]. The proposed technique has boosted the geometric hashing such that the extracted feature is inserted into the hash table exactly once. The first step is to extract features from the FKP image. In the proposed technique, feature vectors have been constructed by determining the key points through two methods called, Scale Invariant Feature Transform (SIFT) [41] and Speeded-Up Robust Feature (SURF) [6] and, then forming descriptors around these key points. SIFT identifies those features that remain stable at different scales. Such key points are identified using the difference of Gaussian (DoG) function. To make the feature robust to image rotation, orientation of the key point is also determined. SURF features are extracted using two major steps, key-point detector and key-point descriptor. Key-point detection is done using the Hessian matrix. The key points are detected at different scales, and the non-maximum suppression is applied in $3 \times 3 \times 3$ neighbourhood to localise the key points. The descriptor is computed by using a rectangular window around each key point. This window is split into $4 \times 4$ sub-regions, and the Haar wavelet responses are extracted from them. The descriptors obtained from all the sub-regions are then concatenated to form a single descriptor. The generated SIFT and SURF feature descriptors are of length 128 and 64, respectively.

**Indexing**: Every feature vector fi extracted using SIFT and SURF is represented by $(x_i, y_i, D_i)$, where xi and yi denote the coordinates and $D_i$ represents the feature vector. The coordinates of the feature vectors have been used for index generation of the hash table. The descriptor $D_i$ has been used for recognition. Three steps have been followed to make the proposed technique robust to translation and rotation. These are (i) mean centring, (ii) feature rotation utilising the principal components, and (iii) normalisation. Sometimes, the features extracted from FKP images of the same subject taken at different times may not have the same coordinate position. Mean centring is done to handle this type of noise. It is accomplished by taking average of all f_i's, denoted by $\bar{f}_i$, and subtracting it from each feature vector. Therefore, the new coordinate position of $f_i' = f_i - \bar{f}_i$ is given by $\left(x_i', y_i'\right)$. In the second step, the feature vectors are rotated in such a way that they all become aligned in a uniform coordinate system. Principal component analysis (PCA) has been used to determine the primary axes of the coordinate system. The coordinates of the features are then rotated in a manner that they become aligned along the determined $X$ and $Y$ axes. Lastly, normalization is done to account for scaling. In this step, the normalised coordinates of every feature vector are computed, and a scaling factor is multiplied to these coordinate values to avoid falling of all the coordinates in one single bin. The hash table is aligned with respect to the normalised coordinate system, and the feature vectors along with their FKP id are stored in the table. The hashing process is shown in Figure 11.8 (image taken from [28]).

**Retrieval**: During the retrieval phase, a query FKP image is shown to the identification system, and the same procedure of feature extraction is carried out for

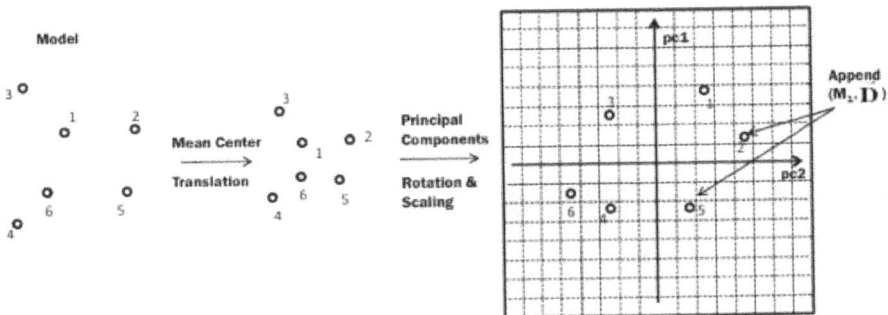

**FIGURE 11.8** Boosted geometric hashing. Hash table showing entries with principal components as a basis.

this image as well. Let us consider for a query FKP image $Q$, $n$ features have been extracted. The extracted features are mapped to a bin for candidate set retrieval. Neighbouring bins of size $k \times k$ are also considered to account for missing or spurious features. Euclidean distance is computed between the features of the retrieved candidate and query. Therefore, corresponding to every feature, a candidate set will be retrieved, making a total of $n$ candidate lists. These lists are concatenated and IDs are sorted in decreasing order of their number of occurrences.

## 11.4.2 RESULTS

The proposed technique has been tested on PolyU Finger-Knuckle Print (FKP) database [23]. The (PolyU)FKP is a publicly available data set containing finger-knuckle print images collected from 660 subjects. Each subject has provided 12 samples which have been collected in two separate sessions with a gap of 25 days. It, thus, comprises 7,920 images in total. Some of the images are shown in Figure 11.9. The first 11 images, out of 12, have been used for indexing, while the remaining one has been used as a query image to the identification system. Correct recognition rates (CRRs) of 96.36% and 99.69% have been reported with SIFT and SURF features, respectively. The technique has achieved 99% hit rate at 10.62% and 94.07% penetration rates when SURF and SIFT features were used, respectively.

The proposed technique has also been tested for robustness to occlusion and rotation in the FKP images. To do so, FKP images have been artificially occluded by 1%, 4%, 9%, 16%, 25%, and 36% as shown in Figure 11.10a. The occluded images have been divided into $2 \times 4$ sub-blocks, and the SIFT and SURF features are extracted from each sub-block. These blocks are compared separately and their matching scores are combined to form a matching decision. The graph between hit rate and penetration rate for various levels of occlusion as shown in Ref. [28] provides two observations. First, SUFT has better features for occlusion as it has around 95% hit rate at 5% penetration rate with low occlusion, whereas SIFT achieves 95% hit rate at around 40% penetration rate when occlusion is very low. Second, SURF is more robust to occlusion, as the hit rate does not change much for 1%–38% of the occlusion. However, with SIFT feature, hit rate decreases from 90% to 30% at 10% penetration rate when occlusion is varied from 1% to 38%.

To test the proposed technique's robustness against rotation, different degrees of rotation (0°, 10°, 50°, 110° and 150°) have been introduced in the FKP images, as shown in Figure 11.10b. The graph between hit rate vs. penetration rate for various

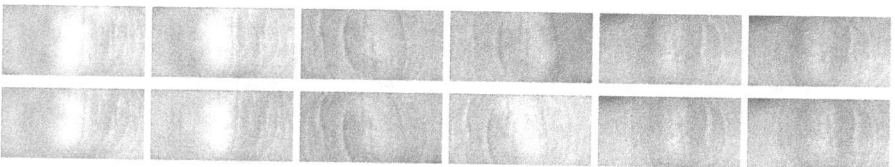

**FIGURE 11.9** Sample FKP images from PolyUFKP database. First row samples are collected in the first session and the second row contains corresponding images collected in the second session.

(a) 1%   (b) 4%   (c) 9%   (d) 16%   (e) 25%   (f) 36%

(a) FKP images showing different levels of occlusion

(a) 0°   (b) 10°   (c) 50°   (d) 110°   (e) 150°

(b) FKP images with different degree of rotation [28]

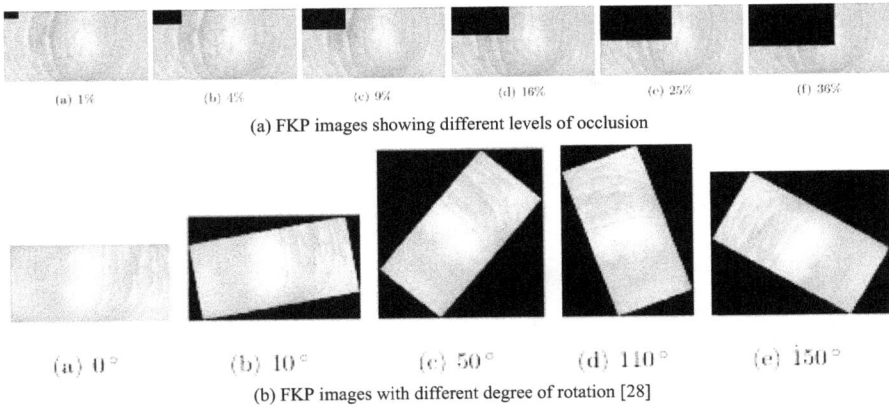

**FIGURE 11.10**   Occlusion and rotation on FKP [28].

degrees of rotation as shown in Ref. [28] shows results as expected. As it is known that SIFT and SURF are both rotation-invariant features, artificially introduced rotation does not deteriorate hit rate at any penetration rate.

## 11.5   INDEXING IRIS IMAGES

Iris is one of the most widely used biometric traits. It consists of rich texture information in the form of colour, minutia, spots, filaments, rifts, *etc.* that makes it unique. It has a small false-matching rate as compared to other available biometric traits [13,15], thus making it a stable biometric trait for authentication. It is an internally protected organ; therefore, it cannot be easily duplicated [30]. Iris database indexing consists of two phases, namely, indexing and retrieval. Indexing refers to associating the extracted features from the iris images to an index. During the retrieval phase, the feature vector for the probe image is generated, and the most similar index is found out. The candidates lying in the similar index are retrieved for comparison with the probe iris image. This results in reducing the search space remarkably.

### 11.5.1   INDEXING OF IRIS DATABASE BASED ON LOCAL FEATURES

The technique proposed in Ref. [32] utilises local features extracted from the iris images for indexing by utilising three transformation methods, namely, Discrete Cosine Transform (DCT) [33], Discrete Wavelet Transform (DWT) [36], and Singular Vector Decomposition (SVD) [19]. Before applying these methods, the iris image is pre-processed by applying segmentation, normalization, and enhancement on the acquired image. The iris segmentation is done using Canny edge detection [3] and the circular Hough transformation. The segmented iris image is then normalised using Daugman's rubber sheet model [31] by converting the iris into a rectangular region. It is then enhanced using the CLAHE approach [56]. The pre-processing steps along with their corresponding outputs have been shown in Figure 11.11. For feature extraction, the pre-processed image is divided into 8 × 8 blocks, and the local

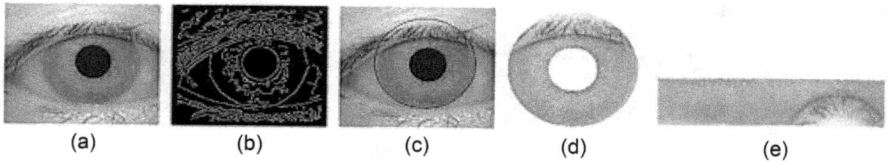

(a)                    (b)                    (c)                    (d)                           (e)

**FIGURE 11.11** Pre-processing of the iris image (a) eye image, (b) edge detection, (c) iris localisation, (d) iris segmentation, and (e) iris normalisation [32].

features are extracted from each block. DWT decomposes the blocks into seven sub-bands and aids in differentiating between the textures. DCT transforms the image space domain to frequency domain, *i.e.* when applied on each sub-band, it transforms them to spectral sub-bands having different importance. SVD is applied to these blocks and important features are selected that are expressed as a series of singular vectors (SVs). Scalable K-means++ has been applied on these features to divide them into distinctive groups leading to the creation of two B-trees. The block diagram of the proposed approach is shown in Figure 11.12.

For indexing, a global key is generated corresponding to every image stored in the database. The key value consists of the group number that contains the image's sub-band features and combined key value. Every group is then sorted in increasing order of the first SV. They are then divided into two bins, where the bins contain features from the first and second SVs, respectively. This divides the database into two B-trees in which traversal is done using the generated global key. The images are stored at the leaf nodes of the B-tree. The B-tree structure is shown in Figure 11.13.

During identification, when a probe iris image is given to the system, the features are extracted from the probe image using the aforementioned procedure. The closest bins to the extracted features are identified and are traversed through the B-tree using the global key. The similar candidates are searched inside the bin using half searching method. This outputs a candidate list for comparison with the probe image.

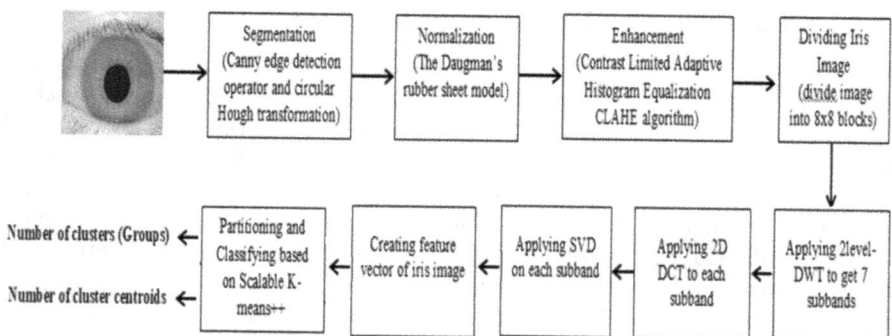

**FIGURE 11.12** Block diagram of the proposed approach [32].

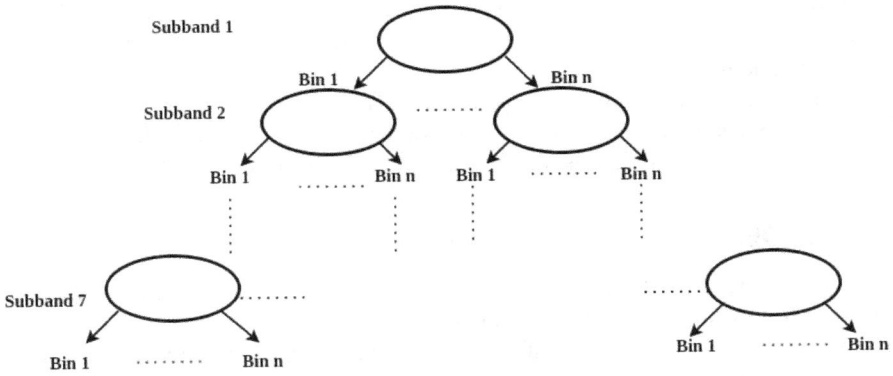

**FIGURE 11.13**  B-tree structure that is used to store the bins [32].

## 11.5.2 Results

The performance of the above-discussed technique has been tested on three databases, viz., CASIA-IrisV3-Interval [43], the BATH University database [55], and the IITK database [47]. CASIA-IrisV3-Interval database contains 54,607 iris images collected from 1,800 real and 1,000 virtual subjects. The images have been captured in two different sessions with a gap of at least one month. The BATH University database contains 2,000 iris images collected from both left and right eye of 50 subjects. The images are in grey-scale format having a resolution of 1,280 × 960. The IITK database consists of 1,800 images collected from the left eye of 600 subjects. Some of the images collected from these databases are shown in Figure 11.14.

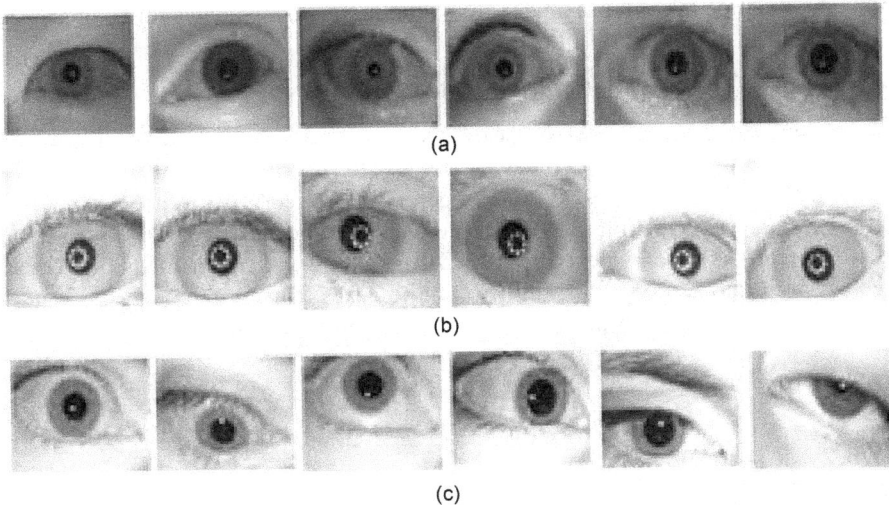

(a)

(b)

(c)

**FIGURE 11.14**  Sample images taken from three databases: (a) BATH, (b) CASIA, and (c) IITK [32].

The proposed technique achieved penetration rates of 0.98%, 0.13%, and 0.12% and bin miss rates of 0.3037%, 0.4226%, and 0.2019% on CASIA-IrisV3-Interval, BATH University database, and IITK database, respectively. The proposed method has also been compared with three methods proposed in Refs. [4,46,58], such as DCT energy histogram and key-point descriptor. The comparison of the same shows that the proposed method has higher penetration rate. This is due to fewer sub-bands used.

## 11.6  INDEXING SIGNATURE IMAGES

Handwritten signature comes under the category of behavioural biometric trait and has been used widely especially for financial transactions [48]. However, it is prone to temporal changes and may get affected by the physical and emotional health of the signatory. A signature is composed of characters that may or may not be readable. It has also been observed that the successive signatures of the same person have a significant difference. Due to these reasons, the signature of a person is analysed as a whole image and not as different characters put together. The handwritten signatures can be categorised as online and offline [16]. An offline signature is captured by scanning or taking a photograph of the signature from a paper. On the other hand, online signature is acquired using an electronic tablet and stylus that also records pen positions, elevation, and pressure. Feature extraction plays an important role in signature identification and verification. The features extracted from online signatures can be classified into two categories called functional and parametric features. The functional features consist of information regarding acceleration, position, force, time, etc. while signing, while the parametric features constitute the parameters calculated from the signals captured from the signing device. Signature identification system tries to establish the identity of the input signature by comparing it with all the signature templates stored in the database. The database is indexed with the goal of reducing the time taken for comparison by finding top-$k$ candidates for comparison.

### 11.6.1  KD-Tree-Based Signature Database Indexing

The technique proposed in Ref. [48] extracts 100 global features for the construction of a feature vector to be used for indexing. Some of the features are total duration of the signature, number of pen-ups, average velocity, velocity correlation, average number of jerks, standard deviation in x- and y-axis, *etc.* The constructed 100-dimensional feature vectors are then indexed using KD-tree [24]. KD-tree partitions the feature vector space into $k$ sub-spaces, thus forming indexes. During retrieval, the feature vector is obtained for the query signature and the range search is invoked to find the most suitable candidates by considering only those candidates that lie within $d$ distance from the query.

### 11.6.2  Results

The discussed technique has been tested on MCYT online signature database [51]. It consists of 50 signatures collected from 330 individuals each. Of the 50 signatures, 25 are original and the remaining 25 are forged. However, the authors have used

**FIGURE 11.15**   Sample signature images from MCYT online signature database.

**TABLE 11.2**
**Comparison of Time Taken for Recognition for MCYT Database with and without Indexing**

| Training Partition (%) | Identification Time (in Seconds) | | Time Reduction (in %) |
|---|---|---|---|
| | Conventional | Indexing | |
| 40 | 0.5213 | 0.0211 | 95.95 |
| 60 | 0.7326 | 0.0235 | 96.79 |
| 80 | 0.9251 | 0.0343 | 96.29 |

only the genuine ones making a database containing 8,250 signatures. Some of the samples from MCYT online signature database are shown in Figure 11.15.

The experimentation has been conducted in three ways, *viz.*, training the model on 40%, 60%, and 80% of the database and then testing on the remaining partition. The identification accuracy (also referred to as Correct Index Power (CIP)) has been reported as 72.59%, 78.03%, and 81.58% when tested on the aforementioned three partitions, respectively. The authors have reported 95.95%, 96.79%, and 96.29% reduction in identification time if indexing is implemented. The time requirement for testing the data set with and without indexing has been shown in Table 11.2. However, it has to be noted that the KD-tree structure depends on the sequence followed to represent the feature vector; therefore, the tree may not always be balanced. A dimensional reduction may further help in the same.

## 11.7   CONCLUSION

Identification is a compute-intensive task that may take longer time to produce results. Indexing is used to fasten the identification process by quickly producing a list of possible candidates who are likely to be similar to the query biometric sample. It is interesting to note that the retrieval takes constant time and produces a short list. As the list is produced in constant time, we can neglect the time spent in retrieval. Also, the time to search through the candidate list is small as the list is expected it to have few elements only. Both these factors contribute to an efficient identification solution. We have seen CGTC-based method for the fingerprint indexing [2], which achieves 95% hit rate at just 3.57% penetration rate. It needs only 7.86% penetration rate to achieve 100% hit rate on FVC 2004 DB1a database [44]. In both scenarios, the system achieves 28 and 13 times speed-up than the non-indexed identification process. For indexing finger-knuckle database, a boosted geometric hashing-based technique is proposed in Ref. [28]. The technique on publicly available (PolyU)FKP data set has achieved 99% hit rate at 10.62% and 94.07% penetration rate when SURF and SIFT features were used, respectively. CRR of the system with SIFT and SURF features is found to be 96.36% and 99.69% on the same database. This technique is robust to occlusion and rotation. To test the same, FKP images are introduced to artificial occlusion of 1%, 4%, 9%, 16%, 25%, and 36%; also for rotation, these images are rotated by 0°, 10°, 50°, 110°, and 150°. It has been observed that not much deterioration is seen in indexing performance. Face is one of the most on-demand biometric traits. There are a large number of facial databases with all kinds of variations such as age, expression, and pose. Many of the face databases have large number of images. PHC was proposed in Ref. [22] for facial image indexing. The technique maps the face images to a hamming space where similar faces could be clustered. The predictability of the hash codes was enhanced by utilising a convolutional neural network. The proposed technique achieved a recognition rate of 83.15% on YouTube Celebrities data set when the length of PHC was 128 bits. It can be observed that with the increase in predictability of the hash codes, the recognition accuracy also improved. Iris is one of the most accurate biometric traits. An indexing system for iris images proposed in Ref. [32] uses three databases, *viz*. CASIA-IrisV3-Interval [43], the BATH University database, and the IITK database [47] to evaluate its performance. The proposed technique has achieved penetration rates of 0.98%, 0.13%, and 0.12% and bin miss rates of 0.3037%, 0.4226%, and 0.2019% on the three databases CASIA-IrisV3-Interval, BATH University database, and IITK database, respectively. A comparison of the proposed method shows that it has higher penetration rate as compared to other methods on iris database. Signature is a very popular biometric trait in offline use. Many organizations in real world widely use this behavioural trait for financial transactions and document authentication. A KD-tree-based indexing technique has been proposed in Ref. [48] for indexing online signature database. Signature images are represented using a 100-dimensional feature vector. These features are indexed using partitioning with KD-tree. Experimentation has been conducted in three different database split, *viz*. training the model on 40%, 60%, and 80% of the database and then testing on the remaining data. The identification accuracy (referred to as CIP) has been reported as 72.59%, 78.03%, and 81.58%, respectively, when tested on the aforementioned three partitions.

## REFERENCES

1. O. Déniz, G. Bueno, J. Salido, and F. D. Torre. Face recognition using histograms of oriented gradients. *Pattern Recognition Letters*, 32(12):1598–1603, 2011.

2. G. Arora, T. Aggarwal, and K. Tiwari. Softmax coaxial Gaussian track code for finger-print indexing. *CoDS-COMAD: ACM India Joint International Conference on Data Science and Management of Data*, Kolkata, pp. 326–329, 2019.
3. P. Bao, L. Zhang, and X. Wu. Canny edge detection enhancement by scale multiplication. *IEEE Transactions on Pattern Analysis and Machine Intelligence*, 27(9):1485–1490, 2005.
4. T. Barbu and M. Luca. Content-based iris indexing and retrieval model using spatial access methods. In *2015 International Symposium on Signals, Circuits and Systems (ISSCS)*, pp. 1–4. IEEE, 2015.
5. S. Barman, S. Chattopadhyay, D. Samanta, S. Bag, and G. Show. An efficient fingerprint matching approach based on minutiae to minutiae distance using indexing with effectively lower time complexity. In *2014 International Conference on Information Technology*, pp. 179–183. IEEE, 2014.
6. H. Bay, T. Tuytelaars, and L. Van Gool. Surf: Speeded up robust features. In *European Conference on Computer Vision*, pp. 404–417. Springer, 2006.
7. N. Bhargava, R. Bhargava, P. Narooka, and M. Cotia. Fingerprint recognition using minutia matching. *International Journal of Computer Trends and Technology*, 3(4):641–643, 2012.
8. V. Bruce and A. Young. Understanding face recognition. *British Journal of Psychology*, 77(3):305–327, 1986.
9. K. Cao and A. K. Jain. Fingerprint indexing and matching: An integrated approach. In *2017 IEEE International Joint Conference on Biometrics (IJCB)*, pp. 437–445. IEEE, 2017.
10. R. Cappelli. Fast and accurate fingerprint indexing based on ridge orientation and frequency. *IEEE Transactions on Systems, Man, and Cybernetics, Part B (Cybernetics)*, 41(6):1511–1521, 2011.
11. R. Cappelli, M. Ferrara, and D. Maltoni. Fingerprint indexing based on minutia cylinder-code. *IEEE Transactions on Pattern Analysis and Machine Intelligence*, 33(5):1051–1057, 2010.
12. L. P. Chew. Constrained Delaunay triangulations. *Algorithmica*, 4(1–4):97–108, 1989.
13. N. Damer, P. Terhörst, A. Braun, and A. Kuijper. Efficient, accurate, and rotation-invariant iris code. *IEEE Signal Processing Letters*, 24(8):1233–1237, 2017.
14. M. Datar, N. Immorlica, P. Indyk, and V. S. Mirrokni. Locality-sensitive hashing scheme based on p-stable distributions. *SoCG04: Annual ACM Symposium on Computational Geometry*, Brooklyn, New York, pp. 253–262, 2004.
15. J. G. Daugman. High confidence visual recognition of persons by a test of statistical independence. *IEEE Transactions on Pattern Analysis and Machine Intelligence*, 15(11):1148–1161, 1993.
16. R. Doroz and P. Porwik. Handwritten signature recognition with adaptive selection of behavioral features. In *Computer Information Systems–Analysis and Technologies*, pp. 128–136. Springer, 2011.
17. A. Gago-Alonso, J. HernaNdez-Palancar, E. RodríGuez-Reina, and A. MuñOz-BriseñO. Indexing and retrieving in fingerprint databases under structural distortions. *Expert Systems with Applications*, 40(8):2858–2871, 2013.
18. A. Gionis, P. Indyk, R. Motwani. Similarity search in high dimensions via hashing. In *Proceedings of the 25th VLDB Conference*, Edinburgh, Scotland, volume 99, pp. 518–529, 1999.
19. G. H. Golub and C. Reinsch. Singular value decomposition and least squares solutions. In *Linear Algebra*, pp. 134–151. Springer, 1971.
20. Y. Gong, S. Lazebnik, A. Gordo, and F. Perronnin. Iterative quantization: A procrustean approach to learning binary codes for large-scale image retrieval. *IEEE Transactions on Pattern Analysis and Machine Intelligence*, 35(12):2916–2929, 2012.

21. P. Gupta, K. Tiwari, and G. Arora. Fingerprint indexing schemes–a survey. *Neurocomputing*, 335:352–365, 2019.

22. R. He, Y. Cai, T. Tan, and L. Davis. Learning predictable binary codes for face indexing. *Pattern Recognition*, 48(10):3160–3168, 2015.

23. C. Hegde, P. D. Shenoy, K. Venugopal, and L. Patnaik. FKP biometrics for human authentication using Gabor wavelets. In *TENCON 2011-2011 IEEE Region 10 Conference*, pp. 1149–1153. IEEE, 2011.

24. W. Hunt, W. R. Mark, and G. Stoll. Fast kd-tree construction with an adaptive error-bounded heuristic. In *2006 IEEE Symposium on Interactive Ray Tracing*, pp. 81–88. IEEE, 2006.

25. O. Iloanusi, A. Gyaourova, and A. Ross. Indexing fingerprints using minutiae quadruplets. In *CVPR 2011 WORKSHOPS*, pp. 127–133. IEEE, 2011.

26. A. K. Jain, P. Flynn, and A. A. Ross. *Handbook of Biometrics*. Springer Science & Business Media, 2007.

27. A. K. Jain and S. Z. Li. *Handbook of Face Recognition*, volume 1. New York: Springer, 2011.

28. U. Jayaraman, A. K. Gupta, and P. Gupta. Boosted geometric hashing based indexing technique for finger-knuckle-print database. *Information Sciences*, 275:30–44, 2014.

29. U. Jayaraman, A. K. Gupta, and P. Gupta. An efficient minutiae based geometric hashing for fingerprint database. *Neurocomputing*, 137:115–126, 2014.

30. R. R. Jha, G. Jaswal, D. Gupta, S. Saini, and A. Nigam. Pixisegnet: Pixel-level iris segmentation network using convolutional encoder–decoder with stacked hourglass bottleneck. *IET Biometrics*, 9(1):11–24, 2019.

31. T. Johar and P. Kaushik. Iris segmentation and normalization using Daugman's rubber sheet model. *International Journal of Scientific and Technical Advancements*, 1(3):2454–1532, 2015.

32. E. T. Khalaf, M. N. Mohammad, and K. Moorthy. Robust partitioning and indexing for iris biometric database based on local features. *IET Biometrics*, 7(6):589–597, 2018.

33. S. A. Khayam. The discrete cosine transform (DCT): Theory and application. *Michigan State University*, 114:1–31, 2003.

34. J. Khodadoust and A. M. Khodadoust. Fingerprint indexing based on expanded Delaunay triangulation. *Expert Systems with Applications*, 81:251–267, 2017.

35. B. Kulis and T. Darrell. Learning to hash with binary reconstructive embeddings. In *Advances in Neural Information Processing Systems*, Vancouver, B.C., Canada, pp. 1042–1050, 2009.

36. M. Lang, H. Guo, J. E. Odegard, C. S. Burrus, and R. O. Wells Jr. Nonlinear processing of a shift-invariant discrete wavelet transform (DWT) for noise reduction. In *Wavelet Applications II*, volume 2491, pp. 640–651. International Society for Optics and Photonics, 1995.

37. S. Lawrence, C. L. Giles, A. C. Tsoi, and A. D. Back. Face recognition: A convolutional neural-network approach. *IEEE Transactions on Neural Networks*, 8(1):98–113, 1997.

38. G. Lin, C. Shen, Q. Shi, A. Van den Hengel, and D. Suter. Fast supervised hashing with decision trees for high-dimensional data. In *Conference on Computer Vision and Pattern Recognition (CVPR)*, Columbus, Ohio, pp. 1963–1970, 2014.

39. W. Liu, J. Wang, R. Ji, Y.-G. Jiang, and S.-F. Chang. Supervised hashing with kernels. In *IEEE Conference on Computer Vision and Pattern Recognition*, Providence, RI, pp. 2074–2081, 2012.

40. M. Liu and P.-T. Yap. Invariant representation of orientation fields for fingerprint indexing. *Pattern Recognition*, 45(7):2532–2542, 2012.

41. D. G. Lowe. Object recognition from local scale-invariant features. In *Proceedings of the Seventh IEEE International Conference on Computer Vision*, volume 2, pp. 1150–1157. IEEE, 1999.

42. J. MacQueen. Some methods for classification and analysis of multivariate observations. In *Proceedings of the Fifth Berkeley Symposium on Mathematical Statistics and Probability*, volume 1, pp. 281–297. Oakland, CA, 1967.

43. M. Mahlouji and A. Noruzi. Human iris segmentation for iris recognition in unconstrained environments. *International Journal of Computer Science Issues (IJCSI)*, 9(1):149, 2012.

44. D. Maio, D. Maltoni, R. Cappelli, J. L. Wayman, and A. K. Jain. FVC2004: Third fingerprint verification competition. In *International Conference on Biometric Authentication*, pp. 1–7. Springer, 2004.

45. D. Maltoni, D. Maio, A. K. Jain, and S. Prabhakar. *Handbook of Fingerprint Recognition*. Springer Science & Business Media, 2009.

46. H. Mehrotra, B. Majhi, and P. Gupta. Robust iris indexing scheme using geometric hashing of sift keypoints. *Journal of Network and Computer Applications*, 33(3): 300–313, 2010.

47. H. Mehrotra, B. G. Srinivas, B. Majhi, and P. Gupta. Indexing iris biometric database using energy histogram of DCT subbands. In *International Conference on Contemporary Computing*, pp. 194–204. Springer, 2009.

48. K. Nagasundara, S. Manjunath, and D. Guru. Indexing of online signatures. *International Journal of Machine Intelligence*, 3(4):289–294, 2011.

49. A. Nigam, K. Tiwari, and P. Gupta. Multiple texture information fusion for finger-knuckle-print authentication system. *Neurocomputing*, 188:190–205, 2016. Advanced Intelligent Computing Methodologies and Applications.

50. M. Norouzi, D. J. Fleet, and R. R. Salakhutdinov. Hamming distance metric learning. In *Advances in Neural Information Processing Systems*, pp. 1061–1069, 2012.

51. J. Ortega-Garcia, J. Fierrez-Aguilar, D. Simon, J. Gonzalez, M. Faundez-Zanuy, V. Espinosa, A. Satue, I. Hernaez, J. J. Igarza, C. Vivaracho, and D. Escudero. MCYT baseline corpus: A bimodal biometric database. *IEE Proceedings-Vision, Image and Signal Processing*, 150(6):395–401, 2003.

52. E. G. Ortiz, A. Wright, and M. Shah. Face recognition in movie trailers via mean sequence sparse representation-based classification. In *Conference on Computer Vision and Pattern Recognition* (CVPR), Portland, Oregon, pp. 3531–3538, 2013.

53. O. M. Parkhi, A. Vedaldi, and A. Zisserman. Deep face recognition. British Machine Vision Association, pp. 1–12, 2015.

54. P. J. Phillips, P. J. Flynn, T. Scruggs, K. W. Bowyer, J. Chang, K. Hoffman, J. Marques, J. Min, and W. Worek. Overview of the face recognition grand challenge. In *2005 IEEE Computer Society Conference on Computer Vision and Pattern Recognition (CVPR'05)*, volume 1, pp. 947–954. IEEE, 2005.

55. N. Popescu-Bodorin and V. E. Balas. Comparing Haar-Hilbert and log-gabor based iris encoders on bath iris image database. In *4th International Workshop on Soft Computing Applications*, pp. 191–196. IEEE, 2010.

56. A. M. Reza. Realization of the contrast limited adaptive histogram equalization (clahe) for real-time image enhancement. *Journal of VLSI Signal Processing Systems for Signal, Image and Video Technology*, 38(1):35–44, 2004.

57. Q. Shi, A. Eriksson, A. Van Den Hengel, and C. Shen. Is face recognition really a compressive sensing problem? In *CVPR 2011*, pp. 553–560. IEEE, 2011.

58. Y. Si, J. Mei, and H. Gao. Novel approaches to improve robustness, accuracy and rapidity of iris recognition systems. *IEEE Transactions on Industrial Informatics*, 8(1):110–117, 2011.

59. C. Strecha, A. Bronstein, M. Bronstein, and P. Fua. Ldahash: Improved matching with smaller descriptors. *IEEE Transactions on Pattern Analysis and Machine Intelligence*, 34(1):66–78, 2011.
60. C. M. Travieso, J. R. Ticay-Rivas, J. C. Briceno, M. del Pozo-Baños, and J. B. Alonso. Hand shape identification on multirange images. *Information Sciences*, 275:45–56, 2014.
61. Y. Wang, J. Hu, and D. Phillips. A fingerprint orientation model based on 2D Fourier expansion (FOMFE) and its application to singular-point detection and fingerprint indexing. *IEEE Transactions on Pattern Analysis and Machine Intelligence*, 29(4): 573–585, 2007.
62. Y. Weiss, A. Torralba, and R. Fergus. Spectral hashing. In *Advances in Neural Information Processing Systems*, Vancouver, B.C., Canada, pp. 1753–1760, 2009.
63. N. Yager and A. Amin. Fingerprint classification: A review. *Pattern Analysis and Applications*, 7(1):77–93, 2004.
64. L. Zhang, L. Zhang, D. Zhang, and Z. Guo. Phase congruency induced local features for finger-knuckle-print recognition. *Pattern Recognition*, 45(7):2522–2531, 2012.
65. W. Zhou, J. Hu, S. Wang, I. Petersen, and M. Bennamoun. Partial fingerprint indexing: A combination of local and reconstructed global features. *Concurrency and Computation: Practice and Experience*, 28(10):2940–2957, 2016.

# 12 Iris Segmentation in the Wild Using Encoder-Decoder-Based Deep Learning Techniques

*Shreshth Saini and Divij Gupta*
IIT Jodhpur

*Ranjeet Ranjan Jha*
IIT Mandi

*Gaurav Jaswal*
IIT Delhi

*Aditya Nigam*
IIT Mandi

## CONTENTS

12.1 Introduction .................................................................................................284
12.2 Deep Learning for Segmentation ................................................................287
12.3 Related Work ...............................................................................................289
    12.3.1 Non-Deep Learning-Based Methodologies ....................................289
    12.3.2 Deep Learning-Based Methodologies .............................................293
12.4 Data Sets and Evaluation Metrics ..............................................................295
    12.4.1 Data sets ...........................................................................................295
    12.4.2 CASIA ..............................................................................................295
        12.4.2.1 UBIris v1 and UBIris v2..................................................295
        12.4.2.2 NICE-I and NICE-II........................................................296
        12.4.2.3 ND-Iris-0405 ...................................................................296
        12.4.2.4 IITD ..................................................................................296
        12.4.2.5 CSIP ..................................................................................297
        12.4.2.6 MICHE-I and MICHE-II..................................................297

12.4.2.7 SBVPI..................................................................................297
12.4.2.8 IRISSEG-CC........................................................................297
12.4.2.9 IRISSEG-EP .......................................................................299
12.4.2.10 MMU1 and MMU2.............................................................300
12.4.2.11 OpenEDS ...........................................................................300
12.4.2.12 iBUG .................................................................................300
12.4.3 Performance Metrics ...........................................................................300
12.4.3.1 Jaccard Index (JI)................................................................301
12.4.3.2 Mean Segmentation Error....................................................301
12.4.3.3 Nice2 Error..........................................................................301
12.5 Experimentation ...............................................................................................302
12.6 Challenges Identified and Further Direction ....................................................303
12.7 Conclusion .......................................................................................................304
Acknowledgements.......................................................................................................305
References.....................................................................................................................305

## 12.1 INTRODUCTION

In today's world, where technology has invaded almost every walk of our life, it has become more important now than ever to safeguard confidential information. While passwords were predominantly used for the task, they have become a thing of the past as they are weak and can be cracked using simple-to-sophisticated techniques by accomplished hackers. In comparison, biometrics provides a much more reliable and challenging environment to crack alternative in comparison to passwords. Biometrics can be broadly categorised into two categories: physiological and behavioural. Physiological identification is based on traits such as face, hand geometry, and iris, whereas behavioural identification is made on the basis of characteristics such as signature and gait. According to Ref. [52], any human behavioural or physiological characteristic may be utilised as a biometric given that it satisfies some specific properties:

1. Universality, i.e., it should be possessed by everyone.
2. Uniqueness, i.e., it should be distinct amongst people.
3. Permanence, i.e., it should remain the same throughout time.
4. Collectability, i.e., it should be easily collectible and quantifiable.
5. Performance, i.e., it should be efficient in the identification of the subject.
6. Acceptability, i.e., it should be acceptable to the people in general.
7. Circumvention, i.e., it should not be easy for the system using it to be fooled.

In Ref. [52], the authors have scored the modalities on the above criteria in majorly three categories: *Low* (⇓), *Medium* (⇕), and *High* (⇑).

The survey has extended or represented differently in many studies such as [19,36,51,94,101]. Table 12.1 represents that although Iris doesn't score 'High' in all the criteria, it is still by far the most suitable biometric modality. Owing to its accuracy and reliability [27], it is used in different biometric applications such as forensics [84] and intelligent unlocking [20].

**TABLE 12.1**

**Comparison of Some Commonly Used Biometrics on the discussed Criteria as in Ref. [52]**

| Biometric | Univer. | Uniq. | Perm. | Collect. | Perf. | Accept. | Circum. |
|---|---|---|---|---|---|---|---|
| Iris | ⇑ | ⇑ | ⇑ | ⇕ | ⇑ | ⇓ | ⇑ |
| Fingerprint | ⇕ | ⇑ | ⇑ | ⇕ | ⇑ | ⇕ | ⇑ |
| Face | ⇑ | ⇓ | ⇕ | ⇑ | ⇓ | ⇑ | ⇓ |
| Gait | ⇕ | ⇓ | ⇓ | ⇑ | ⇓ | ⇑ | ⇕ |
| Hand Geometry | ⇕ | ⇕ | ⇕ | ⇑ | ⇕ | ⇕ | ⇕ |
| Retinal scan | ⇑ | ⇑ | ⇕ | ⇓ | ⇑ | ⇓ | ⇑ |
| Voice print | ⇕ | ⇓ | ⇓ | ⇕ | ⇓ | ⇑ | ⇓ |

The human eye is such that it provides not one but two important biometric traits, namely, the retina and the iris. The iris is an annular structure positioned between the eye parts sclera, which is an off-white colour region, and the pupil, which is the dark region at nearly the centre of the entire eye. The boundary where the iris and pupil meet is called the pupillary boundary, similarly, where the iris and the sclera meet is called the limbic boundary. The iris comprises elastic connective tissues which enrich the region with diverse patterns comprising crypts, ridges, radial and contraction furrows, freckles, and arching ligaments [17]. Unlike the rest of the bio-metric characteristics such as the face or the fingerprint, the iris is guarded by the cornea and aqueous humour which accounts for its high permanence. The iris starts developing in the initial three months of the incubation period through forming and folding of the tissue membranes [39]. Moreover, only the pigmentation of the eye is genetically determined while its intricate structures are independent of genetics, which leads to it having more than 200 distinct features. Owing to its highly complex and random structure, the iris allows for its use as an efficient trait for distinguishing amongst different people. Moreover, it has also been established that the iris pattern amongst twins and even of the same person, the right and the left eyes are different from each other (Figure 12.1).

Using the iris as a biometric trait for human authentication involves the following main processes:

• Acquisition of the image from the subject
• Iris segmentation and pre-processing
• Normalisation of the segmented Iris
• Generation of biometric templates
• Matching of templates and subsequent authentication

For proper image acquisition, the user must stay at a particular distance and look at the camera at a certain angle depending upon the camera specifications, which requires high user cooperation and constrained environment settings. After the image is captured, segmentation of iris is done by extracting it using the set iris

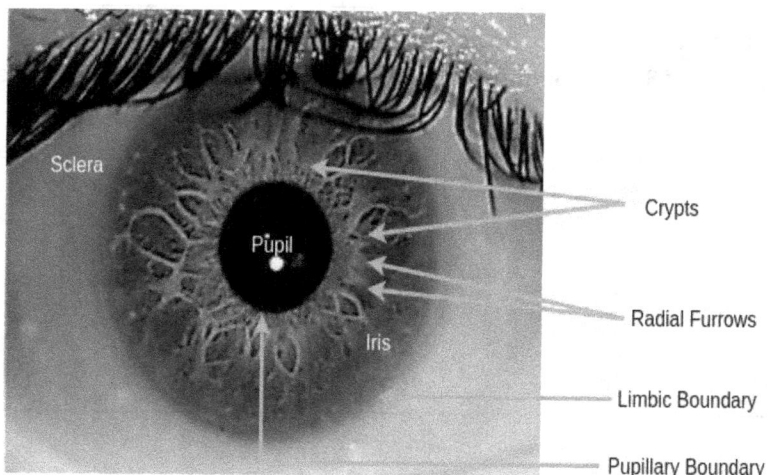

Sclera

Crypts

Pupil

Radial Furrows

Iris

Limbic Boundary

Pupillary Boundary

**FIGURE 12.1**   Anatomy of human eye.

segmentation technique. The next stage is normalisation, wherein the segmented image is transformed into one with a pre-set dimension so as to maintain uniformity in all the segmented images, making it easier to act upon in the further stages. After normalisation, the next stage is feature extraction or biometric template generation. A biometric template is generated by using mathematical functions. The function is defined as such that the template represents the features in the best possible way for an efficient representation. In the last stage, the template is attempted to match with the already existing template of the subject, and authentication is given based on a certain predefined threshold of matching accuracy. Also, if the template is meant to be added into the database against a new subject, then it is added in the 'enrolment' mode of the system, while the former is done in the 'identification' mode of the bio-metric system.

However, that is not always the case, which may lead to the inclusion of various artefacts such as off-angle gaze, eyelash/eyelid occlusion, motion blur, and specular-reflections due to less user cooperation and non-ideal environments. All these will subsequently lead to poor segmentation results, and the error will be propagated and compounded when the information is passed through further in the system, ulti-mately leading to faulty results according to many studies [40,77,81,91]. However, if the segmentation can be done accurately enough, then only the relevant infor-mation, although augmented, can be propagated further, making the system more accurate than before. A majority of iris segmentation methodologies also assume the iris shape to be circular, which is deviated from when the eye is partially closed, which further enhances the need for accurate segmentation techniques in non-ideal environments [26]. Moreover, with the rise in demand for the integration of biometric authentication into our daily lives, non-ideal conditions have to be factored in. In further sections, we discuss the various segmentation techniques which involve both ideal and non-ideal environments.

## 12.2   DEEP LEARNING FOR SEGMENTATION

Segmentation is taken as the first step to process and understand an image; image segmentation is a process when a greyscale or a colour image is broken down into clusters of pixels that contain similar meaningful attributes or homogeneity. Segmentation is titled as the highest domain-independent generalisation of an image. For segmentation, a wider variety and in-depth research has been done, but due to the task-dependent abstraction of pixels, those methods do not always produce good results.

Artificial neural networks were used way back in the 1940s initially, but it was only until the 1990s when research dwelled deep into this field [29]. With the availability of digital data sets and development towards computational power, deep learning made huge progress. The very first fully Convolutional Neural Network (CNN) was developed by LeCun et al. [61]. Following the work [61], many researchers started putting a huge effort into developing variations of CNN models, which could give higher performance. In the previous decade, deep learning has brought forth a revolution in the field of image analysis, computer vision, etc. Deep learning is being used for tasks like recognition, segmentation, detection, classification, and so on. Some most recognisable architectures which are used as base model for many derivative works are VGG [97], ResNet [38], GoogLeNet [99], MobileNet [42], DenseNet [44] (not exhaustive list).

Segmentation, being a necessary step in not just biometric but also in image processing, medical image analysis, vision tasks, augmented reality, etc., uses deep learning-based methods for segmentation tasks that outperform classical approaches. Networks such as fully CNNs for pixel-level classification, encoder-decoder based models, attention modules, generative adversarial models, and recurrent networks have been explored vastly for segmentation. Prior to deep learning, segmentation was done through handcrafted features with rule-based algorithms such as Thresholding [68], K-Means Clustering [31], Watersheds [67], Contours-based methods [55], Graph-Cut approaches [18], and Markov Random Fields [74] (Figure 12.2).

Of all the deep learning-based approaches, encoder-decoder models produce the most promising results in the pipeline of image-to-image translation tasks, i.e. image segmentation. The encoder takes the input image and downsamples it by processing with CNN layers to obtain a compressed high-dimensional representation, which is then sent to the decoder and upsamples the features to map them to the required output, which in our case is a segmentation map. Researchers adopted VGG architecture [97] by removing the fully connected layers of it and then deployed it as encoder while the decoder was developed, such as to mirror the encoder but with upsampling layers. Results obtained for segmentation tasks with such a network were better than other shallow, deep learning models. With modification and replacements in the encoder part research kept developing architectures such as SegNet [12], HRNet [105], and UNet [83], which further helped the community to increase the performance over segmentation tasks. UNet [83] architecture depicted in six was initially developed for medical image segmentation, but later on, it was adopted in other domains as well. Ronneberger et al. [83] introduced novel connections between encoder and deocder layers to facilitate the better parameter updates from the decoder to encoder

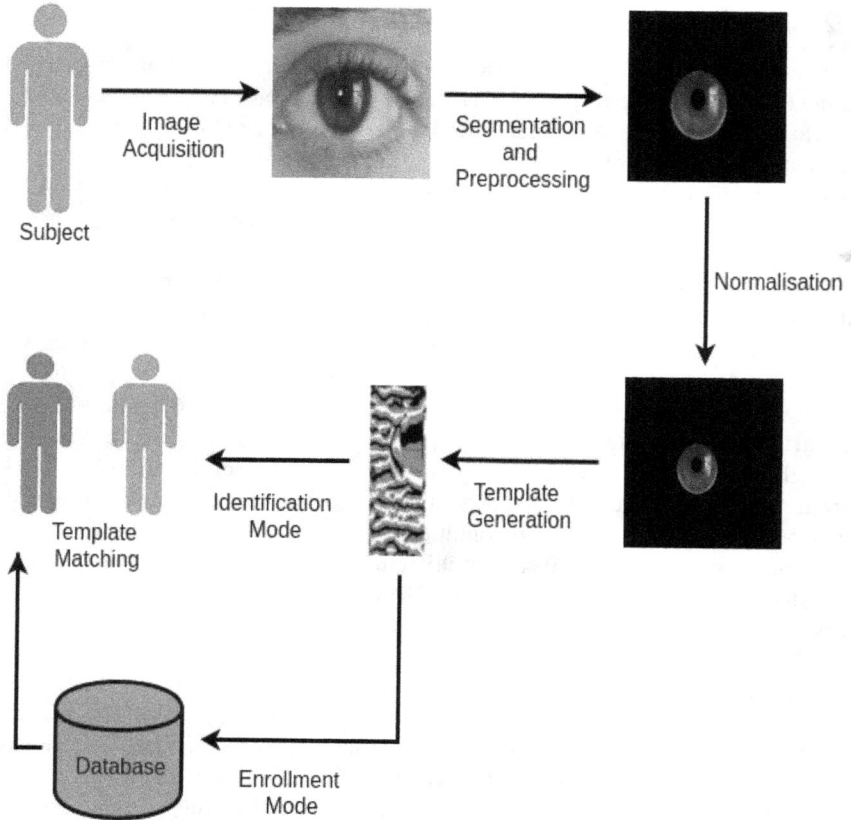

**FIGURE 12.2**  Iris biometric pipeline depicting the identification and the enrolment mode.

layers as an experimental set-up, which accidentally gave them boosted performance while reducing the problem of vanishing gradients. Later on, various modifications were introduced in the UNet, and some noticeable works are nested UNet [109], 3D UNet [24], etc. Some more heavy architectures, which use region proposal networks, for instance, segmentation, are Faster RCNN [82], which was further extended to Mask-RCNN [37]. Many researchers tend to use the atrous-convolution at the lowest dimension of the encoder-decoder model (bottleneck) to increase the receptive field to get the global features better. Chen et al. [21] proposed the famous DeepLab, which uses spatial pyramid pooling, dilated convolution, probabilistic graph model, and deep CNNs for precise boundary detection in the image. The use of the ResNet model [38] as a backbone helps it increase the performance of segmentation tasks. Lastly, General Adversarial Networks (GANs) have shown great potential in segmentation tasks. For example, in the work by Hung et al. [47], they proposed a segmentation network with an FCN as the discriminator, which takes predicted mask and ground truth mask as inputs in addition to the encoder-decoder model for the segmentation stage. Recently, researchers have been focusing on the network-in-network

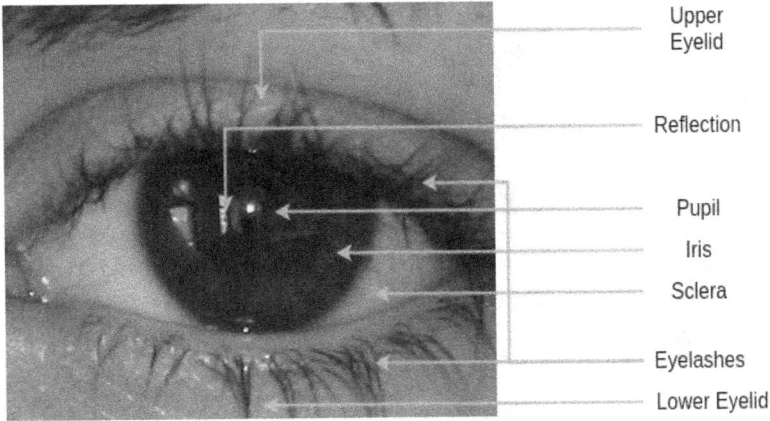

**FIGURE 12.3**    Some of the artefacts in the iris image.

approach to better extract the features from bottleneck for the input image, which may allow the network to look beyond the obstacles in the image for which a more dense and deep network is required. In an attempt to tackle the issue of occlusion, noise, off-angle, and other non-idealities in iris images, a stacked hourglass-based model which sits at the bottleneck of an encoder-decoder model is proposed in Ref. [53]. They proposed a unique training strategy to introduce the optimum number of hourglass modules, which can effectively achieve the task of accurate and precise segmentation without the issue of gradient vanishing (Figure 12.3).

In the area of biometrics, all these deep learning-based methods have been adopted, some with a little while some with significant modifications in them. In the subsequent sections, we discuss in-depth about the related approaches which make use of deep learning for the iris segmentation task.

## 12.3   RELATED WORK

In these subsections, we shall discuss existing methods for the iris segmentation. We cover both classical image processing techniques where algorithms rely on pre-defined rules and in-depth learning-based solutions that exploit the availability of massive paired databases for iris segmentation. There is no doubt that the results of deep learning-based methodologies surpass those of classical approaches. Recently, data-driven deep learning approaches have proven to give exceptional results in the field of biometrics and beyond it. We provide a comprehensive and comparative study amongst the methods.

### 12.3.1   Non-Deep Learning-Based Methodologies

In the iris segmentation, the steps followed in general involve as follows: first, the extraction of the Region of Interest (ROI) from the complete image, followed by the approximation of two circles which separate the iris region from the pupil and the sclera [107]. In this section, we have tried to cover all categories of the iris segmentation and boundary detection. Classical approaches take advantage of either the pixel-based features or the boundary-based features [13,59]. The first noticeable work was done by Daugman [28] back in 1993; his work formed the basis of all the work thereafter. In the eye, pupils along with iris are taken as non-concentric circles; an integrodifferential operator localises the boundary to segment out the iris. The method avoided the use of images, which included any type of occlusions like eyelids, eyelashes, and reflection, etc. Overall, we can say that classical methods followed by Daugman focused on extracting the edges of the pupil and iris to localise the to-be segmented area more precisely. Authors relied on the rule-based algorithmic approaches, which in some sense limited their methods to work on vast variations of images of the iris. We can say that those methods could not perform well on non-ideal images of the iris. Realizing the limitation of methods and instead of using simple edge-detection steps, they started to use more statistical approaches [48]. With the advancements in research, researchers started to model the anatomy of the eye realistically, such as taking the boundaries of the pupil and iris as non-circular. While some of the work did handle the problems of obstruction, specular reflection, and eyelash, etc., their limited feature modelling and extraction approaches were not too vast to cover all the variations in the iris images. In this section, we dive deep into the classical work done so far and provide a comparative analysis.

After Daugman, Wilde [106] in 1997 proposed a new approach in which he used an LED point source in addition to a camera for capturing eye images. He identified iris boundaries by gradient-dependent binary edge map in addition to the circular Hough transform. The paper also presented an in-depth comparative study with Daugman's work. While Wilde's work is considerably complex than Daugman's, the segmentation approach proposed by Wilde was better as it detected the eyelids as well as it worked better with noisy images. In Ref. [15], Boles et al. proposed a circular edge-detection method (Figure 12.4).

The authors of [93] proposed an iris localisation technique, namely, circular sector analysis (CSA), before applying rough entropy for segmentation. Their localisation methods decreased the overall uncertainty in the segmentation mask. Another

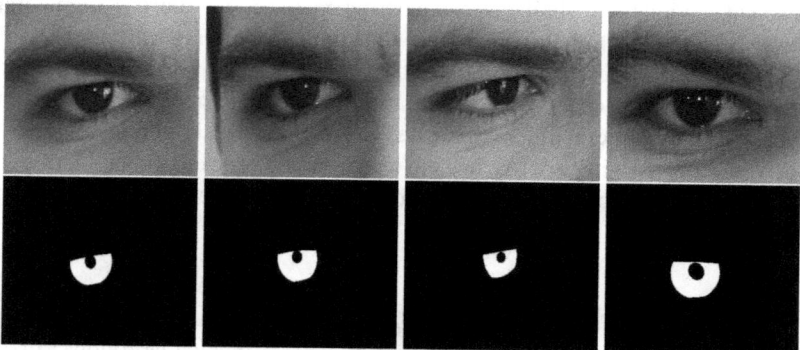

**FIGURE 12.4** Some images and their corresponding groud-truth segmentation masks from the UBIRIS-v2 data set [79].

work [107] proposed the iris localisation by assuming that shapes of the pupil and iris are circular wherein they first localised the pupil through eccentricity-dependent bisection approach, and then for iris, a region totally free from noise was obtained with directional segmentation followed by obtaining the gradients of direction lines to localise iris.

It was not until 2001 when Kong and Zhang in their work [57] incorporated the noisy and occluded images for iris segmentation. They used Hough transform to isolate the iris followed by 1-d Gabor filters for eyelids detection and thresholding to identify specular reflection. Their work gave better results for segmentation as well as in the final recognition task. Lim et al. [62] segmented the iris images by the edge-detection method through finding virtual circles where the pupil was detected first by the centrepoint-detection method. They acquired eye images but from a distance, and to reduce the reflections, they used halogen lamps. Their data set consisted of both eyes, with and without lens and glasses. Daugman, in his work [25], proposed the algorithm where he detected eyelid occlusion while segmenting the iris. Huang et al. [46] applied a median filter prior to canny operator for edge detection. Outer boundary was detected using a voting scheme on the maximum circle, and similarly for an inner boundary, it was identified using a rectangular inter interval. Localised iris was then segmented with the help of an integrodifferential operator. They too handled the eyelid occlusion using thresholding of histogram-based Hough transform. Huang et al. [45] again proposed a novel segmentation technique that also eliminated the noise to improve the results. They localised the iris using a simple filtering step with edge detection and Hough transform; occlusion factors were then eliminated using a Gabor filter.

Dorairaj et al. [32] developed an approach to deal with the off-angle iris image. In this work, he used PCA and global ICA for the encoding of off-angle iris images; while applying PCA/ICA, they first estimated the gazing angle by using Hamming distance followed by a simple integrodifferential operator for segmentation. Daugman in Ref. [26] developed an algorithm to tackle off-angle images similar to that of Ref. [32] with the elimination of occlusion caused by eyelashes. Abiyev et al., in their work [5], came up with the neural network-based method for the iris recognition; a rectangular area of size $10 \times 10$ was used to identify the pupil region. For the removal of noise, they utilised the standard linear Hough transform for eyelids.

The authors of Ref. [48] proposed a multi-stage technique. First, a moving window of circular shape was used for the pupil estimation, following which the estimation of the pupil was done through the standard-deviation peaks in both x as well as y directions, and after that, a median-filter reduced the eyelash effects. In Ref. [3], the authors proposed AdaBoost for eye detection for further segmentation. Reference [80] presents an unsupervised approach where images were modelled as Markov random field. Graph-cut method extracted the texture region, and for the iris segmentation image, intensities were exploited. Roy et al. [88] proposed a non-ideal iris recognition method, in which they used a Mumford-Shah segmentation method. All these classical approaches claim to handle various noises, distortion, and non-ideal iris images, but all being rule-based feature-driven approaches are limited in handling the variation of a non-ideal iris image. In Table 12.2, some classical approaches are compared based on their novelty and performance.

## TABLE 12.2
## Non-Deep Learning-Based Approaches

| Methods | Novelty | Strength | Performance | Data set Used |
|---|---|---|---|---|
| Sardar et al. [93] | CSA and use of rough entropy for Iris localisation | Less computational expensive | E1 error rate: 0.08%, Acc.: 97.12% | IITD, MMU, CASIA |
| Parikh et al. [69] | Colour clustering along with curve fitting | Eyelids detected reduces the chances of error in performance | Acc.: 93.3% (UBIRIS), 94.9% (NICE-II), 92.8% (NICE-I) | NICE-I, NICE-II and UBIRIS v2 |
| Khan et al. [5] | Gradient-based approach for localisation of iris with sclera boundary points | Iris borders were identified using novel gradient approach | Acc.: 98.22%(MMU), 100%(CASIA) | MMU, CASIA |
| Pundlik et al. [80] | Graph-cut approach for iris segmentation | Markov random field removes the eyelashes to boost the performance | None | None |
| Ibrahim et al. [48] | A two-stage hierarchical approach with moving circular window for iris segmentation | Intensity of pupil makes it easier to separate using probability approximation | Acc.: 98.28% (CASIA IrisV3 Lamp), 99.90% (CASIA IrisV1), 99.77% (MMU) | MMU, CASIA |
| Hu et al. [43] | Fusion of three models to extract the iris with the help of Daugman's method | Integral derivative to detect the iris boundary is time-efficient | E1 error rate: 1.75% (MICHE), 1.30% (UBIRIS v2) | MICHE, UBIRIS v2 |
| Huang et al. [4] | Radial suppression after the thresholding of detected edges | For ideal cases of iris images with the assumption of circular iris boundary, radial suppression gives good performance | E1 error rate: 0.32% | CASIA |
| Abate et al. [67] | Novel watershed method in addition to seed selection | Sclera along with eyelash/eyelid are separated using limbus detection approach | None | MICHE |
| Jeong et al. [3] | Eye detection with Adaboost and moving edge detector of shape circular | Detecting the obstructions in the iris images decreases the chances of error in segmentation map | E1, E2 error rates: 2.8%, 14.4%, respectively | UBIRIS.v2 |
| Ibrahim et al. [49] | First derivative-based iris detection with adaptive thresholding | Boundary detection takes lesser time | Acc.: 99.13% (MMU) | MMU v1.0 |
| Patel et al. [70] | Binary-integrated intensity curve with region growing approach | Detection of eyelash and eyelid increases the performance of segmentation task | E1 error rate: 5.14% | CASIA |

## 12.3.2 DEEP LEARNING-BASED METHODOLOGIES

Deep learning-based approaches are the primary source of majority of state-of-the-art solutions. In segmentation, researchers have come up with numerous modifications of simple auto-encoder-based CNNs. UNet, as mentioned earlier, has now become the basis network for segmentation tasks. In the iris segmentation, researchers shifted towards the deep neural network-based approaches gradually as the availability of the iris databases increased, and classical methods were challenged with increasing non-ideality in the iris images. CNN remove pre-/post-processing steps. CNN models were introduced to increase the accuracy of segmentation masks generated over the non-ideal iris images.

Liu et al. [63] developed two modalities with CNN, where a multi-scale CNN along with hierarchical CNN was deployed to detect the iris boundaries in non-ideal cases. In Ref. [59], the authors fed the input iris images to a series of four dense convolution blocks; feature maps extracted from blocks fused with a weighted sum gave coarse as well as fine features to produce the required segmentation mask finally. The authors of Ref. [71] trained two different CNN architectures, which were derived from networks such as Faster RCNN [82] and SSD [64] which localise the circular region of the pupil along with iris. Similarly in Ref. [11], the authors combined two existing CNN networks: the DenseNet [44] and the SegNet [12] for iris segmentation. Another such work which incorporates the features of one CNN architecture into another is Ref. [10]; they developed a CNN model with residual connections in SegNet which allowed the authors to develop deeper network while reducing the chances of vanishing gradient. In Ref. [63], the authors developed two CNNs, the first one was based on hierarchical CNNs and the second one was based on multi-scale CNNs. The authors of Ref. [14] used the data augmentation approach along with the CNN model to virtually increase the non-ideal iris database for segmentation.

In Ref. [104], the authors took the non-ideality to the next level, where the iris images were taken using the mobile images. They developed a lightweight deep CNN as a complete end-to-end segmentation method. Followed by the previous work, Wang et al., in their work [103], came up with a multi-task CNN architecture that also incorporates the attention module for the iris segmentation and boundary localisation. Similarly, in Ref. [54], the authors developed EyeNet, an attention-based CNN for the eye region segmentation.

While numerous works have been published on the iris segmentation with CNN both over ideal and non-ideal images, more or less, they propose a few new additional features or modules towards the existing models. Challenges that the deep learning solutions try to solve are a variety of textural complexity and the shape of iris from person to person, non-ideality such as distortion, non-regular illumination, motion blur, digital noise, and poor image quality. Algorithms are not robust enough to extract the ROI for segmentation. One such work that handles the majority of the above issues while giving state-of-the-art segmentation maps is Ref. [53], wherein the authors proposed a three-stage trained novel deep CNN architecture for the non-ideal iris segmentation. In addition to novel models, they used a combination of multiple loss functions to give precise segmentation maps. Table 12.3 compares some selected deep learning-based methods.

**TABLE 12.3**

**Deep Learning-Based Approaches**

| Methods | Novelty | Strength | Performance | Data set Used |
|---|---|---|---|---|
| Lakra et al. [59] | CNN architecture with patched input of the iris images | Post- and pre-cataract surgery: the CNN-based methods helps in the segmentation of ROI | E1 error rate: 0.98% | IITD, CASIA |
| Liu et al. [63] | CNN architecture trained and deployed an end-to-end method | Give significantly good results over non-ideal iris images | E1 error rate: 0.59% (CASIA v4); 0.90% (UBIRIS v2) | CASIA v4 and UBIRIS v2 |
| Wang et al. [104] | Robust single end-to-end model for segmentation | Worked with mobile images of eye, where data set was generated in nearly wild environment | E1 error rate: 0.72% (CASIA-iris-M1); 0.82% (MICHE-I) | CASIA-Iris-M1, MICHE-I |
| Arsalan et al. [11] | Inspired from DenseNet for the iris segmentation task which does not use any handcrafted features | DensNet architecture allows to increase the network depth which supports in giving improved results | E1 error rate: 0.695% | UBIRIS.v |
| Wang et al. [103] | Multi-task CNN-based model | Use of attention module with novel architecture resulted in increased performance over the segmentation task | E1 error rate: 0.41% (CASIA v4), 0.84% (UBIRIS v2), 0.66% (MICHE-I) | CASIA v4-distance, UBIRIS v2, and MICHE-I |
| Arsalan et al. [9] | Combination of CNN-based model with modified Hough transform for iris segmentation | ROI extracted with HT improves the performance of CNN to produce mask | E1 error rate: 0.82% (UBIRIS v2), 0.345% (MICHE) | UBIRIS v2, MICHE |
| Chen et al. [22] | Dense block-based CNN with drop-out and batch normalisation | Use LabelMe software package to label the occlusion in iris images to work with images takes in non-ideal environment | Acc.: 99.05% (CASIA-interval-v4), 98.84% (IITD), 99.47% (UBIRIS v2) | CASIA-Interval-v4, UBIRIS v2, IITD |

## 12.4  DATA SETS AND EVALUATION METRICS

For any work, it is very important to have pre-set data set(s) and an appropriate evaluation metric(s) to validate the work and the results. Further, this allows to establish fixed guidelines for comparison with other methods and to establish the standard as well as best methods.

### 12.4.1  DATA SETS

Here, we briefly discuss the various data sets [108] used by the researchers in their work.

### 12.4.2  CASIA[1]

CASIA has been compiled by the Chinese Academy of Science – Institute of Automation (CASIA); it is the first freely available iris database for research purposes [8]. To date, it has four versions, CASIA-Iris V1, V2, V3, and V4, wherein each of the data sets has their own subsets [1]. The first version, i.e. CASIA-Iris V1, comprises 756 iris images ($320 \times 280$) that were acquired from 108 subjects using a home-made iris camera. The second version comprises two equal subsets, each comprising 1,200 iris images ($640 \times 480$) acquired through OKI IRISPASS-h device and CASIA-IrisCamV2. As compared to its predecessors, the third version introduced important noise factors and comprised nearly 22,034 images of iris of 700 subjects divided unequally amongst three sets. The Interval set has 2,639 images ($320 \times 280$), the Lamp subset has 16,212 images ($640 \times 480$), and the Twins subset has 3,183 images ($640 \times 480$) collected from 100 pair of twins. The latest version, i.e. CASIA-Iris V4, which is an extended version of CASIA-Iris V3, consisting of the addition of three new subsets. The first subset CASIA-Iris-Distance comprises 2,576 images ($2,352 \times 1,728$), while the second subset CASIA-Iris-Thousand comprises 20,000 images ($640 \times 480$), and the last subset CASIA-Iris-Syn comprises 10,000 ($640 \times 480$) generated images from CASIA-Iris V1. The versions were released in the order in 2002, 2004, 2010, and 2010.

#### 12.4.2.1  UBIris v1[2] and UBIris v2[3]

The UBIris data sets [75,79] were compiled by the Soft Computing and Image Analysis Group (SOCIA), University of Beira Interior, Portugal. V1 comprises nearly 1,877 images from 241 subjects, whereas V2 comprises 11,102 images from 259 subjects. V1 was captured by Nikon E5700 camera in two parts; in the first part, the noise elements were controlled by having the image acquire set-up in a unilluminated room, while in the second part, the images were under normal light which simulated the images captured with minimal active participation and introduced several noise factors such as contrast, focus, and reflections. V2 was acquired using a Canon EOS 5D camera in unconstrained environments, such as on the visible-wavelength

---

[1] Link-http://www.cbsr.ia.ac.cn/english/IrisDatabase.asp
[2] Link-http://iris.di.ubi.pt/ubiris1.html
[3] Link-http://iris.di.ubi.pt/ubiris2.html

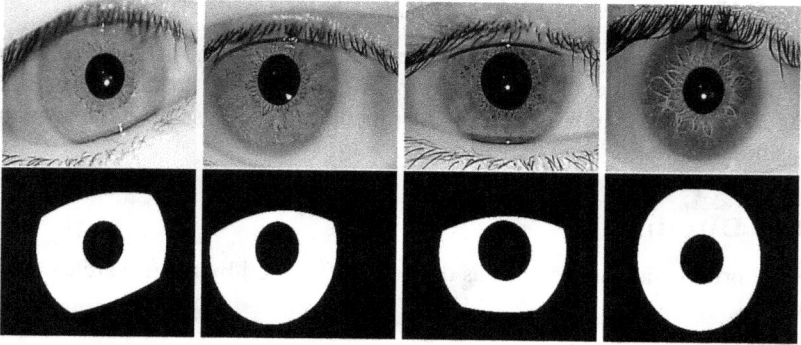

**FIGURE 12.5**   Some images and their corresponding groud-truth segmentation masks from the IITD data set [58].

and on-the-go, which simulated more realistic noise factors as compared to V1. V1 was released in 2004, and V2 was released in 2010. While V1 is available in multiple resolutions such as 800 × 600 pixels and 200 × 150 pixels, V2 is available in 400 × 300.

### 12.4.2.2   NICE-I[4] and NICE-II[5]

Both these data sets [76,78], part of the Noisy Iris Challenge Evaluation, have been distributed by the same group as UBIris. Both the data sets are subsets of UBIris v2. NICE-I was held in 2008, and NICE-II was held in 2010.

### 12.4.2.3   ND-Iris-0405[6]

This data set [16] includes more than 64,979 iris images acquired from nearly 356 subjects taken from 2004 to 2005. Its subset [73] is associated with the iris challenge evaluation, which was organised by the National Institute of Standards and Technology, USA, in 2005. It comprises 2,953 iris images of resolution 480 × 640 acquired from 132 subjects under NIR illumination using an LG EOU 2,200 acquisition system.

### 12.4.2.4   IITD[7]

It was compiled by the Biometrics Research Laboratory of Indian Institute of Technology Delhi, India, and contains 2,240 images of resolution 320 × 240 of nearly 224 subjects acquired using a fully digital CMOS, JPC1000, JIRIS camera. The subjects comprises students and staff at IIT-D itself having age between 14 and 55 years, out of which 48 subjects were female and 176 were male. The data set [58] was published in 2007 (Figure 12.5).

---

4 Link-http://nice1.di.ubi.pt/
5 Link-http://nice2.di.ubi.pt/
6 Link-https://cvrl.nd.edu/projects/data/
7 Link-http://www4.comp.polyu.edu.hk/ csajaykr/IITD/DatabaseI ris.htm

### 12.4.2.5   CSIP[8]

CSIP (Cross-Sensor Iris and Periocular data set) [92] was compiled by the SOCIA group and was acquired using four different mobile devices, i.e. W200 (THL), Xperia Arc S (Sony Ericsson), U8510 (Huawei), and iPhone 4 (Apple). The images were taken by choosing both the front and rear cameras with flash, which led to 10 combinations and their corresponding set-ups. Also, the lighting condition was varied between natural, artificial, and mixed. Owing to all the factors, several noises were incorporated in the data set. The data set comprises 2004 iris images of multiple resolutions from 50 subjects and was released in 2014.

### 12.4.2.6   MICHE-I and MICHE-II[9]

Both MICHE-I and MICHE-II data sets [30] have been created specifically for mobile biometric applications. Part of the Mobile Iris Challenge Evaluation is compiled by the Biometric and Image Processing Lab, University of Salerno, Italy, and it [30] comprises images captured solely from mobile phones in non-restrained environments without the use of any sophisticated equipment to model real-world image acquisition, thereby incorporating various noise factors into the data set. Images were captured using three mobile devices, namely, Samsung Galaxy S4 ($2,322 \times 4,128$), Samsung Galaxy Tab2 ($640 \times 480$), and iPhone5 ($1,536 \times 2,048$), where each captured 1,297, 632, and 1,262 images, respectively. MICHE-II SPECIFICATIONS MICHE-I was released in 2015, and MICHE-II was released in 2016.

### 12.4.2.7   SBVPI[10]

It is distributed by the Faculty of Computer and Information Science, University of Ljubljana, and comprises 1,858 high-resolution ($3,000 \times 1,700$) eye images acquired from 55 subjects. Each subject contributed 32 images, which comprises the person looking at four different gaze-directions, i.e. straight, up, left, and right. As the name suggests, corresponding to each image, there is a separate binary mask for the sclera, pupil, iris, and periocular region. It is a fairly new data set [85,86,102] released in 2018.

### 12.4.2.8   IRISSEG-CC[11]

The data set [7,41] has been compiled by the Halmstad University, wherein the ground truths for the subset of or the whole data set of three other iris data sets are generated. The first is the BioSec Multimodal Biometric Database Baseline [34], which was acquired through an LG IrisAccess EOU3000 close-up infrared iris camera for 3,200 images ($640 \times 480$) from 200 subjects. The IRISSEG-CC comprises ground truth for 75 of them. Next is the CASIA Iris v3 Interval Database [1] which comprises 2,655 iris images ($320 \times 280$) from 249 subjects acquired using a close-up infrared iris camera. The whole ground truth for this one was compiled into the IRISSEG-CC. The last is the MobBIO database [95], which contains 800 iris images ($240 \times 200$)

---

8  Link-http://csip.di.ubi.pt/
9  Link-http://biplab.unisa.it/MICHE/MICHE-II/
10 Link-http://sclera.fri.uni-lj.si/database.html
11 Link-http://islab.hh.se/mediawiki/IrisS egmentationGroundtruth

from 100 subjects through an Asus Eee Pad Transformer TE300T Tablet. There were
two distinct illumination conditions while varying the orientation of the eye with
considerable occlusion. For this, too, the ground truth for all the images was compiled (Table 12.4).

**TABLE 12.4**
**Architecture of the Implemented UNet**

| Bock Name | Layer Name | No. of Filters | Strides | Output Shape |
|---|---|---|---|---|
| - | Input | - | - | $256 \times 256 \times 1$ |
| Encoder | Conv1_1 | 16 | (1,1) | $256 \times 256 \times 16$ |
| | Conv1_2 | 16 | (1,1) | $256 \times 256 \times 16$ |
| | Pool1 | 16 | (2,2) | $128 \times 128 \times 16$ |
| | Conv2_1 | 64 | (1,1) | $128 \times 128 \times 64$ |
| | Conv2_2 | 64 | (1,1) | $128 \times 128 \times 64$ |
| | Pool2 | 64 | (2,2) | $64 \times 64 \times 64$ |
| | Conv3_1 | 128 | (1,1) | $64 \times 64 \times 128$ |
| | Conv3_2 | 128 | (1,1) | $64 \times 64 \times 128$ |
| | Pool3 | 128 | (2,2) | $32 \times 32 \times 128$ |
| | Conv4_1 | 256 | (1,1) | $32 \times 32 \times 256$ |
| | Conv4_2 | 256 | (1,1) | $32 \times 32 \times 256$ |
| | Pool4 | 256 | (2,2) | $16 \times 16 \times 256$ |
| BottleNeck | Conv1_1 | 512 | (1,1) | $16 \times 16 \times 512$ |
| | Conv1_2 | 512 | (1,1) | $16 \times 16 \times 512$ |
| Decoder | Up1 | 256 | (2,2) | $32 \times 32 \times 256$ |
| | Concat1 | - | - | $32 \times 32 \times 512$ |
| | Conv1_1 | 256 | (1,1) | $32 \times 32 \times 256$ |
| | Conv1_2 | 256 | (1,1) | $32 \times 32 \times 256$ |
| | Up2 | 128 | (2,2) | $64 \times 64 \times 128$ |
| | Concat2 | - | - | $64 \times 64 \times 256$ |
| | Conv2_1 | 128 | (1,1) | $64 \times 64 \times 128$ |
| | Conv2_2 | 128 | (1,1) | $64 \times 64 \times 128$ |
| | Up3 | 64 | (2,2) | $128 \times 128 \times 64$ |
| | Concat3 | - | - | $128 \times 128 \times 128$ |
| | Conv3_1 | 64 | (1,1) | $128 \times 128 \times 64$ |
| | Conv3_2 | 64 | (1,1) | $128 \times 128 \times 64$ |
| | Up4 | 16 | (2,2) | $256 \times 256 \times 16$ |
| | Concat4 | - | - | $256 \times 256 \times 32$ |
| | Conv4_1 | 16 | (1,1) | $256 \times 256 \times 16$ |
| | Conv4_2 | 16 | (1,1) | $256 \times 256 \times 16$ |
| - | Output | 1 | (1,1) | $256 \times 256 \times 1$ |

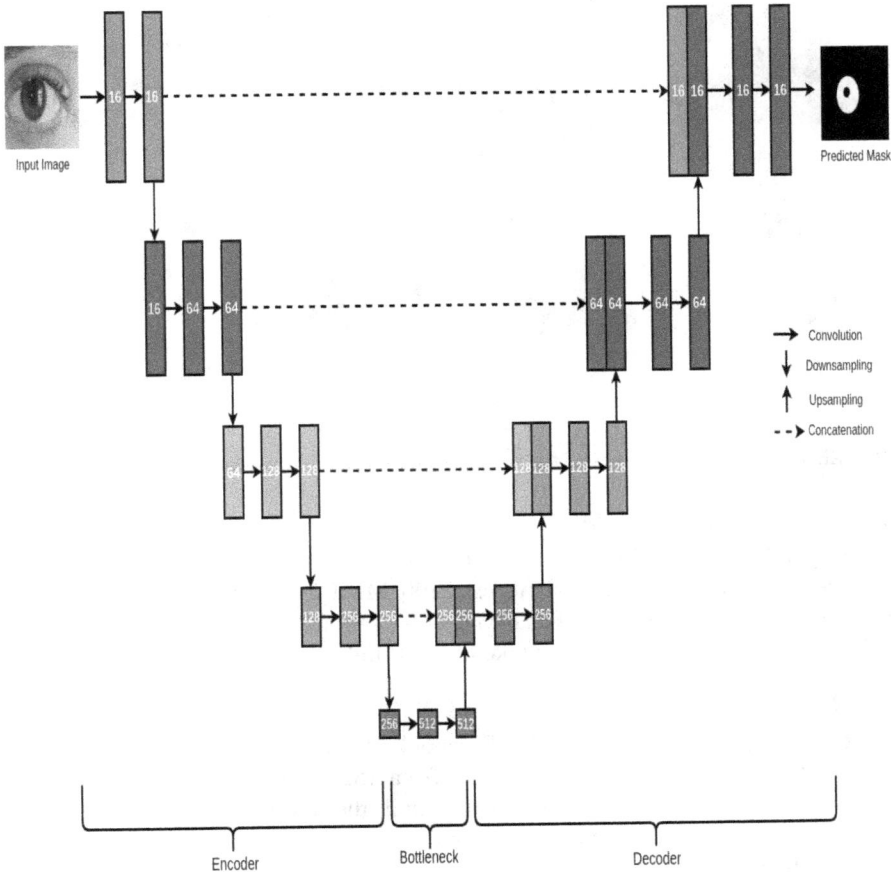

**FIGURE 12.6** UNet [83].

## 12.4.2.9 IRISSEG-EP[12]

Compiled along with IRISSEG-CC [7,41], the data set [41] was compiled by the Multimedia Signal Processing and Security Lab, University of Salzburg. It comprises ground truths of other iris data sets, namely, UBIRIS v2 [79], IIT D [58], Notredame 0405 Iris Image data set [16], and CASIA Iris v4 Interval [1]. The ground truth for the Notredame data set consists of 837 images (640 × 480) whose original images were acquired using LG 2200 close-up near-infrared camera in indoor lighting but with noises such as occlusion, off-angle, and blur. The ground truth for the Casia data set consists of 2,639 images (320 × 280) whose original images were taken using CASIA close-up near-infrared camera (Figure 12.6).

---

[12] Link-http://www.wavelab.at/sources/Hofbauer14b/

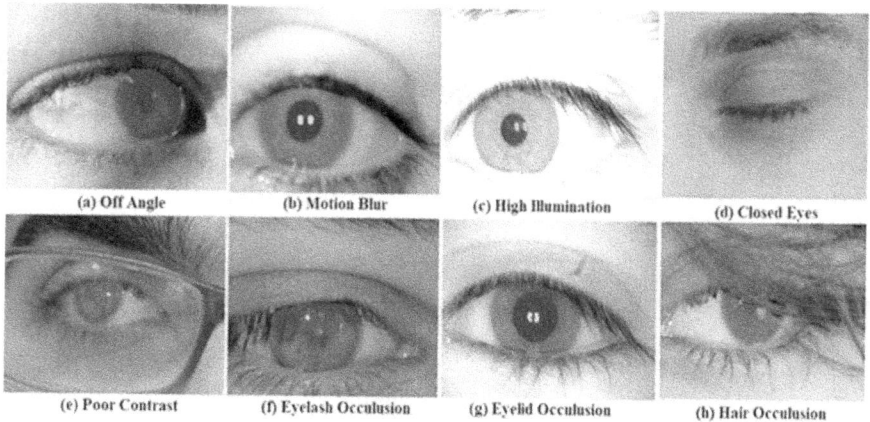

(a) Off Angle          (b) Motion Blur          (c) High Illumination          (d) Closed Eyes

(e) Poor Contrast      (f) Eyelash Occulusion   (g) Eyelid Occulusion         (h) Hair Occulusion

**FIGURE 12.7**   Some examples of the artefacts present in the non-ideal iris images.

### 12.4.2.10   MMU1 and MMU2[13]

Courtesy of Multimedia University, MMU1 [2] comprises about 450 images of nearly 45 subjects captured using LG IrisAccess 2200. MMU2 comprises 995 images from 100 subjects acquired using the Panasonic BM-ET100US camera. Images in the data set are of poor resolution and were taken in NIR lighting.

### 12.4.2.11   OpenEDS[14]

The data set [35] was compiled by Facebook Research comprising 356,649 eye images of resolution 400 × 640 collected from 152 subjects, wherein only 12,759 images have pixel-level annotations of the pupil, the iris, the sclera, and the background. The images were acquired under controlled illumination through a head-mounted display.

### 12.4.2.12   iBUG[15]

This data set was compiled by the Intelligent Behavior Understanding Group for their work [65]. They compiled their own non-ideal iris data set through manual annotation of nearly 4,461 face images picked individually from IMDB [87], HELEN [60], UTDallas Face database [66], 300 VW [96], CVL [72], 300 W [90], and Columbia Gaze database [98] to finally obtain 8,882 iris images (Figure 12.7).

Here, we discussed not only the old and explored data sets but also about some new data sets that are yet to be extensively experimented upon.

### 12.4.3   Performance Metrics

In this section, we will discuss the metrics that are used for quantifying the results of various works.

---

[13]  Link-http://pesona.mmu.edu.my/ ccteo/
[14]  Link-https://research.fb.com/programs/openeds-challenge/
[15]  Link-https://ibug.doc.ic.ac.uk/resources/ ibug-eye-segmentation-dataset/

### 12.4.3.1 Jaccard Index (JI)

It signifies the overlap, i.e. intersection over the combined area, i.e. a union of the segmentation maps for each class. It is calculated over each class and averaged as given by the formula:

$$JI = \frac{1}{N} \sum_{i=1}^{N} \frac{C_{ii}}{GT_i + P_i - C_{ii}} \tag{12.1}$$

Here, $N$ is 2, i.e. binary classes. $C_{ii}$ is the common pixels, i.e. all the pixels having both the ground truth and predicted label as $i$. Here $P_i$ and $GT_i$ are the number of pixels where predicted label is $i$ and the other whose ground truth label is $i$, respectively. The final value is reported after averaging over all the images.

### 12.4.3.2 Mean Segmentation Error

Also termed as E1, it is the overall pixel-wise classification error (PCL) calculated as the exclusive-OR (XOR) ($\oplus$) between the given segmentation map ($Mgt$) and the predicted segmentation map ($M^p$). The equation is given as follows:

$$PCL(M^P, M^{gt}) = \frac{1}{\# \text{ of pixels}} \sum_{i=1}^{\#\text{of pirels}} M^P(\text{pix}_i) \oplus M^{gt}(\text{pix}_i) \tag{12.2}$$

Thereafter, PCL is calculated for all the testing images and averaged to report the overall mean segmentation error, i.e. $E_1$.

### 12.4.3.3 Nice2 Error

Nice2 error or $E_2$ is another measure to evaluate the disparity of the two regions, i.e. non-iris and iris pixels. $E_2^i$, the error for ith image is computed by taking mean of the False-Positive Rate (FPR) along with False-Negative Rate (FNR), which itself is computed at pixel level. Formulas for all stated are given by

$$FNR = \frac{1}{\# \text{ of pixels}} \sum_{i=1}^{\#\text{ of pixels}} [((M^{gt}(\text{pix}_i) \cdot M^P(\text{pix}_i)) \oplus M^{gt}(\text{pix}_i)] \tag{12.3}$$

$$FPR = \frac{1}{\# \text{ of pixels}} \sum_{i=1}^{\#\text{ of pixels}} [((M^{gt}(\text{pix}_i) \cdot M^P(\text{pix}_i)) \oplus M^P(\text{pix}_i)] \tag{12.4}$$

$$E_2^i = \frac{FNR + FPR}{2} \tag{12.5}$$

Thereafter, $E_2^i$ is computed for every test image and averaged to report the final $E_2$. $E_1$ and $E_2$ are bounded between [0, 1], and as they are errors, the closer their values are to "0" the better while the opposite holds for values closer to "1". However, the opposite is true for Intersection Over Union (IOU).

Next, we briefly state some metrics standard in the case of classification; in our case, binary classification. For that, we first define some terms:

- True-Positive ($T_p$): total foreground-pixels classified correctly as iris pixels.
- False-Positive ($F_p$): total pixels incorrectly classified as foreground-pixels.
- True-Negative ($T_n$): total background-pixels classified correctly as non-iris pixels.
- False-Negative ($F_n$): total pixels incorrectly classified as background-pixels.

With the knowledge of the above terminology, the following metrics are defined:

1. **Accuracy**: Fraction of all pixels classified correctly irrespective of the class upon all the pixels in a data set.

$$\text{Accuracy} = \frac{T_p + T_n}{T_p + F_p + T_n + F_p} \tag{12.6}$$

2. **Precision**: Fraction of all positive-class pixels predicted correctly upon all the pixels predicted as positive.

$$P = \frac{T_p}{F_p + T_p} \tag{12.7}$$

3. **Recall**: It is the fraction of all the positive-class pixels classified correctly upon all the positive-class pixels.

$$R = \frac{T_p}{F_n + T_p} \tag{12.8}$$

4. **F-score**: It optimises both the recall and precision as it is the harmonic mean of both.

$$F - \text{score} = \frac{2RP}{P + R} \tag{12.9}$$

## 12.5   EXPERIMENTATION

Here, we briefly discuss an implementation of the UNet [83] done by us on two majorly used data sets, namely, UBIris v2 [79] along with CASIA v4 Interval [1]. We have already discussed the UNet in the above sections whose implementation has been open-sourced by the authors [33].[16] It is a simple encoder-decoder CNN-based architecture wherein novel skip connections joining encoder layers to the decoder layers which provide a global context to the already processed local information for the generation of location-precise maps. For each of the data sets, a similar procedure

---

[16] Link-https://lmb.informatik.uni-freiburg.de/people/ronneber/u-net/

**TABLE 12.5**
**Testing Results on the Implementation of UNet**

| Data set | E1% Error | E2% Error |
|---|---|---|
| **UBIris v2** | 1. **0.54** | 2. **3.12** |
| 3. Casia-Iris v4(Interval) | 4. 0.67 | 5. 1.36 |

as described below was followed. First, the images in the data set were reshaped into $256 \times 256$ and were normalised between [0,1] by dividing it by 255. Next, the data set was divided into a 30% and 70% into testing set and training set, respectively. We coded our model using Keras [23] and trained on the NVIDIA GeForce GTX 1080 Ti for 100 epochs (Table 12.5).

The loss function was Binary-cross-entropy, and E1 and E2 were taken as the validation metrics.

The kernel size was kept $3 \times 3$ with stride 1 for convolution and $3 \times 3$ with stride 2 for both downsampling and upsampling. Each convolution was followed by a ReLu activation [6] and Batch Normalisation [50] for better generalisation. Different optimisers such as SGD [89], RMSProp [100], and Adam [56] were used, and the best results reported below were obtained using Adam. All weights were initialised according to the He et al. [38] initialisation. In accordance with the above nomenclature, the loss function is defined as follows:

$$L_{bce}(M^P, M^{gt}) = \frac{-1}{\#\,of\,pixels} \sum_{i=1}^{\#\,of\,pixels} [M^{gt}(pix_i) * \log(M^P(pix_i))$$
$$+ (1 - M^{gt}(pix_i)) * \log(1 - M^P(pix_i)] \quad (12.10)$$

## 12.6 CHALLENGES IDENTIFIED AND FURTHER DIRECTION

Before discussing about the future work, we again describe the various non-idealities introduced in the data set and also some future non-idealities that may be encountered in the future data sets.

- **Occlusion**: Eyelids, eyelashes, and hair are the leading causes of occlusion in Iris images with massively varying levels of occlusions, sometimes to the tune of 80%–90%.
- **Blurring**: In many cases, either the subject is on the move in unconstrained environments or even in the case of constrained environments, and they may move a little causing the image to blur. Also, in some cases, if the equipment is not correctly set-up, the camera itself may move, adding to the blurriness.
- **Alignment**: In some cases, the subject may move their eye or even their entire head, which causes the iris to appear more oval-shaped and not occupy the centre of the image as intended due to misalignment of the face

and the equipment. Needless to say, in the case of an unconstrained environment, this is highly prevalent.

- **Resolution**: For proper segmentation and subsequent verification, it is always the best to have high-resolution images so that the relevant features are easily extracted and represented. However, this is not always the case as the resolution of the acquired image is solely dependent on the camera equipment, which may vary from a high-end imaging set-up to a mobile camera.
- **Adulteration**: This is an artefact that has the potential to appear in many ways. For example, subjects wearing lenses or glasses alter the natural appearance of the iris. Although indistinguishable to the naked eye, they alter the intensity of the original iris at the pixel level, which may cause substantial hindrance to the segmentation algorithms. In the future, more "artificialness" might be introduced in presently unknown ways.

It has been widely discussed that models based on deep learning are heavily dependent on data characteristics. Iris images acquired in an unregulated condition represent nearly all the characteristics of the real world scenarios. Developed robust and complex deep learning models such as PixISegNet and Iris-DenseNet are now able to process and can be trained on the data sets generated in an unconstrained environment and hence are bound to give excellent performance when deployed to the real world. However, there is still much of room for development where the researchers can improve upon the robustness of the model to look beyond occlusion and identify the features of anatomy nearby iris. A lot of work is yet to be done GANs, a generative model wherein an encoder-decoder may act as an active generator and a separate discriminator to differentiate amongst the predicted map and the ground truth maps may result in the improved performance. Similar to this, the use of attention modules, dictionary learning, recurrent neural network, and many other existing ideas are not thoroughly explored by the researchers. Moreover, most of the models have been trained upon the existing loss functions, as discussed above, leaving room for improvement there too. Hence, there is still a vast space for the development of novel and state-of-the-art techniques and to overcome all the drawbacks of previous works.

## 12.7  CONCLUSION

In the previous decades, numerous researchers have devised several biometric systems based on various biometric traits depending upon the specific need and application of the system such as retina, fingerprints, palm, and voice. However, one common thing with most of the biometric traits is them being invasive or having any form of contact for acquisition, e.g. touching the scanners for fingerprints, hand veins, signing a phrase, or requiring some form of cooperation from the user. Due to this, non-invasive data acquisition and methods for biometric systems have started surfacing more and more. Moreover, with the recent spread of the highly transmissible COVID-19, the need for non-invasive biometrics is bound to increase exponentially. While the face and the iris are suitable non-invasive biometric candidates, the face lacks permanence unlike the iris. In this manner, iris can be said to emerge as the leader in non-invasive biometrics and making its study all the more relevant. In accordance, in this study, we discussed various deep learning and non-deep learning-based methodology concerning the extraction of the iris portion (a crucial step in Iris Biometric). Also, we provided information about the various iris data sets available to the public, along with various metrics to compare the works and help the research community.

## ACKNOWLEDGEMENTS

We acknowledge the journal paper "PixISegNet: Pixel Level Iris Segmentation Network using Convolutional Encoder-Decoder with Stacked Hourglass Bottleneck" published in IET-Biometrics, conference papers "UBSegNet: Unified Biometric Region of Interest Segmentation Network" published in IEEE IAPR Asian Conference on Pattern Recognition (ACPR), "IPSegNet: Deep Convolutional Neural Network Based Segmentation Framework for Iris and Pupil" published in IEEE International Conference on Signal-Image Technology Internet-Based Systems (SITIS), and "HFDSegNet: Holistic and Generalized Finger Dorsal ROI Segmentation Network" published in IEEE International Conference on Pattern Recognition Applications and Methods.

## REFERENCES

1. Casia iris image database, http://biometrics.idealtest.org.
2. Multimedia-university. mmu database [online]. available: http://pesona.mmu.edu.my/ccteo/.
3. A new iris segmentation method for non-ideal iris images.
4. A novel iris segmentation using radial-suppression edge detection. Dec. 2009.
5. R. H. Abiyev and K. Altunkaya. Personal iris recognition using neural network. *International Journal of Security and its Applications*, 37(2):41–50, 2008.
6. A. F. Agarap. Deep learning using rectified linear units (relu). *arXiv preprint arXiv*:1803.08375, 2018.
7. F. Alonso-Fernandez and J. Bigun. Near-infrared and visible-light periocular recognition with gabor features using frequency-adaptive automatic eye detection. *IET Biometrics*, 4(2):74–89, 2015.
8. M. M. Alrifaee, M. M. Abdallah, and B. G. Al Okush. A Short Survey of IRIS Images Databases. *The International Journal of Multimedia and Its Applications*, 9(2):01–14, April 2017.
9. M. Arsalan, H. G. Hong, R. A. Naqvi, M. B. Lee, M.-C. Kim, D. S. Kim, C. S. Kim, and K. R. Park. Deep learning-based iris segmentation for iris recognition in visible light environment. *Symmetry*, 9:263, 2017.
10. M. Arsalan, D. S. Kim, M. B. Lee, M. Owais, and K. R. Park. Fred-net: Fully residual encoder-decoder network for accurate iris segmentation. *Expert Systems with Applications*, 122:217–241, 2019.
11. M. Arsalan, R. A. Naqvi, D. S. Kim, P. H. Nguyen, M. Owais, and K. R. Park. Irisdensenet: Robust iris segmentation using densely connected fully convolutional networks in the images by visible light and near-infrared light camera sensors. *Sensors (Basel, Switzerland)*, 18(5):1501, May 2018.

12. V. Badrinarayanan, A. Kendall, and R. Cipolla. Segnet: A deep convolutional encoder-decoder architecture for image segmentation. *IEEE Transactions on Pattern Analysis and Machine Intelligence*, 39(12):2481–2495, 2017.

13. S. Barpanda, B. Majhi, P. Sa, A. Kumar, and S. Bakshi. Iris feature extraction through wavelet mel-frequency cepstrum coefficients. *Optics Laser Technology*, 110:13–23, March 2018. doi: 10.1016/j.optlastec.2018.03.002.

14. S. Bazrafkan, S. Thavalengal, and P. Corcoran. An end to end deep neural network for iris segmentation in unconstrained scenarios. *Neural Networks*, 106:79–95, 2018.

15. W. W. Boles and B. Boashash. A human identification technique using images of the iris and wavelet transform. *IEEE Transactions on Signal Processing*, 46(4):1185–1188, 1998.

16. K. W. Bowyer and P. J. Flynn. The nd-iris-0405 iris image dataset. *arXiv preprint arXiv:1606.04853*, 2016.

17. K. W. Bowyer, K. Hollingsworth, and P. J. Flynn. Image understanding for iris biometrics: A survey. *Computer Vision and Image Understanding*, 110(2):281–307, 2008.

18. Y. Boykov, O. Veksler, and R. Zabih. Fast approximate energy minimization via graph cuts. *IEEE Transactions on Pattern Analysis and Machine Intelligence*, 23(11): 1222–1239, 2001.

19. T. Burghardt. A brief review of biometric identification. Retrieved August, 6:2010, 2009.

20. J. L. Cambier and J. E. Siedlarz. Portable authentication device and method using iris patterns, March 11 2003. US Patent 6,532,298.

21. L.-C. Chen, G. Papandreou, I. Kokkinos, K. Murphy, and A. Yuille. Deeplab: Semantic image segmentation with deep convolutional nets, atrous convolution, and fully connected CRFS. *IEEE Transactions on Pattern Analysis and Machine Intelligence*, June 2016. doi: 10.1109/TPAMI.2017.2699184.

22. Y. Chen, W. Wang, Z. Zeng, and Y. Wang. An adaptive cnns technology for robust iris segmentation. *IEEE Access*, 7:64517–64532, May 2019. doi: 10.1109/ACCESS.2019.2917153.

23. F. Chollet et al. Keras. https://keras.io, 2015.

24. Ö. Çiçek, A. Abdulkadir, S. S. Lienkamp, T. Brox, and O. Ronneberger. 3d u-net: Learning dense volumetric segmentation from sparse annotation. In S. Ourselin, L. Joskowicz, M. R. Sabuncu, G. Unal, and W. Wells, editors, *Medical Image Computing and Computer-Assisted Intervention - MICCAI 2016*, pp. 424–432, Cham, Springer International Publishing, 2016.

25. J. Daugman. How iris recognition works. *IEEE Transactions on Circuits and Systems for Video Technology*, 14(1):21–30, 2004.

26. J. Daugman. New methods in iris recognition. *IEEE Transactions on Systems, Man, and Cybernetics, Part B (Cybernetics)*, 37(5):1167–1175, 2007.

27. J. Daugman. Information theory and the iriscode. *IEEE Transactions on Information Forensics and Security*, 11(2):400–409, 2015.

28. J. G. Daugman. High confidence visual recognition of persons by a test of statistical independence. *IEEE Transactions on Pattern Analysis and Machine Intelligence*, 15(11):1148–1161, 1993.

29. T. Dean, and D.D. Cox. Neural networks and neuroscience-inspired computer vision. *Current Biology*, 24(18):R921–R929, 2014. doi: 10.1016/j.cub.2014.08.026. PMID: 25247371.

30. M. De Marsico, M. Nappi, D. Riccio, and H. Wechsler. Mobile iris challenge evaluation (miche)-i, biometric iris dataset and protocols. *Pattern Recognition Letters*, 57:17–23, 2015.

31. N. Dhanachandra, K. Manglem, and Y. J. Chanu. Image segmentation using k-means clustering algorithm and subtractive clustering algorithm. *Procedia Computer Science*, 54:764–771, 2015.

32. V. Dorairaj, N. A. Schmid, and G. Fahmy. Performance evaluation of non-ideal iris based recognition system implementing global ICA encoding. In *IEEE International Conference on Image Processing 2005*, Genova, volume 3, pp. III–285, 2005. doi: 10.1109/ICIP.2005.1530384.

33. T. Falk, D. Mai, R. Bensch, Ö. Çiçek, A. Abdulkadir, Y. Marrakchi, A. Böhm, J. Deubner, Z. Jäckel, K. Seiwald, et al. U-net: Deep learning for cell counting, detection, and morphometry. *Nature methods*, 16(1):67–70, 2019.

34. J. Fierrez, J. Ortega-Garcia, D. T. Toledano, and J. Gonzalez-Rodriguez. Biosec base-line corpus: A multimodal biometric database. *Pattern Recognition*, 40(4):1389–1392, 2007.

35. S. J. Garbin, Y. Shen, I. Schuetz, R. Cavin, G. Hughes, and S. S. Talathi. Openeds: Open eye dataset. *arXiv preprint arXiv:*1905.03702, 2019.

36. S. N. Garg, R. Vig, and S. Gupta. A critical study and comparative analysis of multibio-metric systems using iris and fingerprints. *International Journal of Computer Science and Information Security*, 15(1):549, 2017.

37. K. He, G. Gkioxari, P. Dollár, and R. Girshick. Mask r-cnn. In *2017 IEEE International Conference on Computer Vision (ICCV)*, Venice, pp. 2980–2988, 2017. doi: 10.1109/ICCV.2017.322.

38. K. He, X. Zhang, S. Ren, and J. Sun. Deep residual learning for image recognition. In *2016 IEEE Conference on Computer Vision and Pattern Recognition (CVPR)*, Las Vegas, NV, pp. 770–778, 2016. doi: 10.1109/CVPR.2016.90.

39. M. Hill. *Anat2310: Eye Development*. The University of South Wales, 2003. Available from http://anatomy.med.unsw.edu.au/cbl/teach/anat2310/Lecture06Senses(print).pdf

40. H. Hofbauer, F. Alonso-Fernandez, J. Bigun, and A. Uhl. Experimental analysis regard-ing the influence of iris segmentation on the recognition rate. *IET Biometrics*, 5(3): 200–211, 2016.

41. H. Hofbauer, F. Alonso-Fernandez, P. Wild, J. Bigun, and A. Uhl. A ground truth for iris segmentation. In *2014 22nd International Conference on Pattern Recognition*, pp. 527–532. IEEE, 2014.

42. A. G. Howard, M. Zhu, B. Chen, D. Kalenichenko, W. Wang, T. Weyand, M. Andreetto, and H. Adam. Mobilenets: Efficient convolutional neural networks for mobile vision applications. *ArXiv*, abs/1704.04861, 2017.

43. Y. Hu, K. Sirlantzis, and G. Howells. Improving colour iris segmentation using a model selection technique. *Pattern Recognition Letters*, 57:24–32, 2015. Mobile Iris CHallenge Evaluation part I (MICHE I).

44. G. Huang, Z. Liu, L. Van Der Maaten, and K. Q. Weinberger. Densely connected convo-lutional networks. *2017 IEEE Conference on Computer Vision and Pattern Recognition (CVPR)*, Honolulu, HI, pp. 2261–2269, 2017. doi: 10.1109/CVPR.2017.243.

45. J. Huang, Y. Wang, T. Tan, and J. Cui. A new iris segmentation method for recogni-tion. In *Proceedings of the Pattern Recognition, 17th International Conference on (ICPR'04) Volume 3 - Volume 03*, ICPR'04, pp. 554–557. IEEE Computer Society, USA, 2004.

46. Y.-P. Huang, S.-W. Luo, and E.-Y. Chen. An efficient iris recognition system. In *Proceedings. International Conference on Machine Learning and Cybernetics*, Beijing, China, 1(2):450–454, 2002. doi: 10.1109/ICMLC.2002.1176794.

47. W. Hung, Y. Tsai, Y. Liou, Y. Lin, and M. Yang. Adversarial learning for semi-supervised semantic segmentation. Paper presented at *29th British Machine Vision Conference, BMVC 2018; Conference Date: 03-09-2018 Through 06-09-2018*, Newcastle, UK, Jan. 2019.

48. M. T. Ibrahim, T. M. Khan, S. A. Khan, M. A. Khan, and L. Guan. Iris localization using local histogram and other image statistics. *Optics and Lasers in Engineering*, 50(5):645–654, 2012.

49. M. T. Ibrahim, T. Mehmood, M. A. Khan, and L. Guan. A novel and efficient feedback method for pupil and iris localization. In M. Kamel, and A. Campilho, editors, *Image Analysis and Recognition*. ICIAR 2011. Lecture Notes in Computer Science, volume 6754. Berlin, Springer. doi: 10.1007/978-3-642-21596-4_9, 2011.

50. S. Ioffe and C. Szegedy. Batch normalization: Accelerating deep network training by reducing internal covariate shift. *arXiv preprint arXiv:*1502.03167, 2015.

51. A. K. Jain, P. Flynn, and A. A. Ross. *Handbook of Biometrics*. Boston, MA, Springer Science & Business Media, 2007. doi: 10.1007/978-0-387-71041-9.

52. A. K. Jain, A. Ross, and S. Prabhakar. An introduction to biometric recognition. *IEEE Transactions on Circuits and Systems for Video Technology*, 14(1):4–20, 2004.

53. R. R. Jha, G. Jaswal, D. Gupta, S. Saini, and A. Nigam. Pixisegnet: Pixel-level iris segmentation network using convolutional encoder-decoder with stacked hourglass bottleneck. *IET Biometrics*, 9(1):11–24, 2019.

54. P. Kansal and S. Nathan. Eyenet: Attention based convolutional encoder-decoder network for eye region segmentation, October 2019.

55. Tariq M. Khan, Shahzad A. Malik, Shahid A. Khan, Tariq Bashir, & Amir H. Dar. Automatic localization of pupil using eccentricity and iris using gradient based method. *Optics and Lasers in Engineering*, 49(2):177–187, 2011. doi: 10.1016/j.optlaseng.2010.08.020.

56. D. P. Kingma and J. Ba. Adam: A method for stochastic optimization. *arXiv preprint arXiv:1412.*6980, 2014.

57. W. K. Kong and D. Zhang. Accurate iris segmentation based on novel reflection and eyelash detection model. In *Proceedings of 2001 International Symposium on Intelligent Multimedia, Video and Speech Processing. ISIMP 2001 (IEEE Cat. No.01EX489)*, Hong Kong, China, pp. 263–266, 2001. doi: 10.1109/ISIMP.2001.925384.

58. A. Kumar and A. Passi. Comparison and combination of iris matchers for reliable personal authentication. *Pattern Recognition*, 43(3):1016–1026, 2010.

59. A. Lakra, P. Tripathi, R. Keshari, M. Vatsa, and R. Singh. Segdensenet: Iris segmentation for pre- and-post cataract surgery. In *2018 24th International Conference on Pattern Recognition (ICPR)*, Beijing, China, pp. 3150–3155, 2018. doi: 10.1109/ICPR.2018.8545840.

60. V. Le, J. Brandt, Z. Lin, L. Bourdev, and T. S. Huang. Interactive facial feature localization. In *European Conference on Computer Vision*, pp. 679–692. Springer, 2012.

61. Y. Lecun, L. Bottou, Y. Bengio, and P. Haffner. Gradient-based learning applied to document recognition. *Proceedings of the IEEE*, 86(11):2278–2324, 1998.

62. S. Lim, K. Lee, O. Byeon, and T. Kim. Efficient iris recognition through improvement of feature vector and classifier. *ETRI Journal*, 23(2):61–70, 2001.

63. N. Liu, H. Li, M. Zhang, J. Liu, Z. Sun, and T. Tan. Accurate iris segmentation in non-cooperative environments using fully convolutional networks. *2016 International Conference on Biometrics (ICB)*, Halmstad, pp. 1–8, 2016. doi: 10.1109/ICB.2016.7550055.

64. W. Liu, D. Anguelov, D. Erhan, C. Szegedy, S. Reed, C.-Y. Fu, and A. C. Berg. SSD: Single shot multibox detector. In B. Leibe, J. Matas, N. Sebe, and M. Welling, editors, *Computer Vision - ECCV 2016*, pp. 21–37, Cham, Springer International Publishing, 2016.

65. B. Luo, J. Shen, S. Cheng, Y. Wang, and M. Pantic. Shape constrained network for eye segmentation in the wild. In *The IEEE Winter Conference on Applications of Computer Vision* (WACV), Snowmass Village, CO, pp. 1952–1960, 2020. doi: 10.1109/WACV45572.2020.9093483.

66. M. Minear and D. C. Park. A lifespan database of adult facial stimuli. *Behavior Research Methods, Instruments, & Computers*, 36(4):630–633, 2004.

67. L. Najman and M. Schmitt. Watershed of a continuous function. *Signal Processing*, 38(1):99–112, 1994. Mathematical Morphology and its Applications to Signal Processing.

68. N. Otsu. A threshold selection method from gray-level histograms. In *IEEE Transactions on Systems, Man, and Cybernetics*, 9(1):62–66, Jan. 1979, doi: 10.1109/TSMC.1979.4310076.

69. Y. Parikh, U. Chaskar, and H. Khakole. Effective approach for iris localization in nonideal imaging conditions. In *Proceedings of the 2014 IEEE Students' Technology Symposium*, Kharagpur, pp. 239–246, 2014. doi: 10.1109/TechSym.2014.6808054.

70. H. Patel, C. K. Modi, M. Paunwala, and S. Patnaik. Human identification by partial iris segmentation using pupil circle growing based on binary integrated edge intensity curve. *2011 International Conference on Communication Systems and Network Technologies*, Katra, Jammu, pp. 333–338, 2011. doi: 10.1109/CSNT.2011.76.

71. S. M. Patil, R. R. Jha, and A. Nigam. IPSegNet: Deep convolutional neural network based segmentation framework for iris and pupil. In *2017 13th International Conference on Signal-Image Technology Internet-Based Systems (SITIS)*, Jaipur, pp. 184–191, 2017. doi: 10.1109/SITIS.2017.40.

72. P. Peer. Cvl face database. *Computer vision lab., faculty of computer and information science, Slovenia*, University of Ljubljana, Available at http://www. lrv. fri. uni-lj. si/ facedb. html, 2005.

73. P. J. Phillips, K. W. Bowyer, P. J. Flynn, X. Liu, and W. T. Scruggs. The iris challenge evaluation 2005. In *2008 IEEE Second International Conference on Biometrics: Theory, Applications and Systems*, pp. 1–8. IEEE, 2008.

74. N. Plath, M. Toussaint, and S. Nakajima. Multi-class image segmentation using conditional random fields and global classification. In *Proceedings of the 26th Annual International Conference on Machine Learning*, ICML'09, pp. 817–824, New York, NY, USA, Association for Computing Machinery, 2009.

75. H. Proença and L. A. Alexandre. UBIRIS: A noisy iris image database. In *13th International Conference on Image Analysis and Processing – ICIAP 2005*, volume LNCS 3617, pp. 970–977, Cagliari, Italy, Springer, September 2005.

76. H. Proença and L. A. Alexandre. The nice. I: Noisy iris challenge evaluation-part I. In *2007 First IEEE International Conference on Biometrics: Theory, Applications, and Systems*, pp. 1–4. IEEE, 2007.

77. H. Proença and L. A. Alexandre. Introduction to the special issue on the segmentation of visible wavelength iris images captured at-a-distance and on-the-move. *Image and Vision Computing*, 28(2):213–214, 2010.

78. H. Proenca and L. A. Alexandre. Toward covert iris biometric recognition: Experimental results from the nice contests. *IEEE Transactions on Information Forensics and Security*, 7(2):798–808, 2011.

79. H. Proenca, S. Filipe, R. Santos, J. Oliveira, and L. A. Alexandre. The ubiris. v2: A database of visible wavelength iris images captured on-the-move and at-a-distance. *IEEE Transactions on Pattern Analysis and Machine Intelligence*, 32(8):1529–1535, 2009.

80. S. Pundlik, D. Woodard, and S. Birchfield. Non-ideal iris segmentation using graph cuts. *IEEE Computer Society Conference on Computer Vision and Pattern Recognition Workshops*, Anchorage, AK, pp. 1–6, 2008. doi: 10.1109/CVPRW.2008.4563108.

81. A. Radman, K. Jumari, and N. Zainal. Fast and reliable iris segmentation algorithm. *IET Image Processing*, 7(1):42–49, 2013.

82. S. Ren, K. He, R. Girshick, and J. Sun. Faster r-cnn: Towards real-time object detection with region proposal networks. In C. Cortes, N. D. Lawrence, D. D. Lee, M. Sugiyama, and R. Garnett, editors, *Advances in Neural Information Processing Systems*, Vol. 28, pp. 91–99. Curran Associates, Inc., 2015.

83. O. Ronneberger, P. Fischer, and T. Brox. U-net: Convolutional networks for biomedi-cal image segmentation. In N. Navab, J. Hornegger, W. M. Wells, and A. F. Frangi, editors, *Medical Image Computing and Computer-Assisted Intervention - MICCAI 2015*, pp. 234–241, Cham, Springer International Publishing, 2015.

84. A. Ross. Iris as a forensic modality: The path forward. URL http://www. nist. gov/ forensics/upload/Ross-Presentation.pdf, 2009.

85. P. Rot, Ž. Emerši?, V. Struc, and P. Peer. Deep multi-class eye segmentation for ocular biometrics. In 2018 *IEEE International Work Conference on Bioinspired Intelligence (IWOBI)*, San Carlos, pp. 1–8, IEEE, 2018. doi: 10.1109/IWOBI.2018.8464133.

86. P. Rot, M. Vitek, K. Grm, Ž. Emerši?, P. Peer, and V. Štruc. Deep sclera segmentation and recognition. In *Handbook of Vascular Biometrics*, pp. 395–432. Cham, Springer, 2020. doi: 10.1007/978-3-030-27731-4_13.

87. R. Rothe, R. Timofte, and L. Van Gool. Deep expectation of real and apparent age from a single image without facial landmarks. *International Journal of Computer Vision*, 126(2–4):144-157, 2018.

88. K. Roy, P. Bhattacharya, and C. Y. Suen. Towards nonideal iris recognition based on level set method, genetic algorithms and adaptive asymmetrical SVMS. *Engineering Applications of Artificial Intelligence*, 24(3):458–475, April 2011.

89. S. Ruder. An overview of gradient descent optimization algorithms. *arXiv preprint arXiv:*1609.04747, 2016.

90. C. Sagonas, E. Antonakos, G. Tzimiropoulos, S. Zafeiriou, and M. Pantic. 300 faces in-the-wild challenge: Database and results. *Image and Vision Computing*, 47:3–18, 2016.

91. W. Sankowski, K. Grabowski, M. Napieralska, M. Zubert, and A. Napieralski. Reliable algorithm for iris segmentation in eye image. *Image and Vision Computing*, 28(2): 231–237, 2010.

92. G. Santos, E. Grancho, M. V. Bernardo, and P. T. Fiadeiro. Fusing iris and periocular information for cross-sensor recognition. *Pattern Recognition Letters*, 57:52–59, 2015.

93. M. Sardar, S. Mitra, and B. U. Shankar. Iris localization using rough entropy and CSA: A soft computing approach. *Applied Soft Computing*, 67:61–69, 06 2018.

94. M. Schuckers. Some statistical aspects of biometric identification device performance. *Stats Magazine*, 3, 2001.

95. A. F. Sequeira, J. C. Monteiro, A. Rebelo, and H. P. Oliveira. Mobbio: A multimodal database captured with a portable handheld device. In *2014 International Conference on Computer Vision Theory and Applications (VISAPP)*, volume 3, pp. 133–139. IEEE, 2014.

96. J. Shen, S. Zafeiriou, G. G. Chrysos, J. Kossaifi, G. Tzimiropoulos, and M. Pantic. The first facial landmark tracking in-the-wild challenge: Benchmark and results. In *Proceedings of the IEEE International Conference on Computer Vision Workshops* (ICCVW), Santiago, pp. 50–58, 2015. doi: 10.1109/ICCVW.2015.132.

97. K. Simonyan and A. Zisserman. Very deep convolutional networks for large-scale image recognition. *CoRR*, abs/1409.1556, 2015.

98. B. A. Smith, Q. Yin, S. K. Feiner, and S. K. Nayar. Gaze locking: Passive eye con-tact detection for human-object interaction. In *Proceedings of the 26th Annual ACM Symposium on User Interface Software and Technology*, New York, NY, pp. 271–280, 2013.

99. C. Szegedy, W. Liu, Y. Jia, P. Sermanet, S. Reed, D. Anguelov, D. Erhan, V. Vanhoucke, and A. Rabinovich. Going deeper with convolutions. In *2015 IEEE Conference on Computer Vision and Pattern Recognition (CVPR)*, Boston, MA, pp. 1–9, 2015. doi: 10.1109/CVPR.2015.7298594.

100. T. Tieleman and G. Hinton. RMSProp gradient optimization. URL http://www. cs.toronto.edu/tijmen/csc321/slides/lecture_slides_lec6.pdf, 2014.

101. K. Tripathi. A comparative study of biometric technologies with reference to human interface. *International Journal of Computer Applications*, 14(5):10–15, 2011.

102. M. Vitek, P. Rot, V. Štruc, and P. Peer. A comprehensive investigation into sclera biometrics: A novel dataset and performance study. *Neural Computing and Applications*:1–15, 2020. doi: 10.1007/s00521-020-04782-1.

103. C. Wang, J. Muhammad, Y. Wang, Z. He, and Z. Sun. Towards complete and accurate iris segmentation using deep multi-task attention network for non-cooperative iris recognition. *IEEE Transactions on Information Forensics and Security*, 15:2944–2959, 2020.

104. C. Wang, Y. Wang, B. Xu, Y. He, Z. Dong, and Z. Sun. A lightweight multi-label segmentation network for mobile iris biometrics. In *ICASSP 2020 – 2020 IEEE International Conference on Acoustics, Speech and Signal Processing (ICASSP)*, Barcelona, Spain, pp. 1006–1010, 2020. doi: 10.1109/ICASSP40776.2020.9054353.

105. J. Wang, K. Sun, T. Cheng, B. Jiang, C. Deng, Y. Zhao, D. Liu, Y. Mu, M. Tan, X. Wang, W. Liu, and B. Xiao. Deep high-resolution representation learning for visual recognition. *IEEE Transactions on Pattern Analysis and Machine Intelligence*:1–1, 2020. doi: 10.1109/TPAMI.2020.2983686.

106. R. P. Wildes. Iris recognition: An emerging biometric technology. *Proceedings of the IEEE*, 85(9):1348–1363, 1997.

107. A. Witkin, D. Terzopoulos, and M. Kass. Active contour models. *International Journal of Computer Vision*, 1:321–331, 1998.

108. L. A. Zanlorensi, R. Laroca, E. Luz, A. S. Britto Jr, L. S. Oliveira, and D. Menotti. Ocular recognition databases and competitions: A survey. *arXiv preprint arXiv:*1911.09646, 2019.

109. Z. Zhou, M. M. Rahman Siddiquee, N. Tajbakhsh, and J. Liang. Unet++: A nested u-net architecture for medical image segmentation. In D. Stoyanov, Z. Taylor, G. Carneiro, T. Syeda-Mahmood, A. Martel, L. Maier-Hein, J. M. R. Tavares, A. Bradley, J. P. Papa, V. Belagiannis, J. C. Nascimento, Z. Lu, S. Conjeti, M. Moradi, H. Greenspan, and A. Madabhushi, editors, *Deep Learning in Medical Image Analysis and Multimodal Learning for Clinical Decision Support*, pp. 3–11, Cham, Springer International Publishing, 2018.

# 13 PPG-Based Biometric Recognition
## Opportunities with Machine and Deep Learning

*Amit Kaul and Akhil Walia*
EED, NIT Hamirpur

## CONTENTS

13.1 Introduction ............................................................................................. 313
13.2 Photoplethysmogram (PPG) ................................................................... 315
13.3 Literature Review .................................................................................... 315
13.4 Multi-Feature Approach for PPG Biometric .......................................... 319
    13.4.1 Signal Acquisition......................................................................... 323
    13.4.2 Baseline Wander and Noise Removal............................................ 323
    13.4.3 Feature Extraction ......................................................................... 323
        13.4.3.1 Pulse Extraction and Normalisation ............................... 324
        13.4.3.2 First- and Second-Order Derivatives ............................. 325
        13.4.3.3 Autocorrelation ............................................................... 325
13.5 Classification............................................................................................ 326
13.6 Experiments and Results ......................................................................... 327
13.7 Conclusions.............................................................................................. 328
References...................................................................................................... 329

## 13.1 INTRODUCTION

Medical practitioners often monitor the functioning of human body by non-invasively collecting bioelectric signals such as electrocardiogram (ECG), electroencephalogram (EEG), and photoplethysmogram (PPG). These signals are used to examine the health of human beings and diagnose different ailments in them. However, in recent years, in addition to their obvious application in medical science, researchers have extensively used these bioelectric quantities for biometric recognition. Traditionally, the area of biometrics has involved physiological traits like fingerprint, iris, face etc. or behavioural qualities like gait, keystroke, speech, etc. for distinguishing individuals. The USP of biometric lies in the fact that being an integral part of a person, these traits cannot be stolen, shared, or forgotten. The uniqueness and permanence of these traits, especially the physiological ones, are well established and technology linked to them has also matured. Primarily, these tools have been employed in access control and security applications. The systems operate either in verification mode, where one-to-one matching takes place, or identification mode, which involves many-to-one comparison. However, with widespread digitisation, the scope of applications of biometrics has also widened to varied domains ranging from e-commerce to healthcare [1].

With the enhanced popularity and visibility of biometric systems, there has been a rapid upsurge in attempts by the fraudsters to breach these systems and intrude into them. The different points/stages at which biometric system can be intruded are depicted in Figure 13.1.

These can be divided into two main groups: (i) direct attacks involve use of synthetic templates such as pre-recorded speech, face images etc and are at sensor level (attack 1), and (ii) indirect attacks consist of all the remaining errors. Out of these, feature extractor and matcher are bypassed using a Trojan horse in stages 3 and 5, respectively. The stored template is replaced, added, or deleted in attack at stage 6. While weaknesses in the communication channel are exploited by attacks at stages 2, 4, 7, and 8. Indirect attack is basically an insider attack, as intruder should have a significant knowledge about the working of the system.

A solution to the problem of 'direct attack' has been provided by adopting a multimodal scheme, i.e. the use of multiple traits like fingerprint and face or fingerprint and speech in a single system. However, in many of these combinations, also, the liveliness component is still missing and an additional or special hardware needs to be attached to ensure that the samples have been obtained from alive individuals. Another remedy is to use a trait which has vitality property intrinsically embedded in it, and this led the researchers to test the efficacy of bioelectric signals like ECG, EEG, and PPG, etc. for biometric applications. These signals are naturally present in all human beings and also have an in-built trait for vitality check. In the last two decades, a considerable amount of work has been done in this upcoming area. Among them, the acquisition of PPG is most user-friendly and least intrusive. This makes it more suitable for integration with other biometric traits in a multimodal system. This chapter provides a review of PPG-based biometric recognition along with the review of research works applying new age machine learning and deep learning techniques in this area [2].

The remaining chapter contains four more sections with Section 13.2 providing an overview of PPG signal and the associated terminology. The literature related to PPG-based human recognition has been presented in Section 13.3. A scheme for person recognition using PPG and the results obtained are discussed in Section 13.4. Finally, the chapter ends with Section 13.5 in which the conclusions are given.

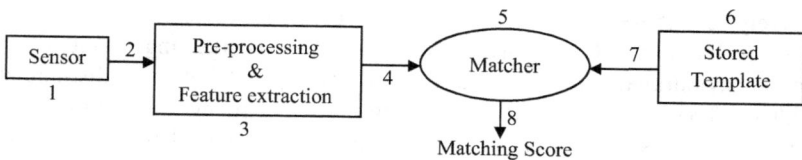

**FIGURE 13.1** Possible attack points in a typical biometric system.

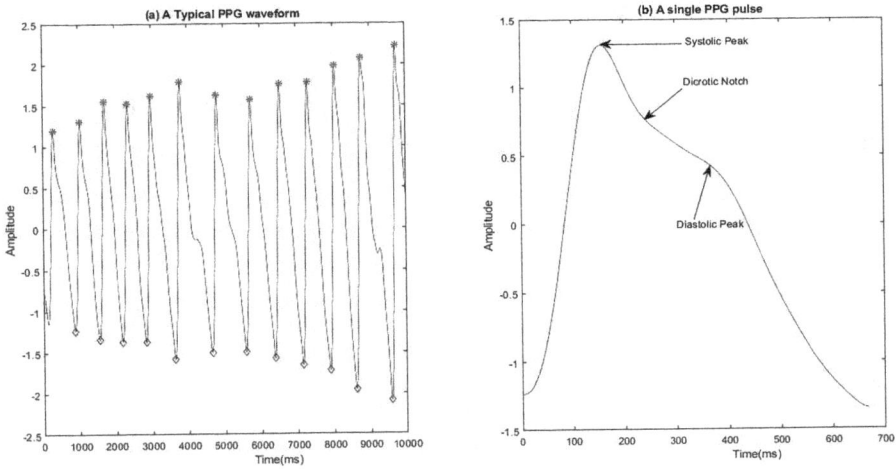

**FIGURE 13.2**    A typical PPG pulse and its characteristic points.

## 13.2   PHOTOPLETHYSMOGRAM (PPG)

Photoplethysmograph, made up of *photo* (light) plus two Greek words *plethysmos* (increasing) and *graph* (write), is an optical instrument through which the changes in the blood volume can be detected and measured. This non-invasive method also known as photoelectric plethysmography was introduced by Alrick Hertzman in late 1930s. The signal acquired by using this instrument is called photoplethysmogram (PPG). A PPG signal is captured using an optical sensor, emitting red and infrared light, placed on the fingertip, earlobe, toe-tip, etc., with the first two acquisition sites being more popular and convenient. PPG signal shows the change in the volume of blood at the site of acquisition (fingertip) as the heart pumps the blood to various extremities. As shown in Figure 13.2, PPG pulse comprises of two waves: one systolic and the other diastolic wave [3].

Predominantly, PPG has been employed for medical applications like measuring saturation of oxygen, blood pressure, cardiac output, etc. However, in the last one-and-half decade, researchers have explored the possibility of utilising this signal for human authentication and identification. A review of literature in this direction is presented in the next section.

## 13.3   LITERATURE REVIEW

Amongst all the medical biometrics, acquisition of PPG is the most convenient, and the possibility of clubbing it with fingerprint and other hand-based biometrics may open a lot many opportunities for automatic authentication in mobile devices, e-healthcare, etc. With these opportunities in mind, a number of studies exploring the utility of PPG in human authentication have been conducted since the beginning of the twenty-first century. The following paragraphs provide a survey of research works conducted in this area since 2003 onwards with the emphasis on techniques used for feature extraction and classification.

Gu et al. in 2003 proposed the use of PPG as a biometric measure [4]. They extracted four features from one PPG pulse – number of peaks in a given pulse, two slopes (one upwards and second downwards), and time duration between the onset of the pulse and the first peak. Euclidean distance was used as a classifier for computing the distance between the input sample and the template vector. Verification accuracy of 94% was found for 17 subjects. In another study, Gu and Zhang used the same feature vector along with fuzzy logic for carrying out the recognition task [5]. They also carried out across session tests, and an accuracy rate of 82.3% was reported by them. Two years later, Bao and his colleagues [6] used the peak-to-peak interval and first-order derivatives of PPG signal along with the Hamming distance to develop an authentication mechanism for mobile health system based on body area sensor networks. In Ref. [7,8], the authors presented a preliminary study in which feature vector was formed from the time intervals between the inflection points in addition to maxima and minima values of PPG signals. These points were extracted from the first- and the second-order derivatives of the PPG. Based on the statistical analysis and the correlation between the intra- and inter-subject datasets for three subjects, the authors concluded that PPG can be employed for biometric recognition. In order to obtain an enhanced PPG signal, rich in discriminatory content, the authors also developed an amplification circuit for signal acquisition. In 2011, Linear Discriminant Analysis (LDA) was used by Spachos and his colleagues for feature extraction from the PPG signal. The classification was carried out using nearest neighbour and majority voting [9]. The experiments were conducted on single session records of two datasets with population size of 14 and 15. Based on the results, the authors concluded that PPG can be used for biometric recognition, but the performance is dependent on the acquisition process. An approach for PPG-based biometric was also put forward by Singh and Gupta [10]. A fixed number of samples (200) between alternate peaks of a time-normalised PPG signal were extracted to form feature vector. The recognition was carried out by computing the correlation between the feature vectors of seven subjects used in this study. Bonisi and fellow researchers also investigated the utility of PPG signal as a biometric measure. A feature vector in the form of template was computed by storing $n$ (16) maximum values of the correlation between every heartbeat signal and the mean heartbeat [11]. Three datasets of short duration and another of 15 minutes were used for evaluation, and an Equal Error Rate (EER) in the range of 5.29%–13.47% was reported by the authors. For improving the accuracy of PPG-based biometric system, Kavsaoğlu and his fellow researchers suggested a method for ranking the 40 time domain features extracted from the PPG signal using its first- and the second-order derivatives [12]. The classification was carried out by k-nearest neighbour for the different combination of features selected based on the ranking algorithm. The entire scheme was evaluated on three datasets of 30 individuals, and an accuracy of around 90% was achieved in majority of the cases. The utility of PPG for continuous authentication application was studied and presented in Ref. [13]. Three heartbeats taken from the processed PPG signal were used to create a template, and subsequently, tests conducted on multiple session data of 10 subjects showed promising results. The possibility of using PPG signal for person authentication was also explored by Lee and Kim [14]. The authors used a twenty-two dimensional feature factor and a feed-forward neural network for the classification task. Accuracy in terms of FAR = 4.2% and FRR = 3.7% was achieved for the evaluation done on PPG samples of 10 subjects. In 2016, Nadzri and Sidek presented their findings related to the feasibility of PPG for biometric recognition of twins considering the variation of gender [15]. They took fiducial points like those corresponding to the systolic peak, dicrotic notch, diastolic peak, etc. to form a sixteen-dimensional feature factor. For four twin couples (eight subjects), accuracy of 94% and 89% was obtained when Radial Basis Function (RBF) and Bayes Network were, respectively, used as classifiers. Almost on the similar lines, a 12-dimensional feature vector consisting of amplitudes and time interval values obtained from two adjacent PPG pulse and LDA was used by Chakraborty and Pal for person identification [16]. The authors reported 100% recognition rate when the performance was checked on a dataset of 15 individuals. Choudhary and Manikandan employed an average of ensemble of PPG pulses for person authentication with PPG signals [17]. The similarity between template and test sample was found using three similarity measures, namely, normalised cross-correlation, wavelet distance measure, and wavelet weighted percent residual difference. The highest accuracy was achieved with the normalised cross-correlation.

In 2016, Jindal et al. put forward one of the first applications of deep neural network for biometric identification based on PPG signals [35]. They used a set of eleven features taken from PPG pulses, which include the values and positions of the characteristic points along with statistical values like mean and standard deviation of the waveform. Classification was done in two stages with a combination of Restricted Boltzmann Machines (RBMs) and Deep Belief Networks (DBN) employed in the second stage. In the same year, Sarkar and his colleagues suggested an approach in which an analytical model-based solution provided a feature vector in terms of Gaussians [18]. The classification was done by Linear Discriminant Analysis (LDA) and Quadratic Discriminant Analysis. This study was conducted on 23 subjects of dataset with 40 session recordings for different emotional states. Later, S. P. M. Namini and S. Rashidi selected thirty superior features through feature selection procedure and then used two comparison combinations for classification [19]. Investigations were done with four classifiers, namely, k-nearest neighbours, Gaussian Mixture Models, Fuzzy k-nearest neighbours, and Parzen window. In the next year, Nima Karimian and his fellow researchers came up with wavelet-based non-fiducial approach for PPG biometric [20]. In one study, they used the Genetic Algorithm for feature selection followed by Support Vector Machine (SVM) and Neural Network for classification. Whereas in another study, the selection of features and dimension reduction was done with a combination of correlation filter based on Kolmogorov-Smirnov and kernel-PCA. Classification comparison was done with Support Vector Machine (SVM), k-nearest neighbour, and Self-Organizing Maps (SOM) [21]. In both works, the identification performance with no-fiducial features was better than the one achieved with fiducial features. E. M. Nowara et al. suggested an anti-spoofing method for face recognition with PPG signals acquired from the videos of the faces [22]. The features employed were the spectra obtained by taking the Fourier transform of the PPG signal, whereas the classification was done with SVM and Random Decision Forest. A study done by V. R. Reddy et al. utilised PPG signal acquired from two sources (finger and face) for the biometric authentication [23]. The instant heart rate computed from signals acquired from both the sources made up the features, and the Pearson correlation coefficient provided the similarity value. Authentication accuracy of 100% was reported for 15 subjects on whom the tests were performed. In 2018, Sidek et al. suggested a method for biometric recognition with peaks of PPG and APG (Acceleration Plethysmogram) as features [24]. The classification done with Bayes Network, Multilayer Perceptron, Sequential Minimal Optimization, and k-nearest neighbours performance showed higher accuracy with APG features for a ten-subject dataset.

In the same year, Yadav et al. conducted investigations for PPG biometric on three datasets [25]. The template was generated by Continuous Wavelet Transform followed by dimension reduction with Direct Linear Discriminant Analysis (DLDA). Pearson distance was employed for matching of templates with the test data. The authors achieved reasonable results for across-session testing and felt that the performance can be improved by exploring more sophisticated feature selection and classification methods. Sancho et al. also published in 2018 the results of their experiments on four widely used public datasets for PPG-based authentication [26]. The templates for the subjects were made through averages of cycles in time and Karhunen-Loève transform domain. The accuracy of across or multiple session tests was found to be lower than the single-session testing. Al-Sidani et al. [27] used 40 features extracted from the PPG signal along with its first- and second-order derivatives for human identification. Template matching was done with k-nearest neighbour. In May 2018, Everson and his colleagues developed a deep learning framework for biometric identification with PPG signals acquired from the wrists of the subjects [36]. A combination of Convolutional Neural Network with long- and short-term memory was employed for modelling and classification of the individuals. Mean accuracy of 96% was achieved for 12 subjects. Luque et al. also used deep learning-based approach for PPG authentication. They carried out feature extraction through the convolutional layers and classification with dense neural net. The tests conducted on two datasets, with population size 43 and 20 respectively, showed reasonably good performance [37,41]. An approach utilising 29 fiducial features and Gradient Boosting Tree for classification was published by Zhao et al. [28]. Accuracy of 90% was reported by them for a dataset of 10 subjects. Patil et al. proposed an image-based alternate method for biometric authentication using PPG images extracted from recorded videos [38]. Continuous Wavelet Transform was used for feature extraction and Deep Neural Networks for classification. An average accuracy of 86.67% was achieved when the evaluation was carried out on database containing samples from 20 individuals. Biswas et al. extended their previous study to provide a more robust PPG authentication system using both Convolutional Neural Network with long- and short-term memory in two-layered combination along with the dense neural net [39]. The modified approach provided an average biometric identification rate of 96%. Hwang and Hatzinakos also suggested a similar deep learning scheme. In addition to within-session tests, investigations were also conducted on across-session data and an accuracy of 78.7% was reported for it [40]. In another work published in 2019, Lee et al. used period setting followed by Discrete Cosine Transform (DCT) for feature extraction [29]. Classification using three classifiers, namely, Decision Trees, k-nearest neighbour, and Random forests resulted in accuracies of 93%, 98%, and 99%, respectively. A technique for PPG biometric using sparse decomposition based on Matching Pursuit for feature extraction and SVM optimised with Crow Search Algorithm for classification has been proposed by Wang and Chen. An accuracy of 97.5% was claimed by the authors for the evaluations carried out for 40 subjects [30]. So as to exploit the discriminatory information provided by different feature extraction techniques, Walia and Kaul employed feature-level and score-level fusion for PPG-based human identification [31]. Features extracted in terms of autocorrelation and first- and second-order derivatives of PPG signal followed by the application of DCT for data reduction were used in the two fusion schemes. Score-level fusion showed better results than other schemes, and an identification rate of 89.7% was achieved with it when the evaluations were done for 38 subjects. Cheng et al. created a feature vector by modelling the PPG waveform with Gaussian function suitably accommodating for the changes in the baseline [32]. The classification was carried out with the help of Probabilistic Neural Network and Random Forest with the latter providing higher accuracy of 98.7% in terms of kappa coefficient. A comparative analysis to study the efficacy of two classifiers, namely, k-nearest neighbour and SVM, was done by Al Sidani et al. The feature vector was the same as the one employed in their earlier works. The performance of k-nearest neighbour was highest in all the cases [33]. In 2020, Khan et al. suggested an approach in which 20 time-and-frequency domain features were extracted from PPG signal processed with Empirical Mode Decomposition and reconstructed from its three intrinsic mode functions. Investigations were carried out with three classifiers (SVMs, K-Nearest Neighbour, and Decision Trees), and a recognition rate of 93.1% was achieved with SVM [34].

All the studies discussed in this chapter have been summarised in Tables 13.1 and 13.2. Table 13.1 covers all approaches where classifiers other than Deep Neural Networks have been employed, whereas Table 13.2 provides an overview of works based upon the utilisation of Deep Learning-based techniques. Overall, based upon this review, it can be stated that in majority of these works, experiments have been conducted on datasets of population size of 50 or less. Most of the studies have been carried out on single-session recordings. However, results depict that the discriminatory content in the PPG signal can be utilised for human recognition. In order to highlight the efficacy of PPG as a trait for person identification and authentication, a multi-feature-based approach has been explained in the following section.

## 13.4   MULTI-FEATURE APPROACH FOR PPG BIOMETRIC

The approach suggested in this chapter is a modified version of the one presented in Ref. [31]. It is a semi-fiducial approach where the individual pulses are extracted from the PPG signal and normalised. Three set of feature vectors are computed from these normalised pulses. The first one is extracted from a synthetic signal created by directly concatenating the normalised PPG pulses. The other two feature vectors are derived from the synthetic signals obtained, respectively, by concatenating the first- and second-order derivatives of the normalised pulses. The recognition is carried out by weighted combination of scores found on matching individual feature vectors with their respective templates. The entire recognition scheme which comprises

**TABLE 13.1**

**Summary of PPG Biometric Studies (Non-Deep Learning Based)**

| S. No. | Source (Research Group) | Features | Classifier | No. of Subjects | Accuracy |
|---|---|---|---|---|---|
| 1 | Y.Y. Gu et al. [4] | Time domain features –slopes, interval, etc. | Euclidean distance | 17 | 94% single session (ss) |
| 2 | Y.Y. Gu and Y.T. Zhang [5] | Time domain features – slopes, interval, etc. | Fuzzy logic | 17 | 94% –ss; 82.3% –multiple sessions (ms) |
| 3 | S. D Bao et al. [6] | Peak-to-peak time interval and first-order derivative | Hamming distance | 12 | 100% (ss) |
| 4 | J. Yao et al. [7,8] | Time intervals from first- and second-order derivatives | Correlation | 3 | 100% (ms) |
| 5 | P. Spachos et al. [9] | LDA | Nearest neighbour and majority voting | 29 (14+15) | FRR/FAR = 0.5%; EER = 25%(ss) |
| 6 | M. Singh and S. Gupta [10] | Fixed number of samples between two alternate peaks | Correlation | 07 | 100% (ss) |
| 7 | A. Bonisi et al. [11] | Maximum cross-correlation values between the mean heartbeat and individual heartbeats | Cross-correlation between templates | 44/14 | EER= [10.06, 8.34, 5.29]% (ss) |
| 8 | A. R. Kavsaoğlu et al. [12] | 40 time intervals and amplitude features from first- and second-order derivatives | k-nearest neighbour | 30 | ≈90% (ms) |
| 9 | J. da Silva Dias et al. [13] | 3 heartbeats | Cross-correlation | 10 (3+7) | FAR = 6%; FRR = 5% |
| 10 | A. Lee and Y. Kim [14] | 22 features comprising angles, area of waveform, etc. | Feed-forward neural network | 10 | FAR = 4.2%; FRR = 3.7% |
| 11 | N.I.M. Nadzri and K. A. Sidek [15] | 16 features based on fiducial points from PPG pulse | RBF, Bayes network | 8 (4 couple of twins) | 94%; 89% (ss) |

(Continued)

**TABLE 13.1 (*Continued*)**

**Summary of PPG Biometric Studies (Non-Deep Learning Based)**

| S. No. | Source (Research Group) | Features | Classifier | No. of Subjects | Accuracy |
|---|---|---|---|---|---|
| 12 | S. Chakraborty and S. Pal [16] | 12 time domain features from two adjacent pulses | LDA | 15 | 100% (ss) |
| 13 | T. Choudhary and M.S. Manikandan [17] | Ensemble average of PPG pulses | Normalised cross-correlation | 24 | FAR = 0.32 FRR = 0.32 |
| 14 | A. Sarkar et al. [18] | Solution of model resulting in sum of Gaussians (Two/Five) | LDA, Quadratic Discriminant Analysis | 23 | 90% 95% (ms) |
| 15 | S. P. M. Namini and S. Rashidi [19] | Two combinations for comparison of 30 superior features obtained through feature selection method | k-Nearest neighbours, Gaussian mixture models, Parzen window, Fuzzy k-nearest neighbours | 30 | EER = 2.17 ± 0.31% (ss) |
| 16 | Nima Karimian et al. [20] | Wavelet transform (DWT)-based features | SVM, SOMs, k-nearest neighbours | 42 | EER = 1.46 ± 2.7%, 1.70 ± 3.4% 1.31±2.6% |
| 17 | Nima Karimian et al. [21] | Wavelet transform-based features. | SVM, Neural Network | 42 | 100% |
| 18 | E. M. Nowara et al. [22] | Fourier transform-based spectrum of PPG signals acquired from face videos | SVM, Random Decision Forest | 50 | 100% |
| 19 | V. R. Reddy et al. [23] | Instant heart rates | Pearson correlation coefficient | 15 | 100% (ss) |
| 20 | K.A. Sidek et al. [24] | Peaks and points from PPG and APG pulses | Bayes network, multilayer perceptron, sequential minimal optimisation, k-nearest neighbours | 10 | 96% |
| 21 | U. Yadav et al. [25] | Wavelet transform (CWT) + Direct Linear Discriminant Analysis based features. | Pearson distance | 42, 32, 34 | 0.46% (ss), 2.11% (ss), 5.88% (ms) |

*(Continued)*

**TABLE 13.1 (*Continued*)**
**Summary of PPG Biometric Studies (Non-Deep Learning Based)**

| S. No. | Source (Research Group) | Features | Classifier | No. of Subjects | Accuracy |
|---|---|---|---|---|---|
| 22 | J. Sancho et al. [26] | Templates generated using cycle or multi-cycle average in time or Karhunen-Loève transform domain | Euclidean, Manhattan distance | 42, 56, 24, 24 | EER=1.0%(ss) 19.1%(ms) |
| 23 | Al-Sidani et al. [27] | 40 features from PPG signal and its first- and second-order derivatives | k-nearest neighbours | 57 (23 + 18 + 16) | 100% (ss) |
| 24 | T. Zhao et al. [28] | 29 fiducial features | Gradient Boosting Trees | 10 | 90% (ss) |
| 25 | S. W. Lee et al. [29] | DCT | Decision Trees, k-nearest neighbours, Random Forest | 42 | 93% 98% 99% |
| 26 | K. Wang and X. Chen [30] | Matching pursuit sparse decomposition | Crow search algorithm + SVM | 40 | 97.5% (ms) |
| 27 | A. Walia and A. Kaul [31] | Autocorrelation, first- and second-order derivatives of PPG | Euclidean distance | 38 | 89.48% (ss) |
| 28 | S. Cheng et al. [32] | Modelling via Gaussian functions | Probabilistic neural network, Random Forest | 31 | 98.67 ± 1.82% |
| 29 | A. Al Sidani et al. [33] | 40 features from PPG signal and its first- and second-order derivatives | K-nearest neighbour, SVM | 23 (57) | 100% (ss) |
| 30 | M. U Khan et al. [34] | 20 time and frequency domain features | SVM, Decision Trees, K-Nearest Neighbours | 20 | 93.1% |

## TABLE 13.2
## Summary of PPG Biometric Studies (Deep Learning Based)

| S. No. | Source (Research Group) | Features | Classifier | No. of Subjects | Accuracy |
|---|---|---|---|---|---|
| 1 | V. Jindal et al. [35] | 11 fiducial features and statistical values from PPG pulses | DBNs, RBMs | 11 | 96% |
| 2 | L. Everson et al. [36] | Convolutional Neural Network + long- and Short-term memory | | 12 | 96% |
| 3 | J. Luque et al. [37] | Convolutional Neural Networks | + Dense Neural Net | 43 (31 + 12) 20 (15 + 5) | AUC-78.7% −83.2% |
| 4 | O. R Patil et al. [38] | Continuous Wavelet Transform | Deep Neural Network | 20 | 86.67% |
| 5 | D. Biswas et al. [39] | Both Convolutional Neural Network and long- and short-term memory with two-layered | + Dense Neural Net | 22 | 96% |
| 6 | D. Y. Hwang and D. Hatzinakos [40] | 2 Convolution Neural Network + 3 long- and Short-term memory | | 20 | 96% (ss) 72% (ms) |

signal acquisition, pre-processing for noise reduction, feature extraction, and classification has been explained in the sub-sections given below. The approach in the block diagram form is also depicted in Figure 13.3.

### 13.4.1 SIGNAL ACQUISITION

The process begins with the capturing or acquisition of the signal for both the creation of template as well as for testing. For this study, a pulse sensor of Biopac MP-36 system has been used for signal acquisition from the index fingertip of individuals. Signals of two minutes duration at a sampling rate of 1,000 Hz were captured. A raw PPG signal during acquisition is susceptible to undesirable changes due to noises such as powerline, respiration, subject movement etc. In order to negate these effects, before feature extraction, the signal is pre-processed using appropriate filters.

### 13.4.2 BASELINE WANDER AND NOISE REMOVAL

In order to denoise the raw PPG signals and remove baseline wander effects, initially, the acquired signal was smoothened using a moving average filter. This was followed by filtering with a Butterworth low-pass filter. Keeping in view the bandwidth of the PPG signal, an 8th order filter with a cut-off frequency of 8 Hz was used.

### 13.4.3 FEATURE EXTRACTION

With the objective of capturing the distinct individual specific characteristic existing in the PPG signal, the features were extracted from the pre-processed signals in

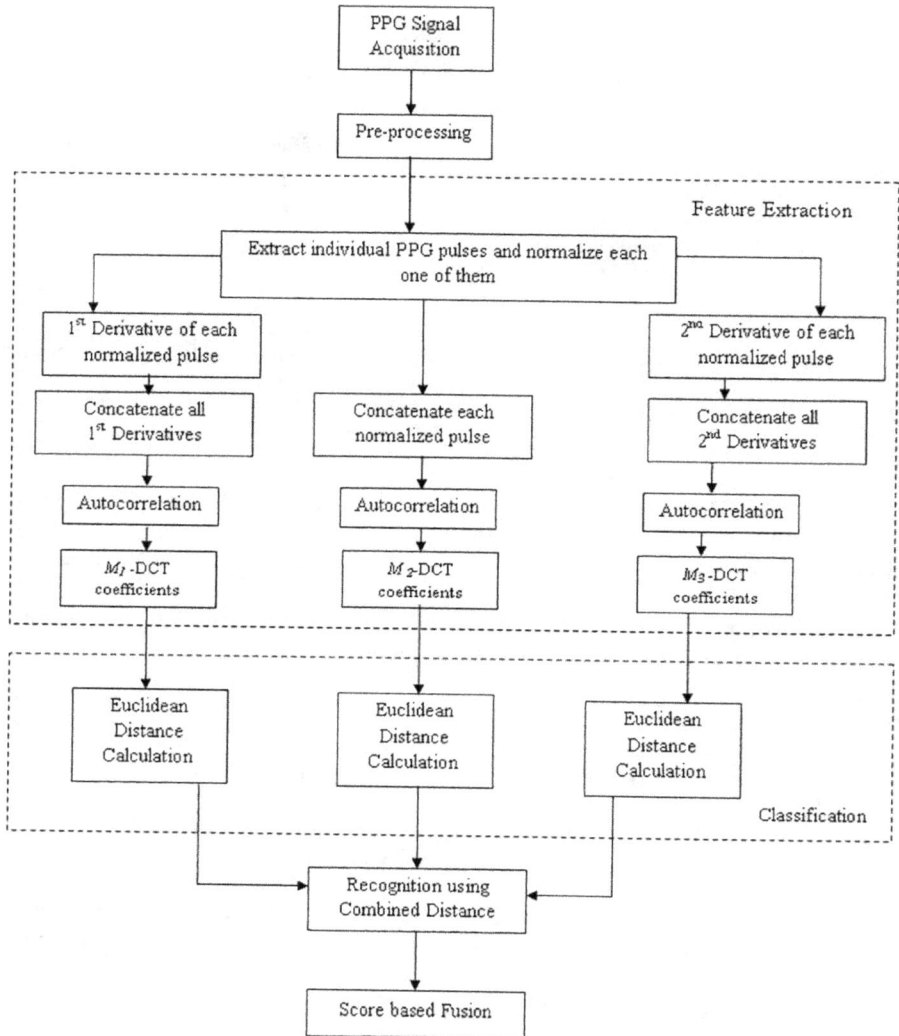

**FIGURE 13.3** Block diagram for multi-feature approach.

three different ways. In each of these three ways, first, the individual pulses were segmented and then normalised.

### 13.4.3.1 Pulse Extraction and Normalisation

For extracting the individual pulses, the peak and the valley points in the pre-processed PPG signal were detected by finding the local maxima and minima. Each pulse was segmented by taking samples between the two adjoining valley points. All the extracted pulses were normalised as per the following expression:

$$n\_ppg = \frac{o\_ppg - \mu(o\_ppg)}{\sigma(o\_ppg)} \qquad (13.1)$$

where $n\_ppg$ is the normalised pulse, $o\_ppg$ is the original pulse, $\mu(o\_ppg)$ is the mean value of the original pulse, and $\sigma(o\_ppg)$ is the standard deviation of the original signal. The normalised signal $n\_ppg$ is further used in creating the synthetic signals, but before that, the first- and second-order derivatives of the normalised pulse were computed.

### 13.4.3.2 First- and Second-Order Derivatives

The derivative of a signal provides information linked to the rate at which the given signal changes. The first derivative and the second derivative can be thought of, respectively, as the velocity and acceleration of the process under consideration. Mathematically, for a discrete signal $x(n)$, the first-order derivative can be computed by first-order difference as given by Equation (13.2).

$$d_1(n) = x(n) - x(n-1) \qquad (13.2)$$

On the same lines, Equation (13.3) can be used to calculate the second-order derivative:

$$d_2(n) = x(n-1) + x(n+1) - 2x(n) \qquad (13.3)$$

where $n = 0, 1, 2, 3, 4 ..... N - 1$.

The first- and second-order derivatives were computed for each normalised PPG pulse using the above expressions. Among these, the second-order derivative, also known as acceleration plethysmogram (APG), has been found to be more effective in detecting the inflection points in comparison to the input PPG signal. As such, these two derivatives also contain certain useful details of the PPG signal.

The normalised PPG pulses, their first- and second-order derivatives, were separately concatenated to form three synthetic signals. These are shown in Figure 13.4 and were subsequently used for feature extraction done with the application of auto-correlation and DCT.

### 13.4.3.3 Autocorrelation

PPG pulses are quasi-periodic in nature and have reasonable degree of similarity which is also true for its two derivatives. In signal processing, correlation is used to find the extent of similarity between two signals (say $x(n)$ and $y(n)$). If the two signals, i.e. $x(n)$ and $y(n)$, are different, it is known as cross-correlation, whereas for $x(n) = y(n)$, it is called auto-correlation. The mathematical expressions for cross-correlation and auto-correlation are, respectively, given by Equations 13.4 and 13.5:

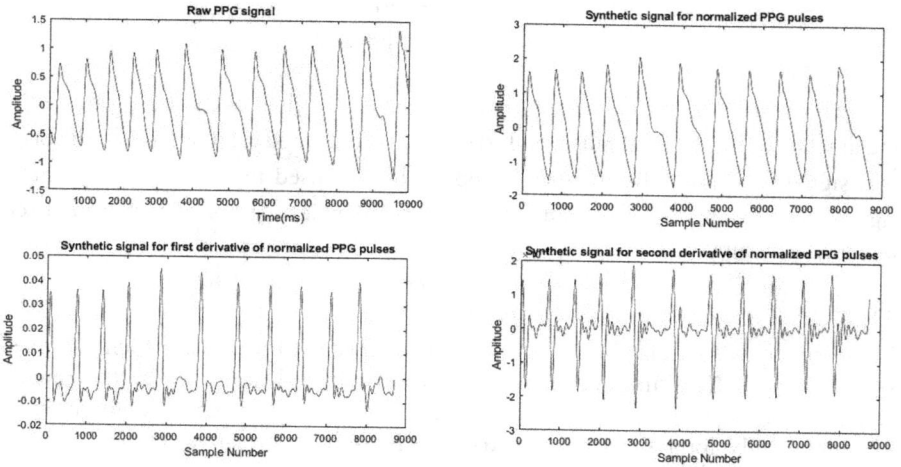

**FIGURE 13.4** Three synthetic signals for feature extraction.

$$r_{xy}(l) = \sum_{n=-\infty}^{\infty} x(n)y(n-l) \tag{13.4}$$

$$r_{xx}(l) = \sum_{n=-\infty}^{\infty} x(n)x(n-l) \tag{13.5}$$

The auto-correlation was separately computed for each of the three signals created from the normalised PPG signal $n\_$ppg, its first derivative $d_1(n)$, and the second derivative $d_2(n)$. For reducing the dimension, DCT was applied to all the auto-correlation sequences. Three feature vectors were formed by retaining the first $M_n$ DCT coefficients from each sequence. In this study, for the normalised pulse and for the two derivatives, the value of $M_n$ was selected to be 18 and 9, respectively. A comparison of feature vectors extracted from two subjects is presented in Figure 13.5.

## 13.5 CLASSIFICATION

Classification was carried out by calculating the Euclidean distance between the stored templates and feature vectors extracted from the test signal. So as to combine the matching distances or scores of the three feature vectors, score normalisation was carried out. The normalisation of matching distances was done using Min-Max normalisation, such that given a set of matching distances $\{d_k\}$, $k = 1, 2, 3...n$, the normalised distances were obtained by utilising the expression mentioned below:

$$s_k = \frac{d_k - \min}{\max - \min} \tag{13.6}$$

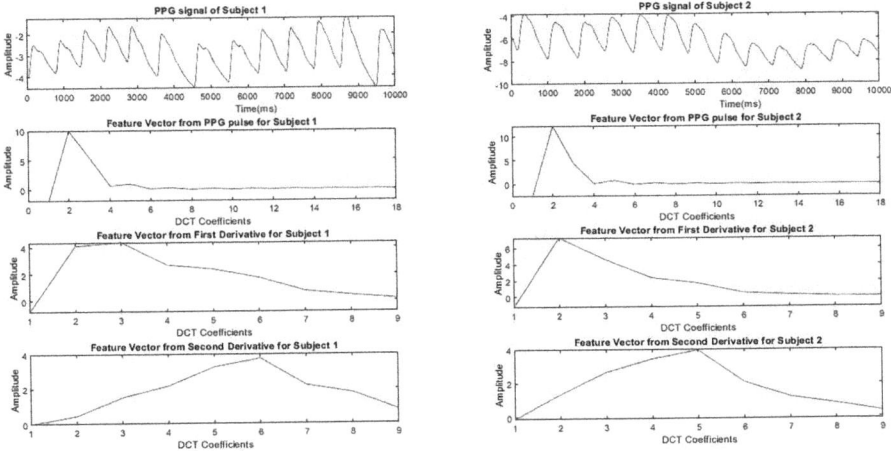

**FIGURE 13.5**   Comparison of feature vectors for two subjects.

where $s_k$ is the matching score after normalisation, and min and max are the minimum and maximum values of $\{d_k\}$, respectively. Score-level fusion by employing simple weighted fusion was adopted for the final identification or verification decision. The combined score was computed as follows:

$$S = w_1 s_1 + w_2 s_2 + w_3 s_3 \qquad (13.7)$$

where $s_1$, $s_2$, and $s_3$ are the normalised scores of matching distances for the corresponding three feature vectors; $w_1$, $w_2$, and $w_3$ are their weights; and $S$ is the combined score. Experiments were conducted to compute identification rate as well as the verification rate.

## 13.6   EXPERIMENTS AND RESULTS

The approach was tested on a dataset consisting of samples from eleven subjects recorded in two sessions. Evaluations were carried out for both within-session and across-session settings. The template for each subject was created from the first part of the recording of first session, and the same template was also used for across-session testing. The feature vectors for the within-session testing were extracted from the last 10 seconds of the input signal. The across-session testing was carried out by extracting the features vectors for the signal samples taken from 10-to 20-second segment of the second set of recordings.

The results achieved for the identification and verification tasks have been listed in Table 13.3. For the within-session experiments, the rank 1 identification rate of 81.82% (9 out of 11) was achieved for the method described in this section. On the other hand, the verification rate for this setting was 90.91%. When the testing was done for across-session setting, the rank 1 identification rate dropped to 54.55% (6 out of 11). The verification performance for across session also dipped to 63.64%. Further analysis of the results for identification accuracy can be done by viewing the

**TABLE 13.3**

**Accuracy for Within- and Across-Session Testing**

| | Session | |
| --- | --- | --- |
| **Mode** | **Within Session (%)** | **Across Session (%)** |
| Identification | 81.82 | 54.55 |
| Verification | 90.91 | 63.64 |

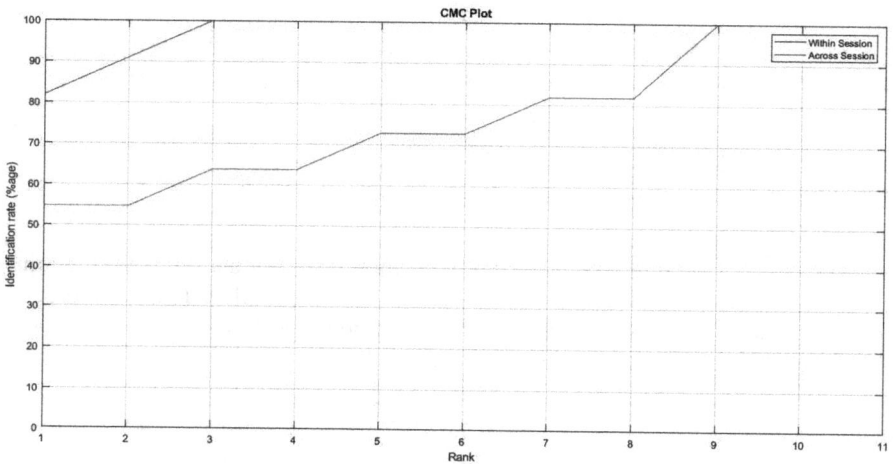

**FIGURE 13.6**   CMC plot for within and across session testing.

Cumulative Match Curve (CMC) plot shown in Figure 13.6. It can be seen that for within session, 100% subjects are recognised by rank 3, and for the across-session scenario, 8 out of the 11 subjects are identified by rank 5. This shows that PPG signal contains content that is characteristic of a particular human being. In this case, the recognition tasks have been performed using simple Euclidean distance classifier. As in the reported literature, machine learning and deep learning techniques have been found to have better classification abilities, so it is expected that using them as classifiers should provide higher accuracy.

## 13.7   CONCLUSIONS

A review of evolution of PPG for biometric application, since its inception in this area, has been presented in this chapter. An approach using multiple feature vectors for PPG biometric has also been explained. Based upon the survey of literature and results obtained for the method described here, it can be stated that PPG signal has a potential to be utilised for the biometric applications. Moreover, the accuracy for the across-session testing is expected to improve by using robust classifiers offered by the new age machine and deep learning techniques. However, the studies published in this area have been carried out on small datasets with less than fifty subjects. In order to establish the efficacy of the methods, experiments need to be carried out on datasets having multiple session recordings and population size of few hundreds. The future work in this direction will be undertaken by keeping these points in mind.

## REFERENCES

1. Chauhan, S., Arora, A.S., &Kaul, A. (2010) A survey of emerging biometric modalities. *Proceedings of the International Conference and Exhibition on Biometrics Technology.*
2. Kaul, A. (2016). *Hybrid Biometric Approach for Human Identification* (Doctoral Thesis). National Institute of Technology Hamirpur, India.
3. Elgendi, M. (2012). Standard terminologies for photoplethysmogram signals. *Current Cardiology Reviews, 8*(3), 215–219.
4. Gu, Y. Y., Zhang, Y., & Zhang, Y. T. (2003, April). A novel biometric approach in human verification by photoplethysmographic signals. *4th International IEEE EMBS Special Topic Conference on Information Technology Applications in Biomedicine,* 2003. (pp. 13–14). IEEE.
5. Gu, Y. Y., & Zhang, Y. T. (2003, October). Photoplethysmographic authentication through fuzzy logic. *IEEE EMBS Asian-Pacific Conference on Biomedical Engineering,* 2003. (pp. 136–137). IEEE.
6. Bao, S. D., Zhang, Y. T., & Shen, L. F. (2006, January). Physiological signal based entity authentication for body area sensor networks and mobile healthcare systems. *2005 27th Annual IEEE Conference on Engineering in Medicine and Biology* (pp. 2455–2458). IEEE.
7. Yao, J., Sun, X., & Wan, Y. (2007, August). A pilot study on using derivatives of photoplethysmographic signals as a biometric identifier. *2007 29th Annual International Conference of the IEEE Engineering in Medicine and Biology Society* (pp. 4576–4579). IEEE.
8. Wan, Y., Sun, X., &Yao, J. (2007, October). Design of a photoplethysmographic sensor for biometric identification. *2007 International Conference on Control, Automation and Systems* (pp. 1897–1900). IEEE.
9. Spachos, P., Gao, J., & Hatzinakos, D. (2011, July). Feasibility study of photoplethysmographic signals for biometric identification. *2011 17th International Conference on Digital Signal Processing (DSP)* (pp. 1–5). IEEE.
10. Singh, M., & Gupta, S. (2012). Correlation studies of PPG finger pulse profiles for Biometric system. *International Journal of Computer Science and Knowledge Management, 5*(1), 1–3.
11. Bonissi, A., Labati, R. D., Perico, L., Sassi, R., Scotti, F., & Sparagino, L. (2013, September). A preliminary study on continuous authentication methods for photoplethysmographic biometrics. *2013 IEEE Workshop on Biometric Measurements and Systems for Security and Medical Applications* (pp. 28–33). IEEE.
12. Kavsaoğlu, A. R., Polat, K., & Bozkurt, M. R. (2014). A novel feature ranking algorithm for biometric recognition with PPG signals. *Computers in Biology and Medicine, 49,* 1–14.
13. da Silva Dias, J., Traore, I., Ferreira, V. G., & David, J. (2015). Exploratory use of PPG signal in continuous authentication. *The Brazilian Symposium on Information and Computational Systems Security.* Florianópolis, Brazil: The Brazilian Symposium on Information and Computational Systems Security.
14. Lee, A., &Kim, Y. (2015, November). Photoplethysmography as a form of biometric authentication. In *2015 IEEE Sensors* (pp. 1–2). IEEE.
15. Nadzri, N. I. M., & Sidek, K. A. (2016). Photoplethysmogram based biometric identification for twins incorporating gender variability. *Journal of Telecommunication, Electronic and Computer Engineering (JTEC), 8*(12), 67–72.

16. Chakraborty, S., & Pal, S. (2016, January). Photoplethysmogram signal based biometric recognition using linear discriminant classifier. *2016 2nd International Conference on Control, Instrumentation, Energy & Communication (CIEC)* (pp. 183–187). IEEE.

17. Choudhary, T., &Manikandan, M. S. (2016, March). Robust photoplethysmographic (PPG) based biometric authentication for wireless body area networks and m-health applications. *2016 Twenty Second National Conference on Communication (NCC)* (pp. 1–6). IEEE.

18. Sarkar, A., Abbott, A. L., &Doerzaph, Z. (2016, September). Biometric authentication using photoplethysmography signals. *2016 IEEE 8th International Conference on Biometrics Theory, Applications and Systems (BTAS)* (pp. 1–7). IEEE.

19. Namini, S. P. M., & Rashidi, S. (2016, October). Implementation of artificial features in improvement of biometrics based PPG. *2016 6th International Conference on Computer and Knowledge Engineering (ICCKE)* (pp. 342–346). IEEE.

20. Karimian, N., Tehranipoor, M., & Forte, D. (2017, February). Non-fiducial PPG-based authentication for healthcare application. *2017 IEEE EMBS International Conference on Biomedical & Health Informatics (BHI)* (pp. 429–432). IEEE.

21. Karimian, N., Guo, Z., Tehranipoor, M., & Forte, D. (2017, March). Human recognition from photoplethysmography (PPG) based on non-fiducial features. *2017 IEEE International Conference on Acoustics, Speech and Signal Processing (ICASSP)* (pp. 4636–4640). IEEE.

22. Nowara, E. M., Sabharwal, A., & Veeraraghavan, A. (2017, May). PPG secure: Biometric presentation attack detection using photopletysmograms. *2017 12th IEEE International Conference on Automatic Face & Gesture Recognition (FG 2017)* (pp. 56–62). IEEE.

23. Reddy, V. R., Deshpande, P., & Pal, A. (2017, November). Simultaneous measurement and correlation of PPG signals taken from two different body parts for enhanced biometric security via two-level authentication. *Proceedings of the 1st ACM Workshop on the Internet of Safe Things* (pp. 32–37).

24. Sidek, K. A., Kamaruddin, N.K., & Ismail, A.F. (2018). The study of PPG and APG signals for biometric recognition. *Journal of Telecommunication, Electronic and Computer Engineering, 10*, 17–20.

25. Yadav, U., Abbas, S. N., & Hatzinakos, D. (2018, February). Evaluation of PPG biometrics for authentication in different states. *2018 International Conference on Biometrics (ICB)* (pp. 277–282). IEEE.

26. Sancho, J., Alesanco, Á., & García, J. (2018). Biometric authentication using the PPG: a long-term feasibility study. *Sensors, 18*(5), 1525.

27. Al-Sidani, A., Ibrahim, B., Cherry, A., & Hajj-Hassan, M. (2018, April). Biometric identification using photoplethysmography signal. *2018 Third International Conference on Electrical and Biomedical Engineering, Clean Energy and Green Computing (EBECEGC)* (pp. 12–15). IEEE.

28. Zhao, T., Wang, Y., Liu, J., & Chen, Y. (2018, October). Your heart won't lie: PPG-based continuous authentication on wrist-worn wearable devices. *Proceedings of the 24th Annual International Conference on Mobile Computing and Networking* (pp. 783–785).

29. Lee, S. W., Woo, D. K., Son, Y. K., & Mah, P. S. (2019). Wearable Bio-Signal (PPG)-based personal authentication method using random forest and period setting considering the feature of PPG signals. *JCP, 14*(4), 283–294.

30. Wang, K., & Chen, X. (2019, October). PPG signal identification method based on CSASVM. In *2019 IEEE 10th International Conference on Software Engineering and Service Science (ICSESS)* (pp. 1–4). IEEE.

31. Walia, A., & Kaul, A. (2019, October). Human recognition via PPG signal using temporal correlation. *2019 5th International Conference on Signal Processing, Computing and Control (ISPCC)* (pp. 144–147). IEEE.

32. Cheng, S., Chou, Y., Liu, J., Gu, Y., & Huang, X. (2019, October). A novel identity authentication method by modeling Photoplethysmograph waveform. *2019 International Conference on Control, Automation and Information Sciences (ICCAIS)* (pp. 1–5). IEEE.
33. Al Sidani, A., Cherry, A., Hajj-Hassan, H., & Hajj-Hassan, M. (2019, October). Comparison between K-nearest neighbor and support vector machine algorithms for PPG biometric identification. *2019 Fifth International Conference on Advances in Biomedical Engineering (ICABME)* (pp. 1–4). IEEE.
34. Khan, M. U., Aziz, S., Naqvi, S. Z. H., Zaib, A., & Maqsood, A. (2020, March). Pattern analysis towards human verification using Photoplethysmograph signals. *2020 International Conference on Emerging Trends in Smart Technologies (ICETST)* (pp. 1–6). IEEE.
35. Jindal, V., Birjandtalab, J., Pouyan, M. B., & Nourani, M. (2016, August). An adaptive deep learning approach for PPG-based identification. *2016 38th Annual International Conference of the IEEE on Engineering in Medicine and Biology Society (EMBC)* (pp. 6401–6404). IEEE.
36. Everson, L., Biswas, D., Panwar, M., Rodopoulos, D., Acharyya, A., Kim, C. H., & Van Helleputte, N. (2018, May). BiometricNet: Deep Learning based biometric identification using wrist-worn PPG. *2018 IEEE International Symposium on Circuits and Systems (ISCAS)* (pp. 1–5). IEEE.
37. Luque, J., Cortes, G., Segura, C., Maravilla, A., Esteban, J., & Fabregat, J. (2018, September). End-to-End Photoplethysmography (PPG) based biometric authentication by using convolutional neural networks. *2018 26th European Signal Processing Conference (EUSIPCO)* (pp. 538–542). IEEE.
38. Patil, O. R., Wang, W., Gao, Y., Xu, W., & Jin, Z. (2018, October). A non-contact PPG biometric system based on deep neural network. *2018 IEEE 9th International Conference on Biometrics Theory, Applications and Systems (BTAS)* (pp. 1–7). IEEE.
39. Biswas, D., Everson, L., Liu, M., Panwar, M., Verhoef, B. E., Patki, S., & Van Helleputte, N. (2019). CorNET: Deep learning framework for PPG-based heart rate estimation and biometric identification in ambulant environment. *IEEE Transactions on Biomedical Circuits and Systems, 13*(2), 282–291.
40. Hwang, D. Y., & Hatzinakos, D. (2019, May). PPG-based personalized verification system. In *2019 IEEE Canadian Conference of Electrical and Computer Engineering (CCECE)* (pp. 1–4). IEEE.
41. Sebastià, G. C. (2018). *End-to-End Photoplethysmography-based Biometric Authentication System by Using Deep Neural Networks* (Doctoral dissertation, Universitat Politècnica de Catalunya. EscolaTècnica Superior d'Enginyeria de Telecomunicació de Barcelona).

# 14 Current Trends of Machine Learning Techniques in Biometrics and its Applications

*B. S. Maaya and T. Asha*
Bangalore Institute of Technology

**CONTENTS**

14.1 Introduction ....................................................................................... 334
    14.1.1 Biometric Systems ................................................................... 334
    14.1.2 Brain Stroke ............................................................................ 334
        14.1.2.1 Risk Factors ............................................................. 337
        14.1.2.2 Blood Pressure ......................................................... 337
        14.1.2.3 Heart Disease ........................................................... 337
        14.1.2.4 Diabetes Mellitus ..................................................... 338
        14.1.2.5 Cholesterol ............................................................... 338
        14.1.2.6 Smoking ................................................................... 338
        14.1.2.7 Alcohol ..................................................................... 339
        14.1.2.8 Other Risk Factors ................................................... 339
    14.1.3 Face Recognition ..................................................................... 339
    14.1.4 Motivation to Machine Learning Techniques ........................... 341
14.2 Related Work ..................................................................................... 341
    14.2.1 Review on Brain Stroke ............................................................ 341
    14.2.2 Review on Face Recognition .................................................... 342
    14.2.3 Brain Stroke Prediction System ............................................... 346
        14.2.3.1 Image Acquisition .................................................... 346
        14.2.3.2 Pre-processing .......................................................... 347
        14.2.3.3 Feature Extraction .................................................... 347
        14.2.3.4 Classification Using Machine Leaning Algorithms .......... 348
        14.2.3.5 Construction of Convolutional Neural Network ............... 352
    14.2.4 Face-Recognition System ........................................................ 352
14.3 Discussion and Results ...................................................................... 353
    14.3.1 Performance of Brain Stroke .................................................... 353
    14.3.2 Performance of Face Recognition ............................................ 355

14.4  Future Scope ..................................................................................... 355
14.5  Conclusion ..................................................................................... 356
References ..................................................................................... 356

## 14.1  INTRODUCTION

The biometrics field is a rapidly growing branch of Information Technology. The innovations are mechanised instruments of distinguishing an individual dependent on their biological and behavioural characteristics. This chapter focuses on the biometric systems, brain stroke classification, and facial recognition using different machine learning (ML) methods.

### 14.1.1  BIOMETRIC SYSTEMS

The biometric framework is a validation system that gives the mechanised distinguishing proof of people dependent on their special physiological or behavioural qualities. Physiological qualities are acquired attributes which are created in the early stage phases of human turn of events. There are few sorts of exceptional physiological or behavioural qualities of people in presence. A portion of the normal biometric methods for distinguishing proof and check include fingerprint acknowledgement, signature elements, keystroke elements, voice acknowledgment, facial acknowledgement, iris examining, retina filtering, hand geometry. The benefits of biometric in social insurance are shown in Figure 14.1.

### 14.1.2  BRAIN STROKE

Stroke is a blood clot or bleeding in the brain that can cause permanent damage affecting mobility, cognition, vision, or communication. Stroke is considered as clinical dire circumstance and can cause long haul neurological harm, complexities,

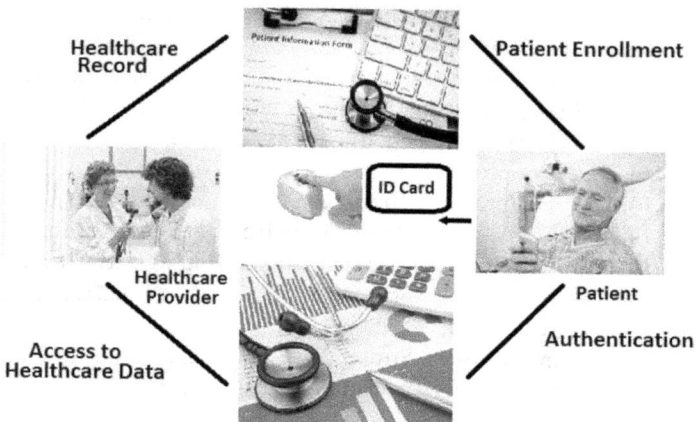

FIGURE 14.1  Biometric-based patient enrolment and authentication system.

and frequently demise [1,2]. Stroke is the third leading cause of death after heart and lung diseases. Most strokes are classified as ischemic having two types: thrombotic and embolic. The blood clot (thrombus) forms during thrombotic stroke in one of the arteries that supplies blood to the brain. An embolic stroke happens when a blood coagulation shapes from the patient mind normally in the patient heart and goes through the patient circulation system to hold up in the smaller cerebrum veins. A haemorrhagic stroke happens when a weak blood vessel bursts and bleeds into the brain. As an explanation of haemorrhagic stroke, the synapses harm as a consequence of the weight from the spilled blood. There are numerous likenesses between these sorts, and it is hard to arrange the cases precisely utilising clinical strategies. Moreover, there are no unmistakable limits between these sorts. This chapter investigated and examined the present examinations on the characterisation of stroke.

One of the main reasons for clot is the fatty deposits that make arteries and lead to a reduced blood flow or other artery conditions. One of the primary methods that is utilised to analyse the coagulation is the computed tomography (CT) examination, which is a test that utilises x beams to take clear, definite photos of the patient's cerebrum [3,4]. CT scan is mainly done immediately after the stroke is suspected. A bleeding in the cerebrum or harm to the mind can be seen utilising cerebrum CT filter. Other brain conditions that cause patients symptoms can be discovered using the brain CT scan. Magnetic resonance imaging (MRI) is the second test that is used to examine brain strokes. MRI depends on magnets and radio waves that are utilised to create photos of the organs and structures in the patient's body. Any adjustments in the mind tissue and harm to synapses from a stroke can be found utilising MRI test. To diagnose a stroke MRI, CT or both can be used [5]. A well-known imaging technique in which x-beams are utilised to mold pictures of cross-segments of the body is CT scan. CT is the choice strategy to distinguish stroke in permitting the patients with suspected extreme stroke. The underlying side effects of the dead tissue, for example, loss of skullcap depiction, obscuration of the lentiform core, loss of separate strip, and hyper-thick centre cerebral supply route, are genuinely swoon on CT. When a patient grumbling of stroke approaches the emergency clinic, then suggested specialists to for CT examination, which will take around 10 minutes. Figure 14.2 shows the CT scan of the normal stroke-free brain, and Figure 14.3 shows as the CT scan of the abnormal stroke lesion brain.

MRI is an innocuous and a convenience free test that traditions an attractive field and radio waves to yield detailed portraits of the body's organs and structures. MRI is amazingly wealthy in data content and expenditures. The image pixel worth can be meticulous as an element of a mass of parameters, including the relaxation time constants T1, T2, and the Proton Density (PD). Figure 14.4 shows the T1, T2, and PD of MRI.

In this examination, we have inspected the connection among ischaemic and haemorrhage stroke and how well that can be dealt with by utilising the modalities CT and MRI. Building up a computer-supported technique which on utilising either CT or MRI would anticipate the rate at which the patient experienced stroke. The main goals of this investigation are to explore the forecasts made by the strategy that will utilise a mix of injury and non-sore issues. Cerebrovascular sicknesses happen by suffering consolidated impacts of hazard factors [6]. It is upgraded by the expanding pace of modifiable hazard factors. An overview of risk factors is given in Section 14.1.1.

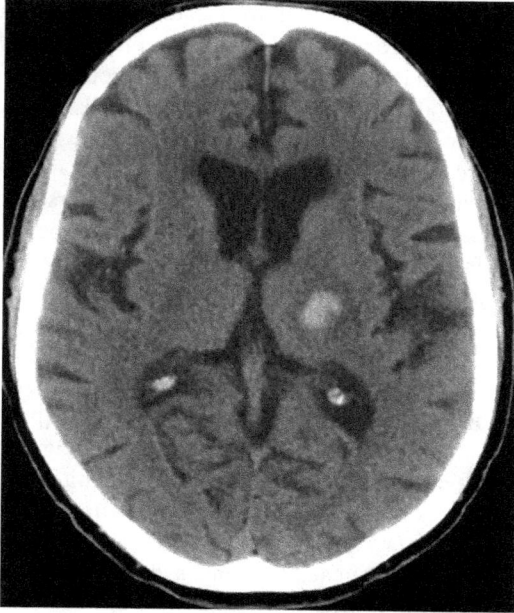

**FIGURE 14.2**   Normal brain CT images.

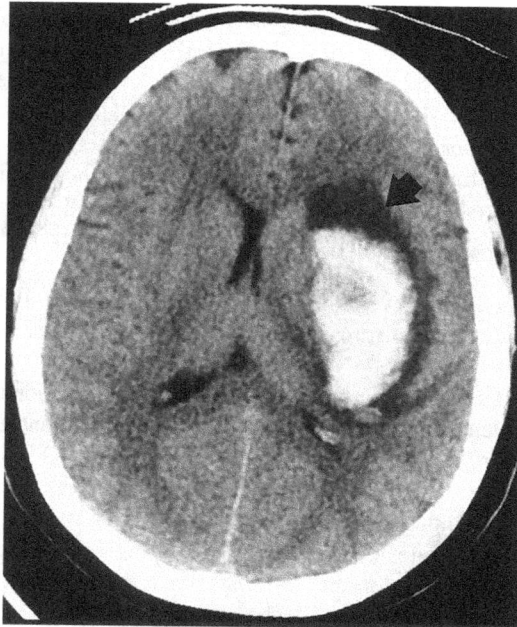

**FIGURE 14.3**   Abnormal brain CT images.

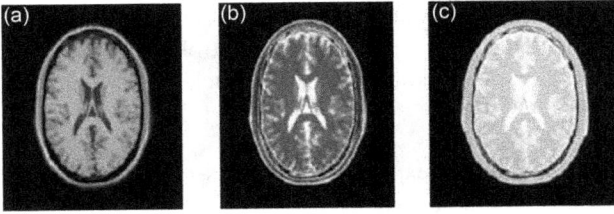

**FIGURE 14.4**   (a) T1 of MRI. (b) T2 of MRI. (c) PD of MRI.

### 14.1.2.1   Risk Factors

A hazard factor is any quality or normal for a person that builds the chance of building up a disease. There exist various hazard factors that improve the danger of stroke, way of life chance components [7] which incorporate eating routine, cigarette smoking propensities, overweight and corpulence, physical latency, liquor utilisation [8], family and hereditary elements, age, sex, sedate use, race, oral prophylactic use, geographic area, season, atmosphere, and financial elements while ailments comprise cardiovascular issue (atrial fibrillation, coronary episode, arrhythmia) [9], pulse [10], diabetes mellitus, cholesterol, mitral valve ailment, raised fibrinogen focus, sickle cell infection, hyper-lipidaemia, transient ischaemic assault, headache, cerebral pains, and headache reciprocal. Hypertension, coronary illness, and diabetes regularly do not cause manifestations in their prior stages. A portion of the normal hazard factors are clarified here.

### 14.1.2.2   Blood Pressure

Circulatory strain is a significant hazard factor in 50%–70% of stroke cases. The drawn-out impacts of expanded weight harm the dividers of supply routes, making them increasingly vulnerable to thickening or narrowing or crack. Stopped-up veins in the cerebrum remove the blood stream to synapses. As hypertension harms supply routes all through the body, it is critical to keep our circulatory strain inside middle of the road reaches to shield our mind from this lethal occasion. About 13% of strokes are haemorrhagic which regularly happen when a vein cracks in or close to the mind. Cracking of the vein causes seeping into the significant tissue in the mind or in space among the cerebrum and skull. Hypertension harms the corridors and can make powerless regions that burst effectively or flimsy spots that top off with blood and inflatable out from the vein divider, aneurysm. Ceaseless hypertension is one of the primary drivers of this sort of stroke. On the off chance that pulse can be decreased through way of life changes and medications, the danger of the event of stroke can be diminished.

### 14.1.2.3   Heart Disease

Coronary illness is a solid hazard factor for ischemic stroke. Harm to the heart may make it almost certain that coagulations will frame inside the heart. These coagulations can make a trip to mind, causing a cardioembolic stroke. Atrial fibrillation can expand our danger of stroke by four to multiple times. Atrial fibrillation upgrades the danger of a blood coagulation shaping inside the offices of heart. This coagulation can go through the circulation system and square the blood gracefully to mind, which eventually prompts stroke. Coronary illness and stroke are likewise related on the grounds that they are the two indications of atherosclerotic sickness in the veins.

### 14.1.2.4 Diabetes Mellitus

People with diabetes have an expanded defencelessness to atherosclerosis and an expanded recurrence of atherogenic hazard factors, especially hypertension, heftiness, and unusual blood lipids. The association among diabetes and stroke is identified in the manner by which body handles blood glucose to make vitality. The greater part of the food we eat is separated into glucose to give vitality. Glucose enters the circulation system and goes to cells all through the body after the food is processed. For the glucose to enter the cells and provide energy, it needs a hormone named insulin. It is the activity of the pancreas to create this insulin to a required extent. For type 1 diabetes, the pancreas doesn't make insulin or it makes too little insulin, or the cells in the muscles, liver, and fat don't utilise insulin in the correct path in type 2 diabetes. At that point, individuals with diabetes end up with an excess of glucose in their blood, while their cells don't get enough vitality. At the appropriate time, this glucose prompts expanded greasy stores or clumps within the vein dividers. These shaped coagulations can limit or square the veins in the cerebrum or neck, preventing oxygen from entering the mind and cause a stroke.

### 14.1.2.5 Cholesterol

As per National Heart, Lung, and Blood Institute, for people over 18 years of age, absolute cholesterol is viewed as high; on the off chance that it is in excess of 200 mg/dL. Low-density lipoproteins (LDL) and high-density lipoproteins (HDL) are the two kinds of lipoproteins that directly affect the cholesterol levels. In the event that the all-out cholesterol is more than 200 or the HDL level is under 40, then the danger of stroke and coronary illness is more. Plaque develops in the supply routes from significant levels of cholesterol and additionally can square blood stream to the cerebrum and cause a stroke. Since cholesterol doesn't break up in the blood all alone, it must be conveyed to and from cells by specific particles named as lipoproteins. Because of its supply route stopping-up properties, LDL cholesterol is frequently alluded to as terrible cholesterol as it can convey cholesterol into the circulatory system and to tissues where our body can store it. This kind of cholesterol can cause plaque to develop. Plaque is a thick, hard material that can obstruct corridors. In the long run, the plaque causes narrowing of the courses or block them completely, causing stroke.

### 14.1.2.6 Smoking

The carbon monoxide we take in from tobacco smoke assembles cholesterol levels in our blood, making it increasingly plausible for hallway dividers to get hurt. The synthetic compounds we breathe in likewise influence the tenacity of our blood and creation of a sort of blood cell called as platelet. This expands the propensity of the blood to frame clumps. These variables increment smokers' danger of creating atherosclerosis whereby conduits become smaller. In the long run, the blood course through the conduits lessens bringing about ischaemic stroke.

### 14.1.2.7 Alcohol

Research shows that drinking a lot of liquor can incredibly expand our danger of having a stroke. This is because liquor adds to various ailments that are hazard factors for stroke. Sensible utilisation of liquor may diminish cardiovascular malady, including stroke. Current epidemiological examinations have indicated a U-moulded bend for the utilisation of liquor and cardiovascular ailment mortality, with low-to-sensible liquor utilisation related to lower overall mortality. In a review examination of stroke considers, a J-formed affiliation bend was suggested for the connection of sensible standard liquor utilisation and ischemic stroke.

### 14.1.2.8 Other Risk Factors

Age, sexual orientation, race, ethnicity, and heredity have been perceived as markers of hazard for stroke sickness. Obesity and heftiness have been connected with more elevated levels of pulse, blood glucose, and atherogenic serum lipids, which are free hazard characteristics for stroke. Hazard factors autonomously increment the likelihood of stroke and may likewise collaborate to expand the likelihood of stroke. Besides, numerous individuals have various marginal heights of hazard trait levels. There are a few research examinations demonstrating the confirmations of utilising physiological parameters as hazard factors for foreseeing the danger of stroke appeared in Table 14.1.

### 14.1.3 FACE RECOGNITION

Face recognition is the mechanism by which a vision system recognises a particular person's face. It has been a pivotal human-PC cooperation device because of its utilisation in security frameworks, get to control, video observation, business regions, and even it is also utilised in interpersonal organisations like Facebook. After the fast improvement of man-made brainpower, face acknowledgment has been stood out because of its nonintrusive nature and as it is a primary strategy for individual recognisable proof for human when it is contrasted with different kinds of biometric methods.

**TABLE 14.1**
**Risk Factor of Stroke**

| | |
|---|---|
| 1 | Age |
| 2 | Sex |
| 3 | Blood pressure |
| 4 | Visuospatial disorder |
| 5 | Dysphasia |
| 6 | Hemianopia |
| 7 | Cerebella signs |
| 8 | Face deficit |
| 9 | Smoking |
| 10 | Married |
| 11 | Gender |

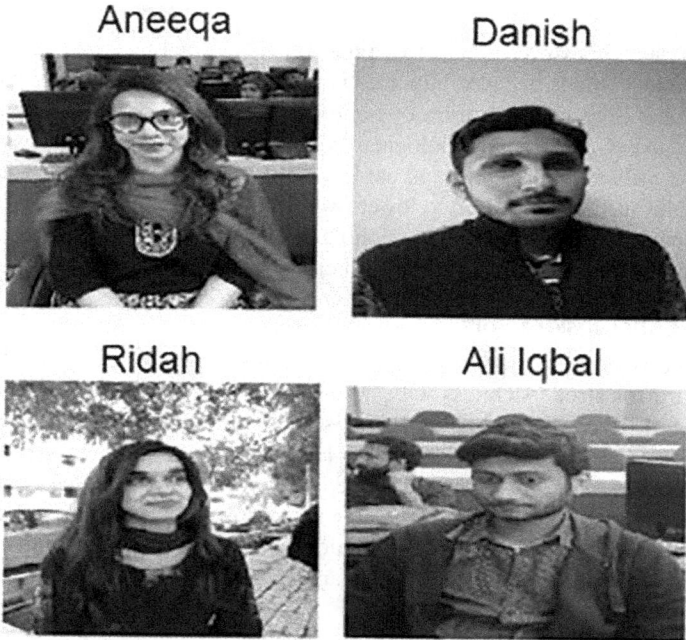

**FIGURE 14.5** Sample images of four subjects from the data set.

Face recognition is one of the most alluring biometric innovations. With the quick advancement of innovation, the precision of face acknowledgment has enormously improved. Numerous strategies for face recognition have been proposed and applied to numerous territories, for example, face-recognisable proof, security, reconnaissance, get to control, and character confirmation [11–14].

Other biometric innovations, for example, fingerprint reader, eye scanner, and voice recogniser include human action and noteworthy delays. To defeat these issues, automatic facial recognition frameworks are generally utilised which doesn't require any human communication for distinguishing the proof [15]. Some model pictures of four subjects from the data set are appeared in Figure 14.5.

In the overview of the face-recognition methods given by [16,17], they classified face-recognition frameworks into three classifications:

1. **Appearance-Based**: this procedure utilises comprehensive surface highlights and they are applied to either entire face or explicit locales of the face picture.
2. **Feature-Based**: this method utilises geometric facial highlights like mouth, eyes, cheeks, and so on, and geometric connection between these highlights.
3. **Hybrid Methods**: as a human being, we are used to for matching face as a whole with holistic approach as well as with the help of features of the face.

### 14.1.4  MOTIVATION TO MACHINE LEARNING TECHNIQUES

ML strategies have been progressively utilised in numerous applications. Specifically, ML has assumed a critical job in improving the presentation of biometric frameworks. With installed ML in biometric frameworks, sometimes tedious tasks such as one-to-one or one-to-many matching tasks can be done automatically and seamlessly. Specifically, Deep learning (DL), a particular ML approach dependent on neural nets made out of numerous layers, has been utilised in various biometrics applications. DL strategies show the capacity to make strong and solid confirmation models that now and again beat the condition of expressions of the human experience frameworks as brought up by certain specialists.

ML can utilise complex calculations to take in highlights from a huge volume of medicinal services information and then use the obtained insights to assist clinical practice. It can be outfitted with learning and self-adjusting capacities to improve its exactness dependent on criticism. ML framework can help doctors by giving state-of-the-art clinical data and help to lessen analytic and restorative blunders that are inescapable in the human clinical practice [18]. Besides, an ML framework separates valuable data from an enormous patient populace to help making constant surmising's for well-being hazard alarm and well-being result forecast [19].

This chapter contains four sections. Section two reviews ML algorithms used in stroke classification and face recognition; Section three describes the proposed methodology; Section four describes the discussion about results analysis for the proposed research work.

## 14.2  RELATED WORK

A considerable amount of studies has been done to develop biometric systems for brain stroke and face recognition using different techniques. This section provides a review of studies that adopted traditional ML and DL approaches in the biometric systems.

### 14.2.1  REVIEW ON BRAIN STROKE

Maier et al. [20] applied nine classification methods, including generalized linear models, random decision forests (RDFs), and CNNs, to order ischaemic stroke and inferred that RDFs and CNNs can give preferred grouping exactness over different strategies. Another study [21] presented a forecast model with DT, ANN, SVM, logistic regression (LR), and ensemble approach generalized boosted model (GBM) to foresee ICU move of stroke patients and inferred that GBM gave the most elevated precision. Kansadub et al. [22] used DTs, naive Bayes, CNN, and ANN to anticipate stroke and reasoned that DT yielded preferable order over different techniques. Sung et al. [23] used kNN, multiple linear regression, and a regression tree model to predict the stroke severity index and exhibited that k-nearest neighbour (kNN) has preferred precision over different models.

Kumar et al. [24] studied the performance of the implemented approach offered higher accuracy for the three-class classification problem which also solved the complexity in segmentation problems. A robust technique for the automatic segmentation of haemorrhage, ischaemic stroke, and tumour lesions from the MRI and CT brain images was contrived by using the Decision Tree characterization model. Snehkunj et al. [25] focused on the feature extraction of the MRI and CT brain images. The abnormalities such as brain haemorrhage and brain tumour were considered into account, which were diagnosed using the same methodology. Various phases were explored such as brain image extraction, transformation, and progression of the MRI or CT images. The accuracy of detecting the abnormalities in the images was enhanced. Ferdian [26] used a robust and accurate segmentation method based on a combination of an atlas-based and active contours segmentation. The experimental analysis revealed an extraordinary correlation with increased accuracy and was well suited for the reliable ventricle segmentation in stroke patients.

Few scientists are working on stroke expectation with ML calculations. Massive research commitments are depicted in this segment. A past report utilised ANN strategy, prepared with six diverse multilayer perceptron calculations to anticipate the mortality of stroke patients which created a precision of 80.7% [27]. Another study utilised SVM, kNN, and ANN to mechanise the discovery of ischaemic stroke, which recommended that SVM has higher expectation precision [28,29]. Amini et al. [30] anticipated stroke rate by utilising k-nearest neighbour and 4.5 decision tree techniques to uncover that C4.5 decision tree strategies yielded a higher exactness rate of 95.42%. Another group [31] used ML methods and SVM to predict stroke thrombolysis result, which indicated that SVM was more accurate. Cheng et al. [32] predicted ischaemic stroke utilising two ANN models that gave an accuracy rate of 79.2% and 95.1%.

Priya et al.'s [33] study predicts the sort of stroke for a patient dependent on classification methodologies. The classes of SVM and ensemble (bagged) gave 91% accuracy with 0.0000 negative predictive value, while ANN prepared with the stochastic gradient descent algorithm outperformed other algorithms, with a higher classification accuracy of 95% with a lower standard deviation of 14.69. Chantamit-O-Pas et al. [34] propose a stroke forecast through DL. The information on clinical area issues couldn't be followed precisely by the conventional prescient models.

Li et al. used generalised linear model, Bayes model, and decision tree model to predict the risk of ischaemic stroke and other thromboembolism of individual with atrial fibrillation [35]. Zhang et al. utilised an assortment of filter-based component choice models to improve the incapable element determination in the existing exploration on stroke hazard recognition [36].

Accurate classification and sensible intercession for high-chance populace can successfully lessen the weight of stroke on families and the society. It is important to consider the review of the grouping model to guarantee the congruity of stroke mediation [37,38]. Al-Maqaleh et al. used decision tree, Naïve Bayesian, and neural system to predict the coronary illness and compared their exhibition in terms of accuracy [39]. Table 14.2 shows the summary of different ML techniques used for brain stroke detection and classification.

## 14.2.2 REVIEW ON FACE RECOGNITION

A great deal of endeavours have been made to create facial recognition frameworks progressively powerful and reliable [40]. Face is considered as one of the most

**TABLE 14.2**

**ML Techniques are Used in Brain Stroke for Different Data sets**

| No | Title | Methods Used | Disadvantages |
|---|---|---|---|
| 1 | Clinical determination of stroke utilising inductive machine learning. | Inductive ML technique and decision tree | Instability for continuous data set. |
| 2 | Intelligent brain haemorrhage diagnosis system | Watershed method, artificial neural network | Over-fitting problem and computational complexity |
| 3 | Automatic CT scan image segmentation to recognise haemorrhage. | Histogram-based centroids initialisation and K-means clustering | Sensitive to noisy or redundancy data |
| 4 | A survival prediction model of rats in haemorrhagic shock using the random forest classifier | Breiman's method and random forest classifier. | Slow prediction |
| 5 | Intelligent diagnosis of brain haemorrhage with neural network system. | Artificial Neural Network and learning tool | Computational burden |
| 6 | Automatic segmentation and classification of brain haemorrhages in CT scans. | Thresholding method, genetic algorithm, Multilayer Neural Network, and K-Nearest Neighbour classification | Convergence rate to obtain better result is low |
| 7 | Pre-segmentation for the computer-aided determination framework. | Pre-segmentation process | Consumes more power |
| 8 | Detection of intracranial haemorrhage using spatial fuzzy c-mean and region-based active contour on brain CT imaging | Fuzzy clustering method, region-based active contour | More number of iterations are required for achieving better detection result |
| 9 | MRI-assisted computer diagnosis of human brain tumour: A survey and a new algorithm | Feedback pulse-coupled neural network, discrete wavelet transform, principal component analysis, and feed-forward back-propagation neural network | More power consumption and software complexity |
| 10 | In a hierarchical classification system, automatic brain haemorrhage segmentation and classification algorithm based on weighted grey-scale histogram. | MDRLSE and synthetic feature selection algorithm | Does not support large data set |

significant biometrics utilised for confirmation and a recognisable proof in a wide assortment of applications. Other biometrics includes fingerprint, iris, signature, ink, and handwritten text [41]. There are different DL approaches such as CNN, Stacked Autoencoder [42], and Deep Belief Network (DBN) [43]. CNN is the mostly used algorithm in image and face recognition. CNN is a sort of artificial neural systems that utilises the convolution procedure to extricate the features from the information to expand the quantity of features. Kim et al.'s [44] study shows that DBN works successfully for the expectation of cardiovascular hazard infection and can be utilised in facial recognition (biometrics) that is transforming into a tremendous structure in the security business.

A convolutional neural system (CNN), one of the most famous deep neural systems in computer vision applications, shows a significant favourable position of automatic facial visual feature extraction [45]. There are two sorts of strategies to train CNN for face recognition: one depends on the characterisation layer [46], and the other depends on metric learning. The primary thought of metric learning for face recognition is boosting interclass fluctuation and limiting intra-class variance. For example, FaceNet [47] utilises triplet misfortune to become familiar with the Euclidean space installing in which all appearances of one personality can be anticipated onto a solitary point. Sphereface [48] proposes angular margin to authorise extra intra-class minimisation and interclass disparity simultaneously. The authors of [49] propose an Added substance Angular Margin Loss work that can successfully upgrade the discriminative intensity of highlight embedding learned through CNNs for face recognition. CNNs prepared on 2D face pictures can viably work for 3D face recognition by adjusting the CNN with 3D facial scans [50]. In addition, the three-dimensional setting is invariant to helping or make-up conditions. The authors of [51] accept some straight amounts as measures and depend on differential geometry to remove important discriminant features from the query faces. Meanwhile, Nicole et al. [52] propose an automatic approach to compute a figure with base-streamlined marker format to be misused in the facial movement catch. Generally, an enormous volume of trained tests are useful to accomplish a high acknowledgment precision.

The structures can be sorted as backbone and assembled networks. A methodical survey on the advancement of the system models and loss functions for deep face recognition is shown in Table 14.3 with accuracy in percentage. The mainstream network architectures, such as Deepface [53], DeepID [54], VGGFace [55], FaceNet [56], and VGGFace2 [57], and other specific architectures like AlexNet, VGGNet, GoogleNet, ResNet, and SENet [58], are presented and broadly utilised as the standard model in face recognition. Different misfortune capacities are sorted into Euclidean-separation-based misfortune, precise/cosine-edge-based misfortune, and softmax misfortune and its varieties.

An examination and investigation on open accessible databases that are at indispensable significance for both model training and testing. Significant face-recognition benchmarks, such as LFW [59], IJBA/ B/C [60], Megaface [61], and MS-Celeb-1M [62], are assessed and analysed, in terms of three perspectives: training methodology, evaluation tasks and metrics, and recognition scenes, which give valuable references to preparing and testing deep face recognition. Some of the key future trends in non-DL and DL are listed in Table 14.4.

The proposed research work focuses on two main objectives: the development of a stroke-prediction system and face recognition in biometrics using machine learning. The design of such system usually has a number of various activities such as data set collection, feature extraction, model selection, and training and finally evaluates

**TABLE 14.3**
**Different CNN Verification Method to Recognize Face with Accuracy [63]**

| Method | Public. Time | Loss | Architecture | Number of Networks | Training Set | Accuracy ± Std (%) |
|---|---|---|---|---|---|---|
| DeepFace | 2014 | Softmax | Alexnet | 3 | Facebook (4.4M, 4K) | 97.35 ± 0.25 |
| DeepID2 | 2014 | Contrastive loss | Alexnet | 25 | CelebFaces+ (0.2M, 10K) | 99.15 ± 0.13 |
| DeepID3 | 2015 | Contrastive loss | VGGNet-10 | 50 | CelebFaces+ (0.2M, 10K) | 99.53 ± 0.10 |
| FaceNet | 2015 | Triplet loss | GoogleNet-24 | 1 | Google (500M, 10M) | 99.63 ± 0.09 |
| Baidu | 2015 | Triplet loss | CNN-9 | 10 | Baidu (1.2M, 18K) | 99.77 |
| VGGface | 2015 | Triplet loss | VGGNet-16 | 1 | VGGface (2.6M, 2.6K) | 98.95 |
| light-CNN | 2015 | Softmax | light CNN | 1 | MS-Celeb-1M (8.4M, 100K) | 98.8 |
| Center loss | 2016 | Center loss | Lenet+-7 | 1 | CASIA-WebFace, CACD2000, Celebrity+ (0.7M, 17K) | 99.28 |
| L-softmax | 2016 | L-softmax | VGGNet-18 | 1 | CASIA-WebFace (0.49M, 10K) | 98.71 |
| Range loss | 2016 | Range loss | VGGNet-16 | 1 | MS-Celeb-1M, CASIA-WebFace (5M, 100K) | 99.52 |
| L2-softmax | 2017 | L2-softmax | ResNet-101 | 1 | MS-Celeb-1M (3.7M, 58K) | 99.78 |
| Normface | 2017 | Contrastive loss | ResNet-28 | 1 | CASIA-WebFace (0.49M, 10K) | 99.19 |
| CoCo loss | 2017 | CoCo loss | - | 1 | MS-Celeb-1M (3M, 80K) | 99.86 |
| vMF loss | 2017 | vMF loss | ResNet-27 | 1 | MS-Celeb-1M (4.6M, 60K) | 99.58 |
| Marginal loss | 2017 | Marginal loss | ResNet-27 | 1 | MS-Celeb-1M (4M, 80K) | 99.48 |
| SphereFace | 2017 | A-softmax | ResNet-64 | 1 | CASIA-WebFace (0.49M, 10K) | 99.42 |
| CCL | 2018 | Centre invariant loss | ResNet-27 | 1 | CASIA-WebFace (0.49M, 10K) | 99.12 |
| AMS loss | 2018 | AMS loss | ResNet-20 | 1 | CASIA-WebFace (0.49M, 10K) | 99.12 |
| Cosface | 2018 | Cosface | ResNet-64 | 1 | CASIA-WebFace (0.49M, 10K) | 99.33 |
| Arcface | 2018 | Arcface | ResNet-100 | 1 | MS-Celeb-1M (3.8M, 85K) | 99.83 |
| Ring loss | 2018 | Ring loss | ResNet-64 | 1 | MS-Celeb-1M (3.5M, 31K) | 99.50 |

**TABLE 14.4**

**Difference between Non-DL and DL**

| Parameter | Non-DL | DL |
| --- | --- | --- |
| Scope | The algorithm needs to be informed on how to make an accurate prediction by providing it with more information. | The algorithm is able to learn that through its own data processing. |
| How does it work | Uses types of automated algorithm which learn to predict future decisions and model function using the data fed to it. | Interprets data features and their relationships using NN which pass the relevant information through several stages of data processing. |
| Management | The various algorithms are directed by the analyst to examine the different variables in the data set. | Once they are implemented, the algorithms are usually self-directed for relevant data analysis. |
| Dependency on data | Great performance on data set which are small to medium sized. | Performs exceptionally on large datasets. |
| Time of execution | Ranges from few minutes to certain hours. | It can take up to one week as the process is long. |
| Output | The output is usually a numerical value like a score or a classification. | The output can be anything from a score, an element, free text, or sound, etc. |

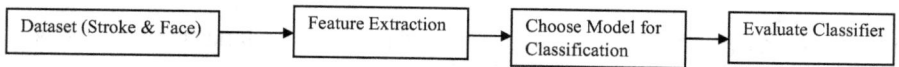

FIGURE 14.6   Block diagram of proposed system.

different classifier efficiencies using metrics. Figure 14.6 shows the design model of the proposed system.

### 14.2.3 BRAIN STROKE PREDICTION SYSTEM

ML classification technology is composed of two models (classification model and evaluation model). The classification model utilises preparing informational index so as to assemble grouping predictive model. Testing informational index is utilised for testing the characterisation productivity. Patient's data set is gathered from medicinal services foundation which has indications of stroke infection. Then, the proposed classification algorithm such as DL, decision tree, artificial neural network, and support vector machine is used to classify and predict whether the patient is suffering from stroke disease or not as shown in Figure 14.7. Then, the performance assessment is done dependent on these algorithms and contrasted using different models, and the precision is estimated.

#### 14.2.3.1  Image Acquisition

Medicinal services associations have increased huge benefit by data mining in the name of big data analysis and decision support system. In this research,

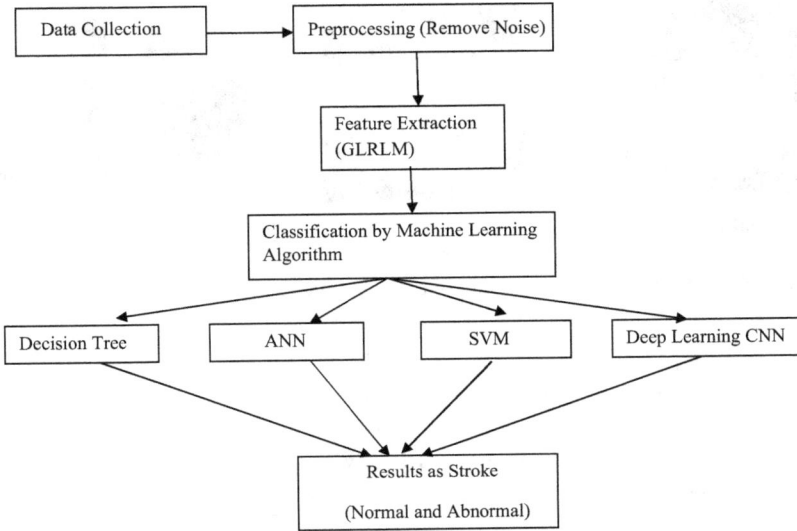

**FIGURE 14.7** Proposed system for stroke classification.

stroke patient database of existing CT images was used in this prospective study. Ongoing CT informational indexes have been gathered from different sources such as clinical focus managing brain diagnosis, Precise Diagnostics Center and KIMS Research Center and Hospital, Bangalore, Karnataka. Every gathered image is stored in a database. All images are having the size $512 \times 512$ reconstruction matrix, 2.3–4 mm slice thickness, X-ray source voltage is 120 kV, and maximum X-ray tube current is 65 mA.

## 14.2.3.2 Pre-processing

### 14.2.3.2.1 Image Cropping and Conversion into Grey-Scale Image

CT cerebrum images have film curios or marks like patient's name, age, date, time, remark, and so on. These names are evacuated utilising the roifill() work in MATLAB. Figure 14.8a shows the original image, and the cropped and grey-scale image is shown in Figure 14.8b.

### 14.2.3.2.2 Skull Extraction

The evacuation of the hard skull encompassing the brain tissue is considered as a test to the cerebrum confinement. The bwareaopen(), imfill(), and imerode() Matlab strategies and numerical activities are utilised to play out the skull evacuation. Figure 14.8c shows the skull-extracted image.

## 14.2.3.3 Feature Extraction

The feature extraction incorporates two distinct techniques: first-order histogram features such as mean, standard deviation, energy, entropy, variance, skewness, and kurtosis, and the other method is the grey-level run length matrix features like

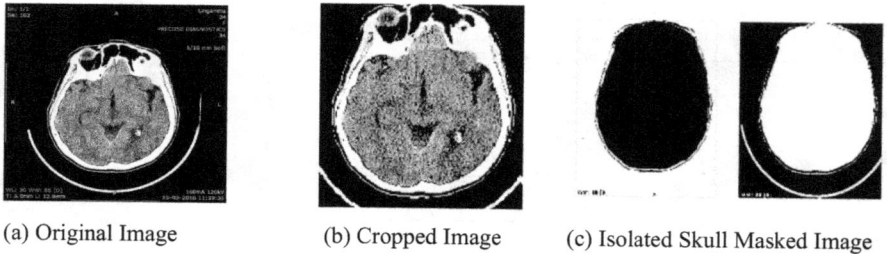

(a) Original Image          (b) Cropped Image          (c) Isolated Skull Masked Image

**FIGURE 14.8**   Steps for pre-processing of original image.

short-run emphasis, long run emphasis, run length non-uniformity, low grey-level run emphasis, run percentage, high grey-level run emphasis, short-run high grey-level emphasis, long-run low grey-level emphasis, short-run low grey-level emphasis, and long-run high grey-level emphasis. In view of these feature vectors, data set is made for characterisation.

### 14.2.3.4   Classification Using Machine Leaning Algorithms

*14.2.3.4.1   Decision Tree*

Decision tree is one of the significant strategies for dealing with high-dimensional information. It would appear as a tree structure. It is exceptionally a basic and simple path for taking care of the data set. Much work has been completed to foresee the hazardous disease utilising decision tree and proved to be progressively proficient. Figure 14.9 represents the decision tree model for predicting stroke diseases.

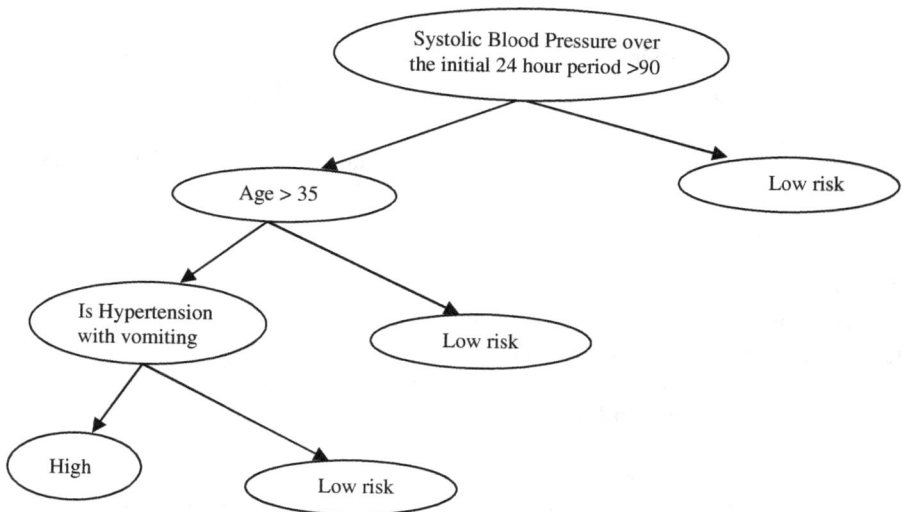

**FIGURE 14.9**   Decision tree.

### 14.2.3.4.2  *Artificial Neural Network*

ANNs [64] can perceive pattern, oversee information, and learn from sample patterns. It is an interconnected system of a gathering of artificial neurons. An artificial neuron can be considered as a computational model which is inspired by the characteristic neurons present in the human brain. These neurons essentially comprise inputs which are further multiplied by a parameter known as weight and then afterward processed by a numerical capacity which decides the actuation of the neuron. After this, there is another capacity that processes the yield of the artificial neuron. In this way, the artificial systems are shaped by combining these artificial neurons to process data.

Backpropagation [65] is a gradient-based algorithm, which has many variants. The most commonly used learning algorithms are Levenberg-Marquardt (LM), Quasi Newton, resilient backpropagation, scaled conjugate gradient, variable learning backpropagation, and scaled conjugate gradient with Powell/Beale restarts. A comparative analysis of the above algorithms has been done and implemented by Levenberg-Marquardt (LM) because of its low Root Mean Squared Error (RMSE) and rapid convergence. Table 14.5 shows the RMSE for the above-mentioned algorithms using Equation (14.1):

$$\text{RMSE} = \sqrt{\frac{1}{Y}\sum_{t=1}^{Y} xt - xdt} \qquad (14.1)$$

where $xt$ is the target value, $xdt$ is the classified value, and $Y$ is the total number of samples.

- **LM Algorithm:**
  The goal of the LM algorithm is to move toward the second-order training speed without figuring the Hessian framework [65]. When the performance function has the form of an aggregate of squares, the Hessian grid can be approximated as follows:

$$H = J^T J \qquad (14.2)$$

$$\text{Gradient}, g = J^T e \qquad (14.3)$$

**TABLE 14.5**

**Comparison of Root Mean Square Error Value with Different Algorithms**

| Algorithms | RMSE Value |
|---|---|
| LM | 0.0171 |
| QN | 0.1335 |
| RBP | 0.0613 |
| SCG | 0.0356 |
| VLBP | 0.0178 |

$J$ is the Jacobian framework that contains first subsidiaries of network error as for parameters, weight, and bias, and **e** is the vector representing network errors. Jacobian grid can be determined through a standard back-propagation technique that is less computational complex than computing Hessian matrix. LM algorithm utilises this estimate to the Hessian grid in the accompanying Newton-like update:

$$x_{k+1} = x_k - \left[ J^T J + \mu I \right]^{-1} J^T e \tag{14.4}$$

When scalar $\mu$ takes the worth zero, it carries on simply like Newton's technique. It goes to be an inclination plummet with a small size when $\mu$ goes to be huge. $\mu$ is diminished after each fruitful advance and is expanded just when a speculative advance would improve the performance work. So, the exhibition capacity will be diminished at every cycle of the algorithm.

The Levenberg algorithm can be summarised as follows:

1.  Do an update as directed by Equation (14.4).
2.  Assess the mistake at the new parameter vector.
3.  In the event that the mistake has expanded thus the update, then at that point, withdraw the progression and increment $\mu$ by a factor of 10 or whatever huge factor. At that point, go to (1) and attempt an update once again.
4.  If the mistake has diminished because of the update, then accept the step and reduction $\mu$ by a factor of 10 or somewhere in the vicinity.

### 14.2.3.4.3   Support Vector Machine

Support vector machine is a broadly utilised supervised ML algorithm for characterisation created by Vapnik, and the present standard manifestation was proposed by Cortes and Vapnik [66]. In the pattern classification, given a lot of sample inputs and the comparing class names, the aim is to restrict the inherent connection among the examples of a similar class, with the goal that when a test information is given, the relating yield class name can be retrieved.

The information focuses are recognized as either positive or negative, and a definitive point is to discover a hyper-plane that isolates the information focuses by a maximal edge. Figure 14.10 shows the two-dimensional situation where the information focuses are directly detachable. The distinguishing proof of the every data point xi is yi, which can take an estimation of +1 or −1.

In many applications, a non-linear classifier provides better accuracy. When we require non-linear separators, a solution is to map the data points into higher dimension (depending on the non-linearity characteristics required) so that the problem is linear in this high dimension. This is the feature space, and the mapping is done by a discriminate function which is defined as follows:

$$f(x) = < w, \ \phi(x) > + b \tag{14.5}$$

$f(x)$ is straight in the feature space characterised by the mapping $\varphi$; however, when seen in the first original space, it is a nonlinear function of $x$ if $\varphi(x)$ is a nonlinear function. This methodology of expressly assessing non-straight highlights doesn't

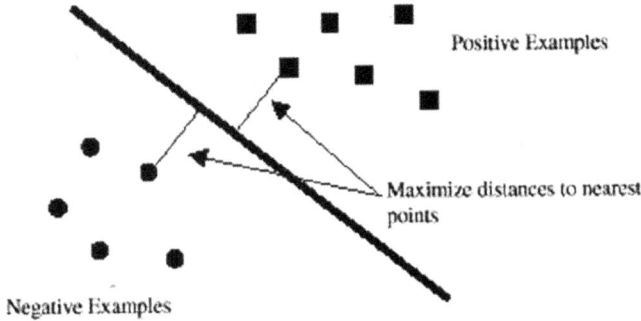

**FIGURE 14.10**   Class separation of SVM classifier.

scale well with the quantity of input features. If monomials of degree $d$ instead of degree 2 monomials are utilised, the dimensionality would be exponential in $d$, bringing about a significant increment in the memory utilisation and the time required to assess the discriminant work.

### 14.2.3.4.4   Deep Learning with CNN

DL is known as various levelled learning. It is a part of ML dependent on a gathering of algorithms that endeavour to show significant level speculations of data by utilising the deep diagram with numerous handling layers and made out of different straight, non-direct change strategies.

Deep neural systems are an exceptional sort of an ANN. The most well-known kind of a deep neural system is a deep convolutional neural network (DCNN). DCNN, while acquiring the properties of a nonexclusive ANN, has likewise its own particular features. To start with, it is deep. A common number of layers is 10–30; however, in outrageous cases, it could surpass 1,000. Second, the neurons are associated with the different neurons share weights. This adequately permits the network to perform convolutions of the input image with the filters inside the CNN. Finally, CNNs commonly utilise an alternate activation function of the neurons when contrasted with traditional ANNs.

Figure 14.11 shows the architecture for a common CNN. One can see that the main layers are the convolution ones which serve the job of producing valuable features for

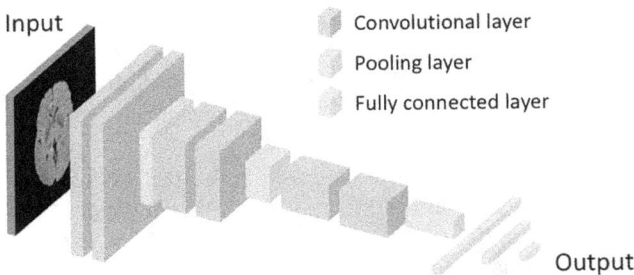

**FIGURE 14.11**   Diagram representing a typical architecture of a convolutional neural network.

classification. Those layers can be thought of as implementing image filters, ranging from basic filter that match edges to those that in the long-run coordinate substantially more confounded shapes such as eyes or tumours. Further from the network input are supposed completely associated layers which use the features separated by the convolutional layers to generate a decision.

### 14.2.3.5   Construction of Convolutional Neural Network

There are two procedures of convolutional neural network to classify stroke, specifically:

- **Training**:
  Training process is the place CNN is being trained with 200 data training of each sort of characterisation.
  - In the training process, CNN comprises two procedures: feedforward and backpropagation. Feedforward checks all the input neuron from the input layer in the hidden layer. Weights from the hidden layer will be sent to the output layer.
  - Backpropagation will follow the error by counting all the weight from the output layer and afterward sent it back to the hidden layer so the neural system acquired new weights with minimum error. These two procedures are finished with 1 EPOCH.
- **Testing**:
  Testing process is the place CNN is being tested with 50 data testing of type classification and contrasted the weights from data testing and weights that have been gotten from the data training. In the testing process, CNN just has the feed-forward procedure.

### 14.2.4   FACE-RECOGNITION SYSTEM

CNN designed for face recognition contains the accompanying layers of structure, which are the input layer, convolution, pooling, and all the associated layers such as yield layer and convolutional layer and the downsampled layer, etc.. In this chapter, the reference to LeNet5 [67] model to accomplish this CNN model set-up. The structure of the model will have LeNetConvPoolLayer, a sum of two layers LeNetConvPoolLayer, and in the third layer convolution in addition to examining layer associated a full association layer, named as hidden layer; this completely connected layer is like the hidden layer in a multi-layer perceptron. The last layer is the output layer, as it is a multi-faceted face classification, so Softmax regression model is utilised, named as LR. Figure 14.12 shows the design of the convolution neural system structure for the face-recognition system.

The input image is applied to the input layer; in this design, a sum of 50 individuals' face have been gathered, and every individual's face number is 15; an aggregate of 750 example samples, the size of each face image is $64 \times 64 = 4,096$, and each image is a grey-scale image. First convolutional and down-sampling layer receives the input image as $64 \times 64$, and the size of the convolution kernel is $5 \times 5$, so the subsequent image size after convolution is $(64 - 5 + 1) \times (64 - 5 + 1) = (60, 60)$.

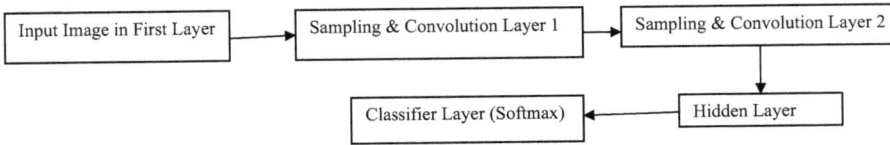

FIGURE 14.12   CNN block diagram for face-recognition system.

After the convolution operation, the image is down-sampled to the maximum, resulting in an image size of 30 × 30.

The input to the second convolution in addition to the sample layer is the output of the principal convolution in addition to the sample layer, so the size of the input image in this layer is 30 × 30. Like the activity of convolution in addition to the sample layer in the first layer, the image is convolution processed first, and the size of the convolution image is 26 × 26. Ensuing image under the most extreme down-sampling activity, the subsequent image size is 13 × 13.

## 14.3   DISCUSSION AND RESULTS

### 14.3.1   PERFORMANCE OF BRAIN STROKE

Throughput of the classifier has been examined based on the error rate. To evaluate the performance metrics in terms of true positive, true negative, false positive, and false negative are used. Validation requires the calculation of statistical parameters like sensitivity, specificity, accuracy, precision, F1 score, and $G$ measure. Mathematically, it is defined as follows:

$$\text{Sensitivity} = \text{TP}/(\text{TP} + \text{FN}) \qquad (14.6)$$

$$\text{Specificity} = \text{TN}/(\text{FP} + \text{TN}) \qquad (14.7)$$

$$\text{Accuracy} = (\text{TP} + \text{TN})/(\text{TP} + \text{TN} + \text{FN} + \text{FP}) \qquad (14.8)$$

$$\text{Precision} = \text{TP}/(\text{TP} + \text{FP}) \qquad (14.9)$$

$$\text{F1 score} = 2.\text{TP}/(2.\text{TP} + \text{FP} + \text{FN}) \qquad (14.10)$$

$$\text{G measure} = \sqrt{(\text{TP}/(\text{TP} + \text{FN})) \times (\text{TP}/(\text{TP} + \text{FP}))} \qquad (14.11)$$

In this research work, the statistical parameters such as sensitivity, specificity, accuracy, precision, F1 score, and $G$-measure are computed to evaluate the performance of the classifiers appeared in Figures 14.13 and 14.14. The number of samples in the training data set was taken as 250, and the number of samples in the testing data set was chosen to be 70. This work has been implemented in MATLAB variant 2018a. A comparison of DT, ANN, and SVM and CNN classifiers has been made, and the results are analysed with CNN yields accuracy with 98.5%. Table 14.6 gives the comparison of statistical parameters' performance.

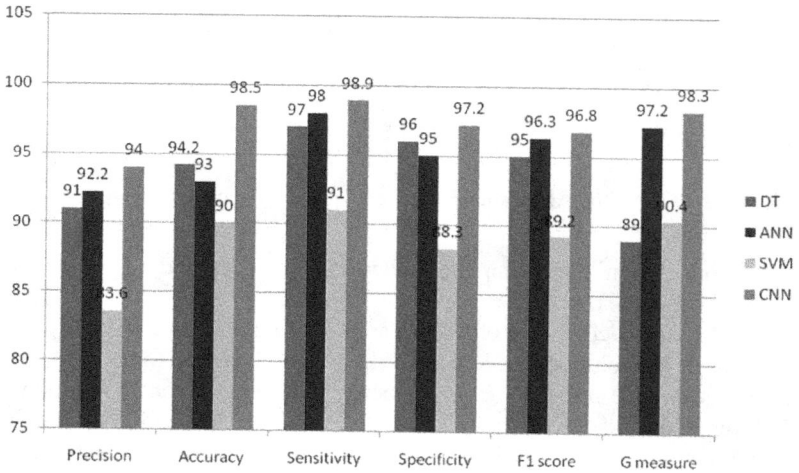

**FIGURE 14.13**   Graphical representation of statistical parameters (%).

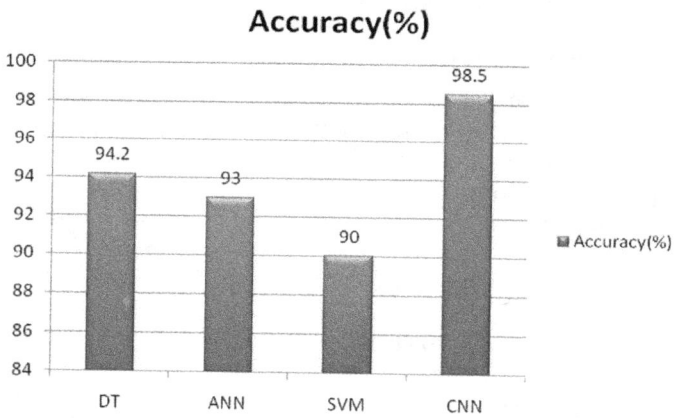

**FIGURE 14.14**   Classification accuracy of brain stroke using different classifiers.

**TABLE 14.6**
**Performance of Statistical Parameters** (%)

| Statistical Parameters | DT | ANN | SVM | CNN |
|---|---|---|---|---|
| Precision | 91 | 92.2 | 83.6 | 94 |
| Accuracy | 94.2 | 93 | 90 | 98.5 |
| Sensitivity | 97 | 98 | 91 | 98.9 |
| Specificity | 96 | 95 | 88.3 | 97.2 |
| F1 score | 95 | 96.3 | 89.2 | 96.8 |
| G measure | 89 | 97.2 | 90.4 | 98.3 |

**TABLE 14.7**
**Training Option for CNN**

| Variables | Value |
|---|---|
| Initial learn rate | 0.001 |
| Momentum | 0.7 |
| Mini batch size | 8 |
| Max epochs | 5 |
| Optimiser | SGDM |

## 14.3.2 PERFORMANCE OF FACE RECOGNITION

The improved data set containing 1,000 images of 50 subjects is separated into training and testing sets for experimentation. Seventy percent of images are utilised for fine-tuning a CNN model, while the remaining 30% images are utilised for execution assessment of the proposed strategy. Determination of proper training alternatives for CNN likewise assumes an essential job in the preparation procedure. The training options discovered reasonable for the proposed technique are appeared in Table 14.7.

The proposed biometric framework dependent on deep face recognition takes an image of a subject or a gathering of subjects in a scene as input, distinguishes faces utilising Viola Jones calculation, and afterward characterises each cropped facial part utilising a trained SqueezeNet model [67]. Empowering exploratory outcomes demonstrating a precision of 98.86% interprets the feasibility of deep face recognition for the biometric framework.

## 14.4 FUTURE SCOPE

- ML algorithms, especially the deep neural system, can improve the expectation of long-term results in ischaemic stroke patients.
- As a drawn-out objective, accuracy medication requests dynamic learning from all biological, biomedical, just as well-being data.
- The deeper systems with initiation modules are enhanced and give higher precision in the biomedical image investigation.
- DL is a promising mediator for various information, serving in disease expectation, prevention, diagnosis, visualisation, and facial recognition and clinical dynamic.
- Currently, increasingly more consideration is being paid to the use of DL in the biomedical data and new utilisations of every blueprint might be discovered in the next future.
- CNNs are most normally utilised in the biomedical image investigation area like facial recognition because of their extraordinary limit in breaking down the spatial data.
- Many DL systems are open source, including ordinarily utilised structures like Torch, Caffe, Theano, MXNet, DMTK, and TensorFlow. Some of them

are structured as eminent-level wrappers for simple use such as Keras, Lasagne, and Blocks.

- ML algorithms can be conveyed with relative straight-forwardness given minimal effort of software apparatuses whenever furnished with a proper establishment of information.

## 14.5  CONCLUSION

Machine intelligence can be projected as one of the significant tools in decision-making in the field of medicine. A machine learning-based approach based on DT, ANN, SVM, and CNN is suggested in this work to predict the possibility of stroke from a group of healthy and stroke patient's data set of age ranging from 30 to 85 years. Based on the outcome for classifying stroke from CT head examined image, convolution neural system can assist nervous system specialist in classifying stroke. The obtained precision likewise relies upon the quantity of gained information for training data set. In this examination, our summed-up strategy can give 98.5% of precision for the classification of stroke. The classification result much relies upon how much images that are being utilised in the training process. More images utilised in preparing process yields the higher precision. Future research can be done utilising different strategies for classifying sub-stroke type also.

With the appearance of huge information and graphical registering, DL has magnificently boosted the conventional computer vision frameworks over the previous decade. Towards this path, we have introduced a CNN-based face-recognition framework which naturally extracts facial features from faces distinguished utilising Viola Jones face detector for face recognition. A huge database containing facial images of 50 subjects was made for training and testing. Empowering trial results demonstrating a precision of 98.86% delineates the viability of deep face recognition for biometric framework. The proposed framework can be utilised in a wide assortment of uses including content-based information recovery, web search by image, observation, criminal distinguishing proof, automated attendance systems, and auto-requirement of limited access to specific regions.

## REFERENCES

1. Lidegaard, Ø., Løkkegaard, E., Jensen, A., Skovlund, C. W., and Keiding, N. "Thrombotic stroke and myocardial infarction with hormonal contraception", *New England Journal of Medicine*, vol. 366, pp. 2257–2266, 2012.
2. Lidegaard, Ø. Milsom, I. A., Geirsson, R. T., and Skjeldestad, F. E. "Hormonal contraception and venous thromboembolism", *Acta obstetricia et gynecologica Scandinavica*, vol. 91, pp. 769–778, 2012.
3. Gierhake D., Weber J., Villringer K., Ebinger M., Audebert H., and Fiebach J., "Mobile CT: technical aspects of prehospital stroke imaging before intravenous thrombolysis", *RoFo: Fortschritte auf dem Gebiete der Rontgenstrahlen und der Nuklearmedizin*, vol. 185, pp. 55–59, 2013.
4. Payabvash, S., Qureshi, M. H., Khan, S. M., Khan, M., Majidi, S., Pawar, S., and Qureshi, A. I. "Differentiating intraparenchymal hemorrhage from contrast extravasation on post-procedural noncontrast CT scan in acute ischemic stroke patients undergoing endovascular treatment", *Neuroradiology*, vol. 56, pp. 737–744, 2014.

5. Lansberg, M. G., Straka, M. M., Kemp, S., Mlynash, M., Wechsler, L. R., Jovin, T. G., Wilder, M. J., Lutsep, H. L., Czartoski, T. J., and Bernstein, R. A., "MRI profile and response to endovascular reperfusion after stroke (DEFUSE 2): a prospective cohort study", *The Lancet Neurology*, vol. 11, pp. 860–867, 2012.

6. Dheeba, C., and Vidhya, S., "A survey on prediction of brain hemorrhage using various techniques", *Indian Journal of Innovations and Developments*, vol 5, No. 6, pp. 1–3, 2016.

7. Subha, P.P., Pillai Geethakumari, S.M., Athira, M. and Nujum, Z.T., "Pattern and risk factors of stroke in the young among stroke patients admitted in medical college hospital, Thiruvananthapuram", *Annals of Indian Academy of Neurology*, vol. 18, p. 20, 2015.

8. Camargo Jr., C.A. "Moderate alcohol consumption and stroke: the epidemiologic evidence", Stroke, vol. 20, pp. 1611–1626, 1989.

9. Benjamin, E. J., Levy, D., Vaziri, S. M., D'Agostino, R. B., Belanger, A. J. and Wolf, P. A., "Independent risk factors for atrial fibrillation in a population-based cohort: the Framingham Heart Study", *Journal of the American Medical Association*, vol. 271, pp. 840–844, 1994. http://dx.doi.org/10.1001/jama.1994.03510350050036.

10. MacMahon, S. and Rodgers, A., "The epidemiological association between blood pressure and stroke: implications for primary and secondary prevention", *Hypertension Research*, vol. 17, No. Supply 1, pp. S23–S32, 1994.

11. https://en.wikipedia.org/wiki/Convolutional_neural_network.

12. Sahani, M., Subudhi, S., and Mohanty, M.N, "Design of face recognition based embedded home security system", KSII transaction on Internet and Information System, 2016.

13. Ramkumar, M., Sivaraman, R., and Veeramuthu, A., "An efficient and fast IBR based security using face recognition algorithm", In: *Proceedings of the 2015 IEEE International Conference on Communications and Signal Processing (ICCSP)*, pp. 1598–1602, Melmaruvathur, India, April 2015.

14. Xu, W., Shen, Y., Bergmann, N., and Hu, W. "Sensor-assisted multi-view face recognition system on smart glass", *IEEE Transaction on Mobile Computing*, vol. 17, pp. 197–210, 2017.

15. Barnouti, N. H., Al-Dabbagh, S. S. M., and Matti, W. E., "Face recognition: a literature review", *International Journal of Applied Information System*, vol. 11, No. 4, pp. 21–31, November 2016.

16. Zhao, W., Chellappa, R., Phillips, P. J. et al., "Face recognition – a literature survey", *ACM Computing Surveys*, vol. 35, No. 4, pp. 399–458, December 2003.

17. http://www.iamwire.com/2017/11/difference-between-machine-learning-and-deep-learning/169100.

18. Ravi, D., Wong, C., Deligianni, F., et al, "Deep learning for health informatics", *IEEE Journal on Biomedical and Healthcare Informatics*, vol. 21, No. 4, pp. 4–21, 2017.

19. Lee, C. S., Nagy, P. G., Weaver, S. J., et al., "Cognitive and system factors contributing to diagnostic errors in radiology", *American Journal of Roentgenology*, pp. 611–617, 2013. doi: 10.2214/AJR.12.10375.

20. Maier, O., Schro¨der, C., Forkert, N. D., Martinetz, T., and Handels, H., "Classifiers for ischemic stroke lesion segmentation: a comparison study", *Journal on PLoS ONE*, vol. 10, No. 12, pp. 1–4, 2015. doi:10.1371/journal.pone.0145118.

21. Alotaibi, N. N., Sasi, S., "Stroke in-patients' transfer to the ICU using ensemble based model", In: *IEEE International Conference on Electrical, Electronics, and Optimization Techniques (ICEEOT)*, Chennai, pp. 2004–2010, 2016. doi: 10.1109/ICEEOT.2016.7755040.

22. Kansadub, T., Thammaboosadee, S., Kiattisin, S., and Jalayondeja, C., "Stroke risk prediction model based on demographic data", In: *8th Biomedical Engineering International Conference*, Pattaya, Thailand, pp. 1–3, 2015. doi: 10.1109/BMEiCON.2015.7399556.

23. Sung. S. F., Hsieh, C. Y., Yang, Y. H., Lin, H. J., Chen, C. H., Chen, Y. W., and Hu. Y. H., "Developing a stroke severity index based on administrative data was feasible using data mining techniques", *Journal on Clinical Epidemiology*, vol. 68, No. 11, pp. 1292–1300, 2015.

24. Kumar, V. and Krishniah, V. J. R. "An automated framework for stroke and hemorrhage detection using decision tree classifier", In: *International Conference on Communication and Electronics Systems (ICCES)*, Coimbatore, India, pp. 1–6, 2016.

25. Snehkunj, R., Jani, A. N., and Jani, N. N. "Brain MRI/CT images feature extraction to enhance abnormalities quantification", *Indian Journal of Science and Technology*, vol. 11, pp. 1–10, 2018. doi:10.17485/ijst/2018/v11i1/120361.

26. Boers, F. A., Beenen, L., Cornelissen, B., et al., "Automated ventricular system segmentation in CT images of deformed brains due to ischemic and subarachnoid hemorrhagic stroke", in *Molecular Imaging, Reconstruction and Analysis of Moving Body Organs, and Stroke Imaging and Treatment*, Europe: Springer, pp. 149–157, 2017. doi:10.3238/arztebl.2017.0226.

27. Shinohara, Y., Yanagihara, T., Abe, K., Yoshimine, T., Fujinaka, T., Chuma, T., Ochi, F., Nagayama, M., Ogawa, A., Suzuki, N., Katayama, Y., Kimura, A., and Minematsu, K., "Cerebral infarction/transient ischemic attack (TIA)", *Journal of Stroke Cerebrovascular Disorder* vol. 20, No. 4, pp. S71–S73, 2011.

28. Rajini N. H., and Bhavani R., "Computer aided detection of ischemic stroke using segmentation and texture features", *Journal of Measurement*, vol. 46, No. 6, pp. 1865–1874, 2013.

29. Sundstro¨m C., "Machine learning algorithms for stroke diagnostics", Master's thesis in biomedical engineering, 2014.

30. Amini, L., Azarpazhouh, R., Farzadfar, M. T., Mousavi, S. A., Jazaieri, F., Khorvash, F., Norouzi, R., and Toghianfar, N., "Prediction and control of stroke by data mining", *International Journal on Preventive Medicine*, vol. 4, No. 2, p. S245, 2013.

31. Bentley, P., Ganesalingam, J., Jones, A. L., Mahady, K., Epton, S., Rinne, P., Sharma, P., Halse, O., Mehta, A., and Rueckert, D., "Prediction of stroke thrombolysis outcome using CT brain machine learning", *Journal on NeuroImage Clinical*, vol. 4, pp. 635–640, 2018.

32. Cheng, C. A., Lin, Y. C., and Chiu, H. W. "Prediction of the prognosis of ischemic stroke patients after intravenous thrombolysis using artificial neural networks", *Journal on Studies in Health Technology and Informatics*, vol. 202, pp. 115–118, 2014.

33. Govindarajan, P., Soundarapandian, R. K., Gandomi, Amir H., et al., "Classification of stroke disease using machine learning algorithms", Journal on Neural Computing and Applications, *Special Issue on Intelligent Biomedical Data Analysis and Processing*, Issue 3, London: Springer, pp. 603–878, 2019.

34. Chantamit-O-Pas, P., Goyal, M, "Prediction of stroke using deep learning model", In: *Neural information processing ICONIP, Lecture notes in computer science* 10638, 2017.

35. Li, X., Liu, H., Du, X., et al. "Integrated machine learning approaches for predicting ischemic stroke and thromboembolism in atrial fibrillation", In: *AMIA Annual Symposium Proceedings*, Europe, pp. 799–807, 2017.

36. Zhang, Y., Zhou, Y., Zhang, D., et al., "A stroke risk detection: improving hybrid feature selection method", *Journal of Medical Internet Research*, vol. 21, issue. 4, pp. 1–17, 2019. doi: 10.2196/12437.

37. Wang, X., Fu, Q, Song, F, et al., "Prevalence of atrial fibrillation in different socioeconomic regions of China and its association with stroke: results from a national stroke screening survey", *International Journal of Cardiology*, vol. 71, pp. 92–97, 2018.

38. Gurovich, Y., Hanani, Y., and Bar, Omri, e. a. "Identifying facial phenotypes of genetic disorders using deep learning", *Nature Medicine*, vol. 25, pp. 60–64, 2019.

39. Al-Maqaleh, B. M., and Abdullah, A. M. G. "Intelligent predictive system using classification techniques for stroke disease diagnosis", *International Journal of Computer Science and Engineering*, vol. 6, No. 6, pp. 145–151, 2017.

40. M. J. Khan, K. Khurshid, and F. Shafait, "A spatio-spectral hybrid convolutional architecture for hyperspectral document authentication", In: *2019 15th IAPR International Conference on Document Analysis and Recognition (ICDAR)*, Sydney, Australia, 2019.

41. Yousaf, A., Khan, M. J., Javed, N., et al., "Size invariant handwritten character recognition using single layer feedforward backpropagation neural networks", In: *2nd International Conference on Computing, Mathematics and Engineering Technologies (iCoMET)*, Sukkur, Pakistan, pp. 1–7, 2019.

42. Saleem, M. S., Khan, M. J., Khurshid, K., Hanif, M. S., "Crowd density estimation in still images using multiple local features and boosting regression ensemble", *Journal on Neural Computing & Applications*, pp. 1–10, 2019.

43. Shelhamer, E., Long, J., and Darrell, T., "Fully convolutional networks for semantic segmentation", *IEEE Transaction on Pattern Analysis of Machine Intelligence*, vol. 39, No. 4, pp. 640–651, 2017.

44. Kim, J. M. S., Kang, U., and Lee, Y., "Statistics and deep belief network-based cardiovascular risk prediction", *Journal on Healthcare Informatics Research*, vol. 23, No. 3, pp. 169–175, 2017. doi: 10.4258/hir.2017.23.3.169.

45. Shin, Y., and Balasingham, I., "Comparison of hand-craft feature based SVM and CNN based deep learning framework for automatic polyp classification", In: *Proceedings of the 2017 39th Annual International Conference of the IEEE Engineering in Medicine and Biology Society (EMBC)*, Jeju Island, Korea, pp. 3277–3280, July 2017.

46. Taigman, Y., Yang, M., Ranzato, M., and Wolf, L. "Deepface: closing the gap to human-level performance in face verification", In: *Proceedings of the IEEE Conference on Computer Vision and Pattern Recognition*, Columbus, OH, pp. 1701–1708, June 2014.

47. Tornincasa, S., Vezzetti, E., Moos, S., Violante, M. G., Marcolin, F., Dagnes, N., Ulrich, L., and Tregnaghi, G. F., "3D facial action units and expression recognition using a crisp logic", *Journal on Computer Aided Design and Application*, vol. 16.

48. Liu, W., Wen, Y., Yu, Z., Li, M., Raj, B., and Song, L., "Sphereface: deep hypersphere embedding for face recognition", In: *Proceedings of the IEEE Conference on Computer Vision and Pattern Recognition*, Honolulu, HI, USA, pp. 212–220, July 2017.

49. Deng, J., Guo, J., Xue, N., Zafeiriou, S., "Arcface: additive angular margin loss for deep face recognition", In: *Proceedings of the IEEE Conference on Computer Vision and Pattern Recognition*, Long Beach, CA, USA, pp. 4690–4699, June 2019.

50. Dagnes, N., Marcolin, F., Vezzetti, E., Sarhan, F. R., Dakpé, S., Marin, F., Nonis, F., and Mansour, K. B, "Optimal marker set assessment for motion capture of 3D mimic facial movements", *Journals on Biomechanics*, vol. 93, pp. 86–93, 2019.

51. Schroff, F., and Kalenichenko, D, Philbin, J. "Facenet: a unified embedding for face recognition and clustering", In: *Proceedings of the IEEE Conference on Computer Vision and Pattern Recognition*, Boston, MA, USA, pp. 815–823, June 2015.

52. Kim, D., Hernandez, M., Choi, J., Medioni, G., "Deep 3D face identification", In: *Proceedings of the 2017 IEEE International Joint Conference on Biometrics (IJCB)*, Denver, CO, USA, pp. 133–142, 2017.

53. Whitelam, C., Allen, K., Cheney, J., Grother, P., Taborsky, E., Blanton, A., Maze, B., Adams, J., Miller, T., and Kalka, N., Iarpa Janus "Benchmark-b face dataset", In *CVPR Workshops*, USA, pp. 592–600, 2017.

54. Sun, Y., Liang, D., Wang, X., and Tang, X., "Deepid3: face recognition with very deep neural networks", *arXiv preprint arXiv*:1502.00873, 2015.
55. Parkhi, O. M., Vedaldi, A., Zisserman, A., et al., "Deep face recognition", In: Proceedings of the British Machine Vision Conference (BMVC), Swanesh, UK: BMVA Press, pp. 41.1–41.12, September 2015.
56. Schroff, F., Kalenichenko, D., and Philbin, J., "Facenet: a unified embedding for face recognition and clustering", In: *Conference on Computer Vision and Pattern Recognition*, Boston, MA, pp. 815–823, 2015.
57. Cao, Q., Shen, L., Xie, W., Parkhi, O. M., and Zisserman, A., "VGGface2: a dataset for recognising faces across pose and age", *arXiv preprint arXiv*:1710.08092, 2017.
58. He, L., Li, H., Zhang, Q., and Sun, Z., "Dynamic feature learning for partial face recognition", In: *The IEEE Conference on Computer Vision and Pattern Recognition (CVPR)*, China, pp. 7054–7063, June 2018.
59. Hu, J., Shen, L., and Sun, G., "Squeeze-and-excitation networks", *arXiv preprint arXiv*:1709.01507, 2017.
60. Klare, B. F. Klein, B., Taborsky, E., Blanton, A., Cheney, J., Allen, K., Grother, P., Mah, A., and Jain, A. K., Pushing the "frontiers of unconstrained face detection and recognition: Iarpa Janus benchmark", In: *Conference on Computer Vision and Pattern Recognition*, Boston, MA, pp. 1931–1939, June 8–10, 2015.
61. Kang, B.-N., Kim, Y., and Kim, D. "Pairwise relational networks for face recognition", In: *The European Conference on Computer Vision (ECCV)*, Vol.11, Cham, Switzerland: Springer, September 2018. https://doi.org/10.1007/978-3-030-01216-8_39.
62. Hu, G., Yang, Y., Yi, D., Kittler, J., Christmas, W., Li, S. Z., and Hospedales, T., "When face recognition meets with deep learning: an evaluation of convolutional neural networks for face recognition", In: *International Conference on Computer Vision workshops*, Santiago, pp. 142–150, 2015. doi: 10.1109/ICCVW.2015.58.
63. Wang, M., and Deng, W. "Deep face recognition: a survey", *arXiv*:1804.06655v8[cs.cv] 12 February 2019.
64. Colak, C., Karaman, E., and Turtay, M. G., "Application of knowledge discovery process on the prediction of stroke", *Journal on Computer Methods and Programs in Biomedicine*, vol. 119, No. 3, pp. 181–185, 2015.
65. More, J. J. in Watson, G. A., *"Numerical Analysis"*, Lecture Notes in Mathematics 630, Springer Verlag, Germany, pp. 105–116, 1997.
66. Cortes, C., and Vapnik, V., "Support-vector networks", *Machine Learning*, vol. 20, No. 3, pp. 273–297, 2017.
67. Wang, J. and Li, Z., "Research on Face Recognition Based on CNN", In: *2nd International Symposium on Resource Exploration and Environmental Science, IOP Conference Series of Earth and Environmental Science*, vol 170, Red Hook, NY, pp. 1–6, 2018. doi: 10.1088/1755-1315/170/3/032110.

# Index

Aadhaar 24
acceptability 134, 258, 284
across-session 146
AdaBoost 291, 292
adversarial loss 57, 68
AlexNet 113, 211, 345
alignment 33, 156, 260, 303
anatomy 135, 207, 286
anti-spoofing 47, 242
area under the curve 89
artificial neural network 143, 349
Attack Presentation Classification Error Rate
        108, 123
attribute 25K dataset 87
autocorrelation 139, 325
autoencoder 55, 163, 343
auxiliary data 157

batch normalization 63, 116, 294
behavioural 133, 134, 135
bi-directional associative memory 163
binarisation 7, 61
binary code 158, 260, 280
bin salting 159
bioelectric signals 134
biometric authentication 32
biometric cryptosystems 26, 27, 31, 33, 45, 156
biometric data privacy 156
biometric encryption 26, 31
biometric identity 156
biometric recognition 1, 135
biometric system 25, 27, 43
biometric templates 25, 27, 32, 156
biometric traits 25, 31, 34, 133
biometric verification 25
Biopac 145
BioPhasoring 159
BioHashing 34, 38, 158, 163
biological 88, 258, 334
biometric standards 176
blood pressure 337
bona fide presentation 108
Bona fide Presentation Classification Error Rate
        109, 123
brain stroke 334, 353

cancelable biometrics 26, 33, 155, 157
CASIA 165, 275, 295
Celeb-DF Dataset 94
Celeb-Faces Attributes dataset 87

charge coupled device 5
circumvention 134, 258, 284
competitive code 2
contactless 116, 210
convolutional neural networks 92, 116, 161, 344
correlation 163, 276
cumulative match characteristics 220
CycleGAN 84

DCT 142, 143
decision tree 348
decoder 61, 287
deep convolutional feature 3
deep convolutional neural networks 4, 156, 351
DeepFake 81, 84, 86
DeepFakeTIMIT Dataset 83, 94
DeepID 344, 355
deep learning 1, 82, 341, 351
deep residual network 81
DET curves 123, 124, 126
DFFD Dataset 85, 89
dictionary learning 56
Discrete cosine transform 4, 139, 273, 318
Discrete Wavelet Transform 191, 273, 343
discriminability 36, 157, 160
diversity 157, 163, 169
Dorsal Hand Vein 1, 3, 8
downsampling 212, 303, 353
DET curves 123, 126

ear 4
ECG Biometric 137
edge-detection 244
electrocardiogram 134, 141, 142, 161
electroencephalogram 134, 161
epoch 68, 245, 352
equal error rate 39, 109, 316
esoteric biometrics 135
Euclidean distance 139, 216, 267

face 1, 4, 44, 81, 100, 165
face detection 83, 263
face forensics 86, 90
fake biometric 242
false accept rate 28, 169
false match rate 146, 148, 150
false-negative 247, 301, 353
false-positive 302, 353
false rejection rate 28, 94
feature extraction 10, 30, 141

feature information 161, 189
feature selection 164, 318
feedforward network 146, 274, 290
finger dorsal 163
finger knuckle image 209
finger knuckle print 260, 270
fingernail 208, 217
fingerprint 4, 33, 45, 52, 67, 116, 165
F1 score 354
forensics 29, 136, 234, 264
filtering 259, 291, 323
fusion 18, 43, 109, 161
fuzzy commitment 156
fuzzy logic 316
fuzzy vaults 156
FVC2002 165

Gabor filter 4, 10, 20, 88, 160, 200, 281
gait 4, 29, 133, 161, 258
gallery images 64, 263
GAN 78, 85, 241, 288
gaussian filter 247, 250
gaussian mixture model 139, 317
gender 82, 317, 339
gender classification 164, 175, 179
genetic algorithm 37, 343
Genuine Acceptance Rate 218
genuine distribution 40
genuine score 40, 216
genuine user 39, 134
group sparse representation 4

Haar Wavelet 208, 271
hand geometry 135, 258, 285, 334
hashing 16
heart disease 337
hidden layer 93, 138, 146, 353
hough transform 273, 290
hybrid indexing 268
hybrid methods 340
hyper-parameter 53, 64, 77, 116, 160, 213, 265
hyperspectral 3, 8, 14
human computer interaction 176

identical twins 206
identification 17, 83, 106
identification rate 15, 145
IITD 30, 41, 292
illumination 115, 185, 293
image classification 10, 61, 211
imposter score 216
impostor 106, 169
inception-v3 111, 328
independent component analysis 181, 208
indexing 37, 257, 271, 273
index-of-max 160
information fusion 213

inter-class distance 262
inter-class similarity 29, 30
internet of things 155
interoperability 260
intra-class 18, 30, 110, 344
intra-class variation 43, 61
Iris 30, 134, 273
Iris Recognition 30, 291

Jaccard Index 301

Kalman filter 139
Kerberos 32
key binding 31
keystroke 33, 133, 313
keystroke dynamics 133
K-Means 54, 267, 287, 343
knuckle biometric 163
knuckleprint 135, 163, 257

laser 112
latent fingerprint 51, 56, 57
layered LSTM 83
LDA-1NN 180
likelihood ratio 39, 190
linear discriminant analysis 4, 180, 318
liveliness 25, 314
liveness 2, 134
local binary pattern 3, 10, 88, 177, 244
local derivative pattern 3, 10
local phase quantization 177
logistic regression 3, 89, 214
long short-term memory 83

majority voting 316
marginal 214, 339
marginal loss 345
match score 54, 65, 220
max-pooling 120, 161, 246
max rule 213
median filtering 7, 160, 291
Min-Max 58, 220
min rule 213
minutiae points 160, 265
MobileNet 107, 126, 285
multi-biometric 32, 161
multi-instance 48, 161
multimodal 1, 4, 180, 207, 297
multimodal database 180
multimodal identification 209, 17, 222
multimodal verification 17, 220
multi-sensor 63, 66
multi-spectral 104, 115, 175, 183

naive Bayes 341
near-infrared 2
neutral network 160

NIR sensor 182
NIR spectrum 180, 185
NIST 60, 64, 75, 89
non-ideal 177, 286, 300
non-invertibility. 27, 36, 157
non-universality 43
normalisation 26, 212, 271, 285

occlusion 87, 178, 260
ocular 175, 177, 179
one to many 28
one to one 28

palmprint 1, 25, 165, 206
Particle Swarm Optimisation 214
Parzen window 317, 321
password 25, 32, 106, 156, 284
pattern recognition 145, 260
*penetration rate* 260, 270, 272
performance 68, 94, 120
permanence 25, 133, 284, 305
physiological 25, 260, 284, 339
physiological trait 313
PolyU 165, 272, 296
presentation attack 24, 28, 04, 108
principal component analysis 145, 161, 180, 271
privacy 25, 33, 155, 170
probe 39, 65, 157
product rule 213, 232

quadratic discriminant analysis 317
quadruplets 267
quality analysis 64
quantization 144, 163

radial basis function 139, 181, 317
random forest 180, 245, 319, 343
random mapping 160, 165
random noise 158, 165
random projection 36, 158
rank level fusion 223, 225
reconstruction loss 58, 68
reflection 111, 185, 190
ridge 51, 66, 246

sampling 142, 207, 353
score level fusion 4, 120, 139, 220, 319
Score Max 213
score normalisation 326
self-organizing maps 317
signal-to-noise-ratio 194
signature 135, 165, 276, 334
similarity score 237, 259
singular vector decomposition 273
smart cards 255
soft biometrics 183, 134
Softmax 12, 30, 91, 344

speaker verification 161
speech 133, 161, 243, 313
speed 43, 261, 349
Speeded up Robust Feature 4, 97, 270
spoof attacks 100
spoofing 2, 83, 234, 242
stacked autoencoder 161, 166, 343
stacked hourglass 289, 305
statistical models 138
stroke 161, 333
subspace 158, 190
sum of gaussian 321
sum rule 108, 213, 220
super resolution 181
support vector machine 139, 178, 346
surveillance 170, 176, 260
system design 161, 213

tanh 63, 93
template 8, 23, 26, 145, 155
template protection 23, 31, 156
template security 43, 45
temporal 88, 112, 264, 276
temporal domain 57, 87
tensor 112, 117
testing 140, 177
testing set 178, 190
test sample 18, 128, 190, 317
texture 2, 141, 242, 266
thermogram 206
threshold 7, 10, 37, 96, 128
thresholding 158, 291
throughput 353
tokanised 36, 45, 157, 171
token 39, 106, 156
tongue print 135
training set 92, 112, 262, 303
transformation 26, 279, 342
triplet 267, 344
triplet loss 345
true-negative 248, 302, 353
true-positive 220, 247, 302, 353
true positive identification rate 220

ubiquitous 152
unimodal biometric 3, 32
unimodal identification 17, 220
unimodal verification 206, 216
uniqueness 25, 133, 284, 313
universality 134, 257
user-specific 169

verification 17, 25, 140, 176, 209
VGG 126, 179, 243
virtual 271, 295
voice 33, 135, 161, 234, 285
vulnerability 24, 197, 199, 240

Wavelet 178, 191, 209
weighted fusion 223, 227
weighted layers 211, 212
Weighted method 215
weighted sum rule 108
weights 120, 137, 214

XOR 301
XORed 32, 160

zero-padded 64
zero-pole 53, 56

For Product Safety Concerns and Information please contact our EU
representative  GPSR@taylorandfrancis.com
Taylor & Francis Verlag GmbH, Kaufingerstraße 24, 80331 München, Germany

www.ingramcontent.com/pod-product-compliance
Lightning Source LLC
Chambersburg PA
CBHW052011230326
41598CB00078B/2465